19th International School on Condensed Matter Physics (ISCMP): Advances in Nanostructured Condensed Matter: Research and Innovations 2016

Journal of Physics: Conference Series Volume 794

Varna, Bulgaria
28 August – 2 September 2016

ISBN: 978-1-5108-3695-2
ISSN: 1742-6588

Printed from e-media with permission by:

Curran Associates, Inc.
57 Morehouse Lane
Red Hook, NY 12571

Some format issues inherent in the e-media version may also appear in this print version.

Printed by Curran Associates, Inc. (2017)

For permission requests, please contact the Institute of Physics
at the address below.

Institute of Physics
Dirac House, Temple Back
Bristol BS1 6BE UK

Phone: 44 1 17 929 7481
Fax: 44 1 17 920 0979

techtracking@iop.org

Additional copies of this publication are available from:

Curran Associates, Inc.
57 Morehouse Lane
Red Hook, NY 12571 USA
Phone: 845-758-0400
Fax: 845-758-2633
Email: curran@proceedings.com
Web: www.proceedings.com

Table of contents

Volume 794

19th International School on Condensed Matter Physics (ISCMP): Advances in Nanostructured Condensed Matter: Research and Innovations

28 August to 2 September 2016, Varna, Bulgaria

Accepted papers received: 15 December 2016
Published online: 22 February 2017

Preface

Papers

Invited lectures

Oral presentations

Posters

IOP Conf. Series: Journal of Physics: Conf. Series **794** (2017) 011001 doi:10.1088/1742-6596/794/1/011001

PREFACE

We are pleased to introduce the Proceedings of the 19th International School on Condensed Matter Physics *"Advances in Nanostructured Condensed Matter: Research and Innovations"* (19th ISCMP). The school was held from August 28th till September 2nd, 2016 in Varna, Bulgaria. It was organized by the Institute of Solid State Physics of the Bulgarian Academy of Sciences (ISSP-BAS), and took place at one of the fine resorts on the Bulgarian Black Sea "Saints Constantine and Helena".

The aim of this international school is to bring together top experimentalists and theoreticians, with interests in interdisciplinary areas, with the younger generation of scientists, in order to discuss current research and to communicate new forefront ideas. This year special focus was given to discussions on membrane biophysics and quantum information, also not forgotten were some traditionally covered areas, such as characterization of nanostructured materials. Participants from 12 countries presented 28 invited lectures, 12 short oral talks and 44 posters. The hope of the organizing committee is that the 19th ISCMP provided enough opportunities for direct scientific contacts, interesting discussions and interactive exchange of ideas between the participants. The nice weather certainly helped a lot in this respect. The editors would like to thank all authors for their high-quality contributions and the members of the international program committee for their commitment.

The papers submitted for publication in the Proceedings were refereed according to the publishing standards of the Journal of Physics: Conference Series. The Editorial Committee members are very grateful to the Journal's staff for the continuous fruitful relations and for giving us the opportunity to present the work from the 19th ISCMP.

Prof. DSc Hassan Chamati, Assist. Prof. Dr. Alexander A. Donkov, Assoc. Prof. Dr. Julia Genova, and Assoc. Prof. Dr. Emilia Pecheva

 doi:10.1088/1742-6596/794/1/011002

Bulgarian Academy of Sciences

GEORGI NADJAKOV INSTITUTE OF SOLID STATE PHYSICS

SOFIA, BULGARIA

19th INTERNATIONAL SCHOOL ON CONDENSED MATTER PHYSICS

Advances in Nanostructured Condensed Matter: Research and Innovations

August 28th - September 2nd, 2016
Varna, Bulgaria

Organising Committee

Chairman

Prof. Hassan Chamati, Director, Georgi Nadjakov Institute of Solid State Physics, Bulgarian Academy of Sciences, Sofia, Bulgaria (director@issp.bas.bg).

Honorary Chairman:
Academician Alexander G. Petrov, Georgi Nadjakov Institute of Solid State Physics, Bulgarian Academy of Sciences, Sofia, Bulgaria (agpetrov@issp.bas.bg)

Local Organising Committee

Victoria Vitkova (scientific secretary, victoria@issp.bas.bg), Emilia Dimova, Alexander Donkov, Julia Genova, Albena Paskaleva, Emilia Pecheva

International Programme Committee

D. Alexandrov (Thunder Bay), O. Batischev (Moscow), H. Chamati (Sofia), V. Čelebonović (Belgrade), D. Dimova-Malinovska (Sofia), A. Iglič (Ljubljana), D. Nesheva (Sofia), A. Nesrullajev (Mugla), A. G. Petrov (Sofia), Stephen Reynolds (Dundee), N. Tonchev (Sofia), Ph. Vanderbemden (Liege).

Local Scientific Council

Bulgarian Academy of Sciences: S. Vodenicharov (President), O. Stoilova (Nanosciences, New materials and Technologies)
Georgi Nadjakov Institute of Solid State Physics: H. Chamati, A. G. Petrov,D. Nesheva, N. Tonchev, V. Vitkova
Central Laboratory of Solar Energy and New Energy Sources: D. Dimova-Malinovska
Sofia University - Faculty of Physics: V. Donchev, E. Valcheva (Heads of Departments).
Plovdiv University Paisii Hilendarski - Faculty of Physics and Engineering Technologies: T. Yovcheva (Head of Department).

IOP Conf. Series: Journal of Physics: Conf. Series **794** (2017) 011003 doi:10.1088/1742-6596/794/1/011003

Peer review statement

All papers published in this volume of *Journal of Physics: Conference Series* have been peer reviewed through processes administered by the proceedings Editors. Reviews were conducted by expert referees to the professional and scientific standards expected of a proceedings journal published by IOP Publishing.

IOP Conf. Series: Journal of Physics: Conf. Series **794** (2017) 012001 doi:10.1088/1742-6596/794/1/012001

Nanostructured ZnO - its challenging properties and potential for device applications

D Dimova-Malinovska
Central Laboratory of Solar Energy and New Energy Sources, Bulgarian Academy of Sciences, bulv. Tzarigradsko chaussee, 72, 1784 Sofia, Bulgaria

E-mail: dmalinovska@yahoo.co.uk

Abstract. Nanostructured ZnO possessing interesting structural and optical properties offers challenging opportunities for innovative applications. In this lecture the review of the optical and structural properties of ZnO nanostructured layers is presented. It is shown that they have a direct impact on the parameters of devices involving ZnO. An analysis of current trends in the photovoltaic (PV) field shows that improved light harvesting and efficiency of solar cells can be obtained by implementing nanostructured ZnO layers to process advanced solar cell structures. Because of amenability to doping, high chemical stability, sensitivity to different adsorbed gases, nontoxicity and low cost ZnO attracted much attention for application as gas sensors. The sensitivity of nano-grain ZnO gas elements is comparatively high because of the grain-size effect. Application of nanostructured ZnO for gas sensors and for increasing of light harvesting in solar cells is demonstrated.

1. Introduction

ZnO is a semiconductor with large optical band gap 3.0-3.4 eV and resistivity between 1.10^{-3} and 1.10^{12} Ω cm depending on doping and the methods of deposition [1]. Recently, ZnO films have attracted high interest in the scientific society. They offer various advantages: i) They consist of cheap and abundant elements, ii) They can be produced by large scale coating, iii) They allow tailoring of the ultraviolet absorption, iv) They have high stabilities in a hydrogen plasma, v) They can be deposited at low growth temperatures.

ZnO is a compound occurring in the nature as the mineral zincite discovered in 1810 by Bruce in Franklin (New Jersey, USA) [1]. Regarding the historical application and development of the investigaton of ZnO several periods could be pointed, as presented in table 1. In the bronze era ZnO has been produced as co-product of copper ore smelling and has been used for healing of wounds and its application in the pharmacy industry continues till now. In the middle age ZnO usage for production of brass, as a white paint in the vulcanization process, in the porcelain enamels production has started. This period can be regarded as "prehistorical". The next period between 1900 -1960 is related to the beginning of the first scientific research and dissemination of the results - the first scientific paper with results of ZnO study was published in 1912 [2], it is devoted to investigation of the electrical properties of ZnO cylindrical rods. Study mainly of the structural and electrical properties mainly of sintered polycrystalline ZnO samples has been performed [3-7]. Fabrication of ZnO thin films by evaporation has been object of study, as well [4, 5]. In 1955 the effect of internal negative photo-effect has been reported by Borissov M and Kunev S [6]. This period named "Basic research period", can be regarded as „incubation" period which "opens the door" for the next period

Table 1. Periods of development of the researches devoted to ZnO

Year	Development steps
Bronze Age	1. **Prehistorical Period**: Zn - Byproduct of copper ore smelting; used for healing of wound.
Middle age	2. **Middle age period: First industrial applications**: Application in the production of brass (before its replacing by Zn in Cu-Zn alloy).
1800	White paint, in rubber for activation of the vulcanization process, in porcelain enamels-till the middle of the 19th century.
1810	Mineral zincite discovered by Bruce in Franklin (New Jersey, USA).
1900-1960	3. **Basic research period - incubation period; Study of the sintered polycrystalline samples mainly.**
1912	First published scientific paper: Somerville A A 1912 *Phys. Rev. I* **34** 311 [2].
1930	Study of dependence of the electrical properties and stoichiometry, Wagner C and Schottky W 1930 *Z. Phys. Chem. B* **11** 163 [3].
1935	Polycrystalline plates, single crystalline needles and evaporated thin films, Fritsh O 1935 *Ann. Phys.* **5** 375 [4].
1951	ZnO thin films by Zn oxidation, Schalimova K V *Compt Rendus Russ. Acad. Sci.***78** 1127[5].
1954	Negative internal photoeffect in ZnO: Borissov M, Kunev S. 1954 *Compt Rendus Bulg. Acad. Sci.* **7** 21; Borissov, Kunev S. 1955 *Z. Phys.Chem.* **205** 56[6].
1959	Review paper: Heiland G, Mollwo E 1959 *Solis State Phys.* **8** 191 **[7].**
1960-2005	4. **Development of new ZnO thin film technologies and intensive basic research for applications.**
1960	Piezoelectrical properties: Hutson A R 1960 *Phys. Rev. Lett.* **4** 505 [8].
1962	Application as gas sensor: Seiama T, Kato A, Fujiishi K and Nagatani M 1962 *Anal. Chem.* **34** 1502 [9].
1971	Application for varistors: Matsuoka M 1971 *Jpn.J.Appl.Phys.***10** 736 [10].
1976	First electronic application as thin layer for surface acoustic wave devices (SAW): Hickernell 1976 *Proc.IEEE* **64** 631 [11].
1996	Application as phosphor: Vanheusdalen K, Seager C H, Warren W L Tallant D R and Voigt J A 1996 *Appl. Phys. Lett.* **68** 403 [12]
1999	Application as TCO in thin film a-Si:H solar cells: Rech B and Wagner H *Appl. Phys. A* **69** 155 [15].
1999	In flat plate panel: Katayama M 1999 *Thin Solid Film* **341** 140 [13].
1999	ZnO nano-wire growth and application in a-Si:H solar cell: Paporte T and Lincot D 1999 *Appl. Phys. Lett.* **75** 3817 [16], Kopenkampt R, Boedecker K, Lux-Steiner M C, Poschenrieder M, Zenia F, Levy-Clement C 2000 *Appl. Phys. Lett.* **77** 2575 [19].
2002	In OLED: Hung L S and Chen C H 2002 *Mater. Sci. Eng.* R **39** 143 [14].
2004	In thin film transistor: Nomura K, Ohta H, Takagi A, Kamiya T, Hirano M, and Hosono H 2004 *Nature* **432** 488 [18].
21st Century	**5. Nanostructured ZnO: Demonstration and large industrial nanostructured ZnO applications period ????**

"Development of ZnO thin film technologies and intensive based researches for application". In this period (1960-2005) the potential for different applications of ZnO has been demonstrated which has passed through a scientific and technical renaissance in the last few years. Zinc oxide has been recognized as one of the most important semiconductor materials for piezoelectricity [8], gas sensing [9], electronics [10] surface acoustic wave devices [11], optoelectronics [12-14], solar cells [15-16], bio-applications [17], transparent thin film transistor [18], etc.

The first publications on ZnO nanowires growing [16, 19, 20] together with the study of the polycrystalline ZnO films with nanosized grains [15, 21] have stimulated the development of different technologies for their fabrication and enlargement of the area of research. Nanostructured ZnO possessing interesting structural and optical properties offers challenging opportunities for innovative applications. Will it be the next period of "Demonstration and large industrial ZnO applications?

An analysis of current trends in the PV field shows that improved light harvesting and efficiency can be obtained by implementing nanostructured ZnO layers to process advanced solar cell structures. Because of amenability to doping, high chemical stability, sensitivity to different adsorbed gases, nontoxicity and low cost, ZnO attracted much attention for application as gas sensors. In this contribution a review of the optical and structural properties of ZnO nanostructured layers fabricated by RF magnetron sputtering and electrochemical growth is presented. It is shown that they have a direct impact on the parameters of devices involving ZnO. Application of nanostructured ZnO for gas sensors and for increasing of light harvesting in solar cells is demonstrated, as well.

2. Optical, structural properties and potential for application of polycrystalline ZnO thin films in gas sensors

2.1. Nano-polycrystalline ZnO thin films deposited by RF magnetron sputtering

Polycrystalline ZnO thin films can be deposited by different physical and chemical methods: thermal evaporation in vacuum, oxidation of Zn, ZnS, ZnSe thin films, MOCVD, e-beam evaporation, DC and RF magnetron sputtering, ALD, spray pyrolysis, pulsed laser deposition, etc. [1 and ref. there]. The choice of the method of deposition should be chosen regarding the goal of their application, the properties requested for this application and, as well as, the cost of the method of their production should be limiting factor.

For some applications the ZnO films have to be doped. Doping with the III group elements (Al, Ga, In) leads to decreasing of the resistivity and good transparency of the thin films [20, 21]. Doping of ZnO polycrystalline thin films with Er, Ta, Co deposited by magnetron sputtering results in increasing of the sensitivity of the films to different gases, such as NH_3, CO_2, as well [22, 23]. The results of study of their structural properties by X-Ray Diffraction (XRD) spectroscopy [21-23] demonstrate that the deposited films are polycrystalline. Reflections corresponding to the (002), (110), (102), (110), (103), (112) and (004) planes of wurtzite ZnO are observed as shown in figure 1 a, b and c [20-23].

The value of the coefficient of absorption, α, of the samples is obtained from the spectral dependence of transmission and reflection. The value of the optical gap, E_g, is calculated using the dependences (shown in figure 1 d, e and f) of $(\alpha h\nu)^2$ *vs.* $h\nu$ by the formula [24]:

$$\alpha (h\nu) = B\,[(h\nu - Eg)^{1/2} / h\nu] \tag{1}$$

The coefficient B was assumed as a parameter for the band tail states [25, 26]. According to [10], a decrease in B implies an increase in the disorder of the structure and a consequent increase in the conduction band tail width. In our case, a decrease in B is observed with doping and with T_s decreasing, which is accompanied by a deterioration of the crystalline structure of the ZnO films.

In the lower energy range where α varies exponentially with photon energy the spectral dependence of the absorption edge depends on energy, $h\nu$, by the Urbach formula [25]:

$$\alpha (h\nu) = \alpha_0 \exp\,[(h\nu - E_1)/E_0] \tag{2}$$

where α_0 is the Urbach absorption at the edge (E_l), and E_0 is the Urbach energy width, which is believed to be a function of the structural disorder.

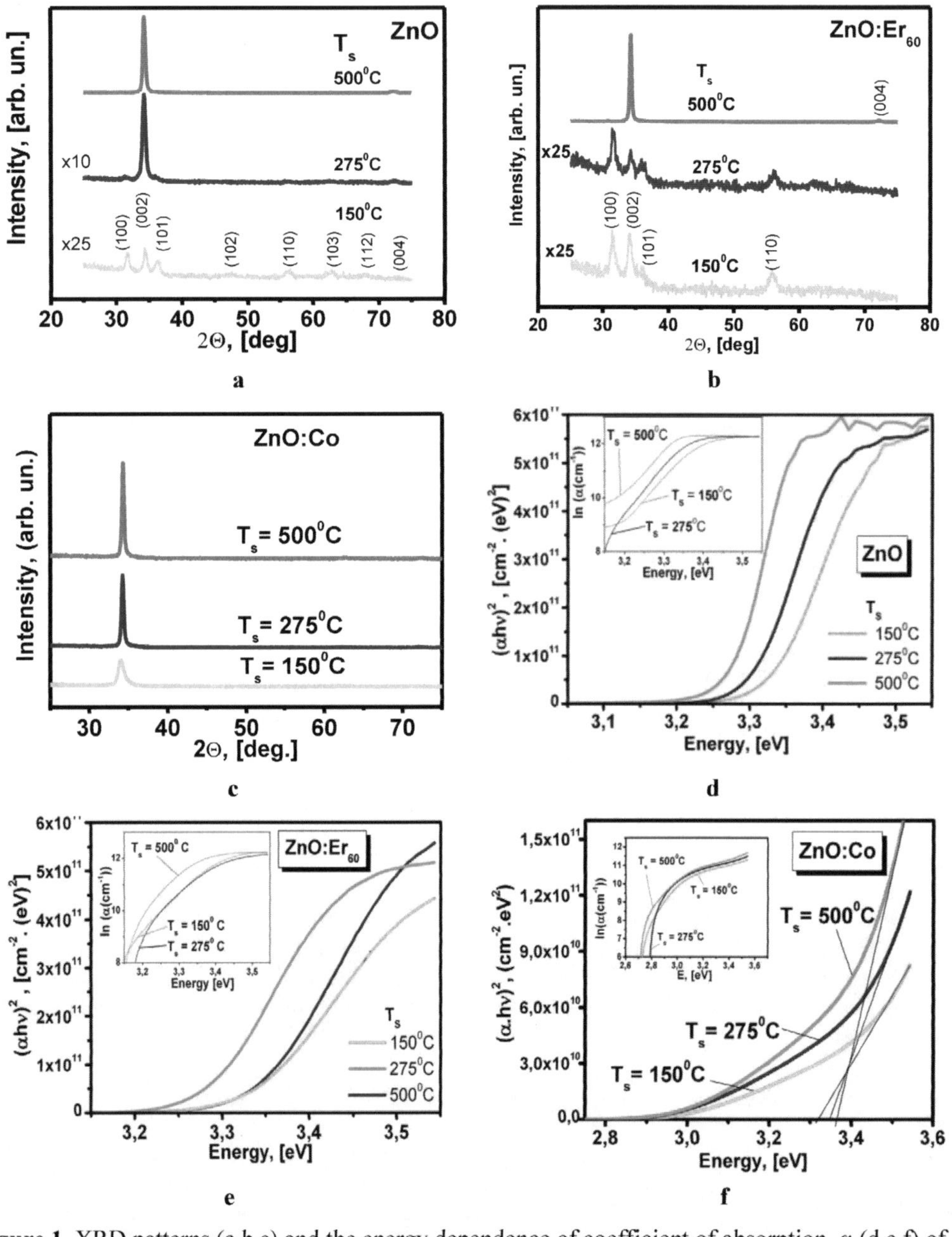

Figure 1. XRD patterns (a,b,c) and the energy dependence of coefficient of absorption, α,(d,e,f) of different series of ZnO films, deposited at different substrate temperatures T_s: ZnO (a and d), ZnO:Er (0.2at.%) (b and e) and ZnO:Co (4-5 at%) (c and f). The insert shows a plot of lnα vs.$h\nu$ [22, 23].

The exponential dependence of the absorption on $h\nu$ in the Urbach region ($h\nu < Eg$) is due to the perturbation of the parabolic density of the states at the band edge – increasing structural disorder results in an increase in E_0 [21-27].

Table 2 contains the data obtained from the analyses of XRD and optical spectra for undoped and doped ZnO samples with Al, Er, H, Ta and Co [21-27]. The (002) peak shows deviations from 34.44^0, which is the value for ZnO powder. Larger deviations were observed for ZnO films deposited at lower T_s and those doped with Al, Er and Ta. They indicate that the interplanar spacing changed relative to that of ZnO powder, which is probably due to factors such as lattice strain and interstitial defects due to entrancing defect and the doping as well [28]. The values of the stress, σ, derived from the XRD data were calculated using the following equation [29]:

$$\sigma = -233\,(c_{film} - c_{bulk})/c_{bulk}, \tag{3}$$

where c_{film} and c_{bulk} are lattice parameter of c axis for the films (evaluated using the Fullprof software from XRD spectra [30]) and of the bulk hexagonal ZnO (c_{bulk} = 0.5206Å) [29 and ref. there in].

Figures 2 a and 2 b display the relation between the values of Urbach energy (a) and the coefficient B^2 (b) in dependence on the stresses, σ, respectively, for all of the ZnO films under investigation. The changes of the structural properties of magnetron sputtered ZnO thin films (undoped and doped with

Table 2. The values of 2θ of (002)XRD peak and its FWHM , the average grain sizes, D, the strain, Δc, the stress, σ, the optical band gap, E_g, the Urbach energy, E_0, and the coefficient B for different series of ZnO films [27]. The concentration of the dopants is given in the brackets.

Samples description	T_s, ^{0}C	2θ, deg	FWHM deg	D, nm	Δc_{002}, %	σ, GPa	E_g, eV	E_0, meV	B^2, (cm^2.eV)
ZnO	150	34.30	0.52	16.0	0.34	-0.81	3.33	63	3.90E12
ZnO	275	34.40	0.47	17.7	0.08	-0.18	3.30	59	4.70E12
ZnO	500	34.40	0.33	25.2	0.08	-0.18	3.27	60	6.30E12
ZnO:Al (1 at.%)	150	34.20	0.62	13.6	0.65	-1.52	3.36	100	2.60E12
ZnO:Al (2 at.%)	275	34.15	0.61	13.4	0.80	-1.79	3.41	120	7.60E11
ZnO:Al (1 at.%)	500	34.27	0.62	13.4	0.40	-0.98	3.33	91	2.26E12
ZnO:Er (0.2at.%)	150	34.10	0.62	13.4	0.92	-2.15	3.33	72	2.47E12
ZnO:Er (0.2at%)	275	34.30	0.62	13.4	0.34	-0.81	3.34	70	3.29E12
ZnO:Er (0.2at%)	500	34.20	0.57	14.5	0.65	-1.52	3.28	63	3.15E12
ZnO:Er (0.4at%)	150	34.16	0.59	14.0	0.80	-1.79	3.34	65	3.14E12
ZnO:Er (0.4at%)	500	34.11	0.53	15.7	0.88	-2.06	3.30	52	3.15E12
ZnO:H (2 at.%)	150	34.42	0.43	19.3	0	0	3.31	70	5.30E12
ZnO:H (2 at.%)	275	34.37	0.48	17.3	0.15	-0.36	3.31	56	4.25E12
ZnO:H (1 at.%)	400	34.26	0.47	17.7	0.46	-1.07	3.27	56	7.10E12
ZnO:Ta (1 at.%)	150	33.81	0.79	10.5	1.77	-4.12	3.32	78	3.56E12
ZnO:Ta (1 at%)	275	33.86	0.40	20.8	1.61	-3.76	3.31	80	3.01E12
ZnO:Ta (1 at.%)	500	34.02	0.29	28.6	1.15	-2.68	3.31	73	4.50E12
ZnO:Ta (2 at.%)	150	33.09	1.21	6.80	3.92	-9.13	3.44	100	2.00E12
ZnO:Ta (2 at.%)	275	33.47	0.88	9.42	2.77	-6.44	3.44	95	3.77E11
ZnO:Ta (3 at.%)	500	34.07	0.45	18.5	0.52	-2.33	3.37	91	1.57E12
ZnO:Co (5at.%)	150	34.09	0.81	10.2	0.50	-2.24	3.28	118	9.7E11
ZnO:Co (4.8at.%)	275	34.33	0.28	30.0	0.22	-0.98	3.31	114	5.54E11
ZnO:Co (4.8at.%)	500	34.33	0.25	33.0	0.14	-0.63	3.33	104	1.10E11

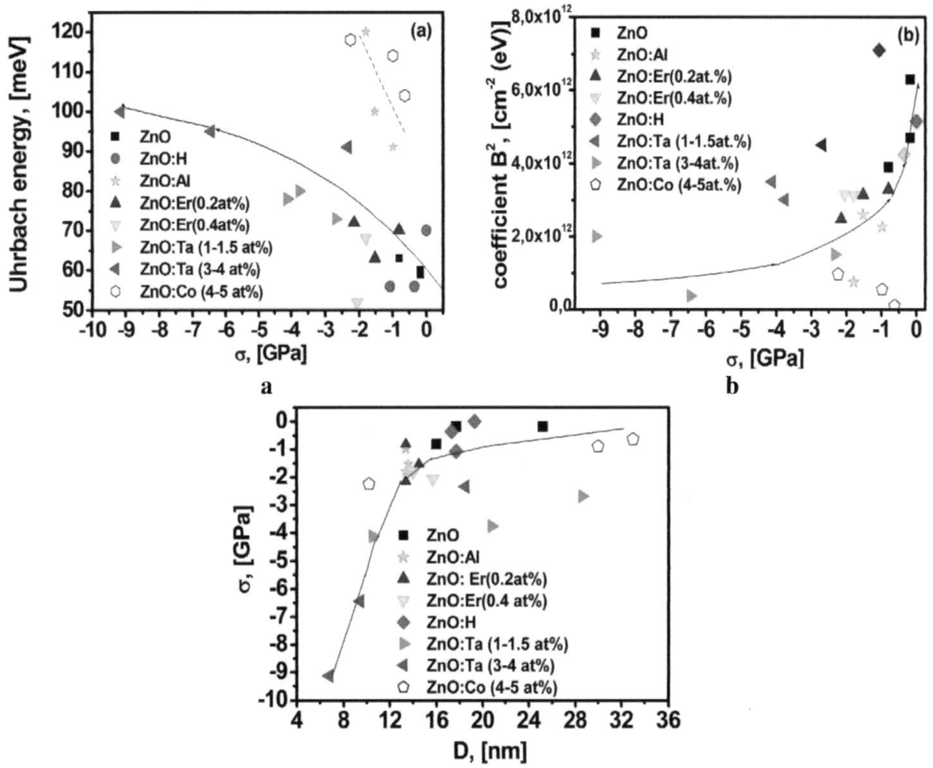

Figure 2. Dependence of the Urbach energy (a) and coefficient B^2 (b) on the stress and dependence of the stress on the grain size in ZnO, ZnO:Al, ZnO:Er, ZnO:H, ZnO:Ta and ZnO:Co films deposited under different conditions [27]. The lines are guided by eye.

Al, Er, Ta and Co, and deposited in H_2 presence) are compared with changes in the values of B and Urbach energy tail, E_0, calculated from the optical properties. The values of the stress, σ, are negative - this demonstrates the presence of tensile stress in the films. It is seen that there is a correlation between the stresses and the structural disorder in the films – the increase of the stress results in an increase of the Urbach band tail width and a decrease of the coefficient B. A correlation between the values of the stress and the band tails has been observed - increasing stress is accompanied by increasing of value of the band tails width. The correlation between the average grain size, D, (calculated by applying the Scherer equation to the Full Width of the Half Maximum (FWHM) of the (002) peak [31]) and the stress, shown in figure 2 c, demonstrates that the value of the stress decreases with increasing D. A similar result has been reported in [28]. The increase of the structural disorder due to the doping or to the lower deposition temperature is accompanied by decreasing of the average grain size and increasing of the stress and the strain in the ZnO films, which cause a perturbation of the parabolic density of the states at the band edges. The results can be explained having in mind that accommodation of the impurities in the lattice results in localised strains, reflected in a perturbation of the parabolic density of the states at the band edges.

2.2 Application of ZnO nano-polycrystalline films in gas sensors

ZnO is one of the first oxide materials, which have attracted the attention for their potential application as gas sensors due to their chemical sensitivity to different gases, high chemical stability, non-toxicity and low cost [23, 32-36]. The material properties are determined mostly by its structure and surface morphology, which strongly depend on the preparation method. In the last few years, ZnO has been an object of intensive investigation, particularly in its low dimensional structures – thin layers with nanometer crystallites, nano-wires, nano-rods or nano-tubes, nano-balls and others [36 -38].

The sensitivity of the ZnO films to NH_3 was determined by two approaches - as the ratio of the resistivity measured before and after exposure to NH_3 and by the frequency shifts of the quartz crystal microbalance (QCM) after gas exposure. The variation in the sensitivity (the ratio of the resistivity) of undoped and H doped ZnO films with NH_3 concentration is shown in figure 3 [23]. Initial trend of saturation at higher NH_3 concentrations (>18 000 ppm) is also observed. It is seen that the sensitivity of the ZnO:H films is higher than that of the ZnO films. The ZnO:H film deposited at T_s = 400^{0}C exhibited a higher sensitivity than that deposited at 150^{0}C, for NH_3 concentrations < 18,000 ppm.

It is well accepted that the sensitivity of semiconductor gas sensors is attributed to the chemisorption of oxygen and OH^- on the oxide surface and the subsequent reaction between adsorbed species and the tested gas, which causes the resistance change [39]. It is known that atmospheric oxygen molecules and water vapor are adsorbed on the surface of the ZnO in the forms of O^{2-} and OH^-, thereby reducing the electrical conductivity. When a ZnO sensor is exposed to a reductive atmosphere, the test gas like NH_3 reacts with the surface species, the trapped electrons are returned to the conduction band of the ZnO, causing an increase in the conductivity of the sensor. The different effective surface area of the films could be one of the reasons for their different sensitivity. However, the effective surface area estimated from the grain size (calculated form XRD spectra) is almost the same for all set of samples studies here. Thus, the results presented here require another explanation. The adsorbed hydrogen and OH^- species in the ZnO:H films during the deposition could explain the higher response of samples ZnO:H as compared to that of the pure ZnO films and the ZnO:H sample deposited at higher T_s. One of the possible reactions on the substrate during the process of sputtering in an $Ar+H_2$ atmosphere is:

$$ZnO + H_{2gas} \leftrightarrow Zn + H_2O_{gas} \quad (4)$$

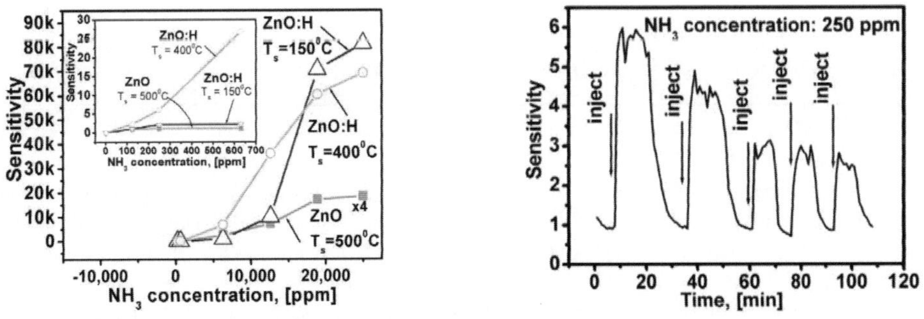

a b

Figure 3. (a) A sensitivity of ZnO:H and ZnO films toward NH_3 at room temperature. The insert is the sensitivity at lower NH_3 concentration. (b) Transient response of a ZnO:H film (T_s=400^{0}C) for a 250 ppm NH_3 concentration [23].

The thermodynamic analyses [39] show, that this reaction will be shifted to the right with increasing of the temperature. According to the equation (4) the deposition at 150^{0}C would results in higher concentration of H and at 400^{0}C – in higher water vapor presence in ZnO:H films. Possibly the presence of OH^- would accelerate the surface reaction between the reducing gas with NH_4OH as a product, additionally to the NH_3 reaction with adsorbed oxygen species. Thus, the samples with higher OH^- concentrations (deposited at T_s = 400^{0}C in $Ar+H_2$) would be more sensitive to a lower NH_3 concentration.

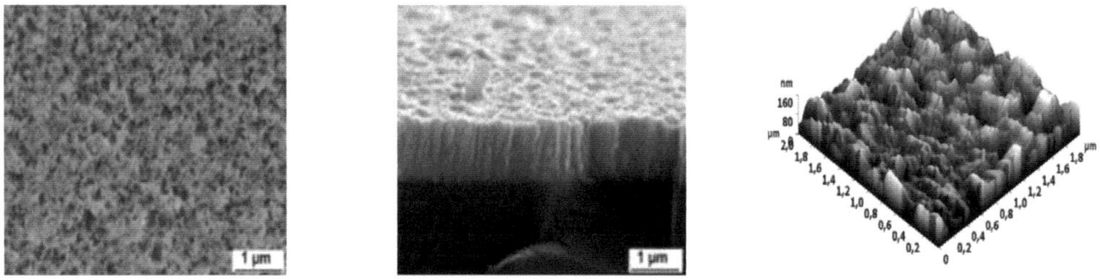

a b c

Figure 4. SEM pictures of ZnO film – top surface (a) and cross section (b) and AFM image (2 x2 μm) of ZnO film deposited by RF magnetron sputtering of ZnO target in an Ar atmosphere (c) [40].

The ability of the undoped ZnO thin films and those doped with Er, Ta and Co to sorb ammonia vapor is demonstrated by the QCM method [40]. For this purpose, the resonance frequency shift response was measured. The sorption process was investigated over an aqueous solution of ammonia with 100 ppm concentration. The change in the frequency, Δf, of the QCM, as a function of its mass loading is detected. Surface morphology and cross section images of ZnO film used for this purpose (deposited by RF magnetron sputtering in Ar atmosphere) are shown in figure 4. The column structure of the film is clearly seen with size of the nanocrystallites of about 100 nm. Figure 5 shows the frequency change, Δf, as function of the exposure time over a 100 ppm ammonia solution for different samples. The highest rate of the change in frequency, S_p, is registered in the first two minutes, for all investigated samples. After that, the rate slowly decreases and the frequency approaches to a constant value. The rate of change for the first two minutes is calculated in order to estimate the sensitivity of each kind of ZnO layer. The S_p value and the total sorbed mass, Δm, was evaluated using the relationship between Δf and Δm for AT- cut of quartz crystal given by the Sauerbrey equation [41]:

$$\Delta f = - (2.26 \times 10^{-6} f^2 \Delta m)/A, \tag{5}$$

where f (MHz) is the frequency of the QMB before exposure to ammonia, Δm (g) is the mass of the sorbed gas on the surface of the ZnO films and A (cm^2) is the surface area of the electrodes.

The total sorbed mass on the samples is calculated for the whole period until saturation is achieved. The data are presented in table 3. The lowest rate of sorption and the smallest total sorbed mass (4.5 Hz/min and 4.86 ng, respectively) are registered for ZnO. In the case of ZnO films doped with Ta, and Co, values of 8.5 Hz/min and 8.76 ng and 5.65 Hz/min and 12.78 ng for the sorption rate and mass, respectively, are obtained. It has to be noted that the largest rate of change of the frequency – 13.2 Hz/min is measured in the case of sorption on the ZnO:Er layers, which corresponds to a total sorbed mass of 32.20 Higher sensitivity was observed for Er doped ZnO films, compared to the undoped and ZnO:Co and ZnO:Ta films. The differences in the grain size in the films could be one of the reasons for the different behaviour of the QCM sensors. The higher sorbed mass is detected for the sample with smaller grain size which possesses larger effective surface area leading to higher concentration of the adsorption centres.

3. Properties and potential for application of electrochemically grown nanostructured ZnO

ZnO nano-rods, nanowires, nano-dots represent important class of nanostructures which has attracted the attention of the scientific society due to their very interesting and unique properties with potential for various device applications. In particular the nano-materials with suitable size of their nano-structures could improve the light harvesting by solar cells and thus their efficiency. Different methods of deposition of nanostructured (NS) ZnO layers have been applied, such as chemical vapour deposition, metal-organic CVD, aqueous solution growth, electro-chemical deposition (EChD), etc.

Table 3. Values of the optical band gap, E_g, the average grain size, *D*, the rate of sorption, S_p, and the sorbed mass, Δm, over 100 ppm NH_3 for different kinds of samples. The valued of *D* are calculated using values of 2Θ (given in the brackets) and the corresponding FWHM [40].

Samples	E_g , eV	D, Å, (2θ)	S_p, Hz/min	Δm, ng
ZnO:Er	3.30	95 (002)	13.2	32.20
ZnO:Co	3.35	105 (002)	5.65	12.78
ZnO:Ta	3.46	131 (015)	8.5	8.76
ZnO	3.30	177 (002)	4.50	4.86

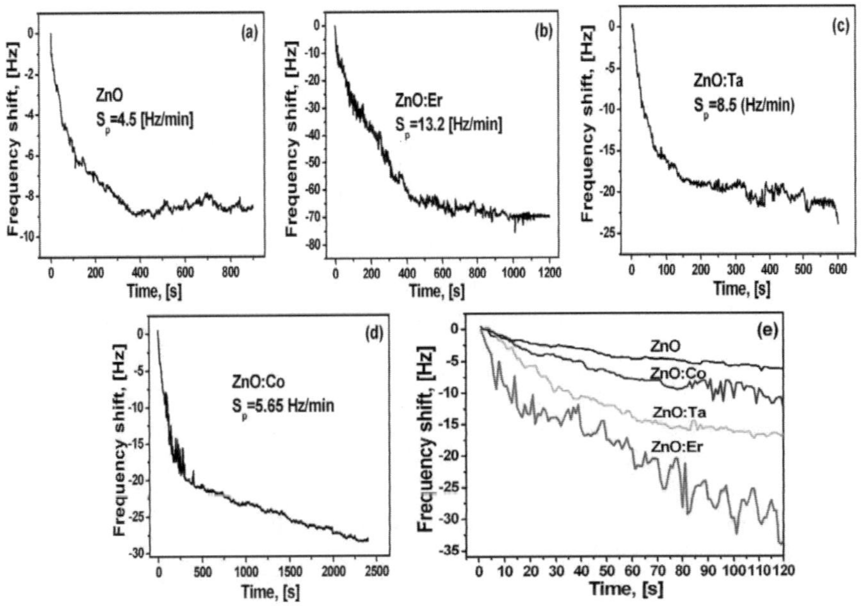

Figure 5. Time-frequency shift characteristics of quartz crystal microbalance with: (a) ZnO, (b) ZnO:Er, (c) ZnO:Ta, (d) ZnO:Co films; (e) presents the frequency shifts for the all samples for 120 sec. over a 100 ppm NH_3 solution.over a 100 ppm NH_3 solution [40].

3.1. Correlation between the surface roughness of NS ZnO and haze ratio in optical reflection

The method of EChD has some advantages - it is a low temperature method the ZnO nanostructures can be deposited on very large area and it is applicable in industrial conditions. ZnO nanostructured

Figure 6. SEM plane views of SnO_2:F film (a) and ZnO layers; samples: 2 - (b), 6 - (c), 7 - (d) and 9 - (e) [43]. The markers correspond to 500 nm.

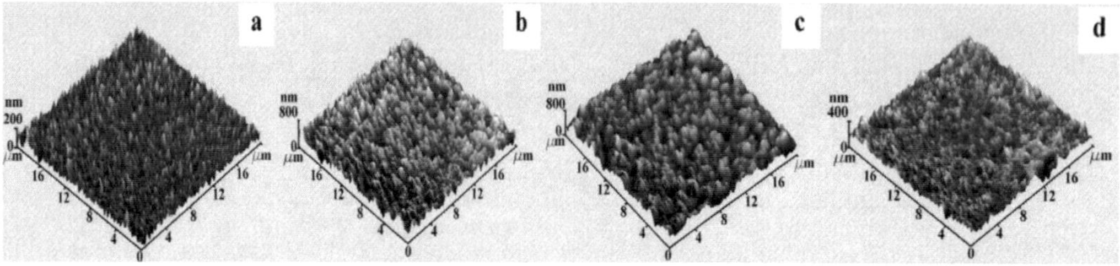

Figure 7. AFM images of SnO_2:F film (a) and ZnO layers - samples: 6 - (b), 7 - (c) and 9 – (d) [42].

films were deposited by an electrochemical process from acid aqueous solution of $ZnCl_2$ (5. 10^{-3} M) and KCl (0.1 M) with pH 4.0 at temperature of 60^0C in flowing air and suspension containing ZnO powder as precursors using a three-electrode electrochemical cell and saturated calomel electrodes (SCE) as reference electrode described earlier in [42,43]. It is supposed that ZnO powder suspension could serve as a source of seeding particles. For comparison a layer without adding of the ZnO suspension was prepared as well (sample 2). Duration of the ZnO deposition was 60 min. The prepared ZnO films were 0.55 – 1,57 μm thick. Different regimes were explored – sample 2 – without drops, (dr.), sample 3– 1x dr., sample 4– 2x dr., sample 5– 2x dr. + H_2O_2, samples 6 and 6'– 3x dr., samples 7 and 7'– 3x dr. + annealing at 300 °C, sample 8– 3x dr. + H_2O_2 and sample 9– 3x dr. + Ni seeding layer. In the case of sample 9 thin Ni seeding layer (1-2 nm thick) was chemically deposited on the surface of the glass/SnO_2:F substrate before ZnO EChD growth. The surface morphology was studied by SEM (figure 6) and AFM (figure 7) and the thickness of the film was determinate by AFM. The crystalline structure was studied by X-Ray Diffraction spectrometry (XRD) (figure 8). The XRD patterns demonstrate that the deposited ZnO films are polycrystalline with reflection corresponding to the (100), (002), (101), (102) and (103) planes of wurtzite structure with c-axis perpendicular to the substrate. The most intensive is the XRD peak of the (002) plane. From the FWHM of the (002) peak, $\Delta 2\theta$, the average size of the grains, D, is estimated. The value of the optical band, E_g, is calculated using transmittance and reflectance spectra [43]. The calculated values are presented in table 4.

The ZnO layer grown without adding of ZnO suspension (sample 2) consists of grains with size of about 56 nm (very inhomogeneous grains size, D) and has an average surface roughness of ~82 nm (figure 7b). It is seen from figures 6 and 7 that the grain size in the layers increases with adding of different quantity of the suspension which is in agreement with the obtained from the XRD data (table 4). Annealing of the film at 300°C in air for 1 hour results in changes in the morphology – the

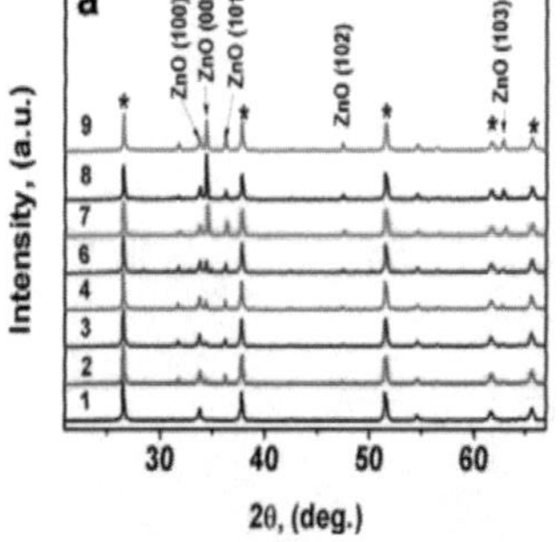

Figure 8. XRD spectra of the ZnO layers deposited at different conditions [43].

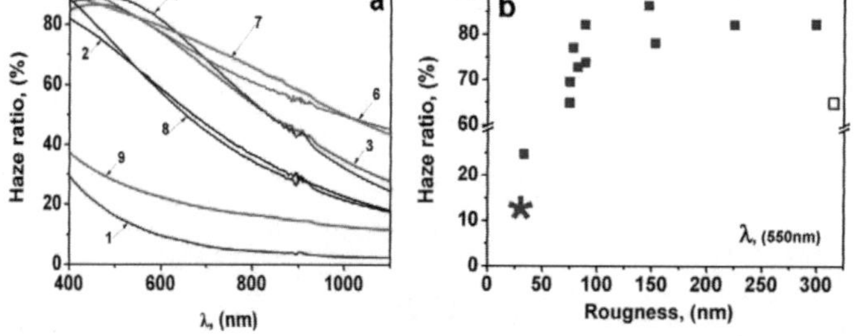

Figure 9. Spectral dependences of haze ratio in reflection (a) and its dependence (measured at 550 nm) on the surface roughness (b) of ZnO layers deposited at different conditions. The signs* corresponds to the value of the SnO_2:F substrate [43].

Table 4. Structural and optical characteristics of the electrochemically deposited ZnO films: the optical band gap, E_g; the position of the (002) peak in the XRD spectra, 2θ; the FWHM of (002) peak, $\Delta 2\theta$; the values of the stress, σ; the average grains size, D; the thickness, d; the average roughness, R; the haze ratio. For comparison the corresponding data for the SnO_2:F films are given, too [43].

Sample	Eg, eV	2θ, deg.	FWHM deg.	σ, GPa	D, nm	d, nm	R, nm	Haze, (%) 400nm	Haze, (%) 550nm	Haze, (%) 1000nm
1: SnO_2:F	3.75	34.39	0.14	-0.25	59	910	29	30	13	3
2: w/t drops	3.39	34.39	0.15	-0.24	56	550	315	82	65	23
3: 1 x drops	3.37	34.41	0.16	-0.12	52	580	89	85	82	35
3':1x drops	3.38	34.38	0.14	-0.32	59	620	89	86	74	24
4: 2x drops	3.38	34.37	0.10	-0.38	83	700	147	88	86	34
5:2xdr+H_2O_2	3.46	34.39	0.12	-0.25	69	1570	82	90	73	29
6: 3x drops	3.42	34.39	0.11	-0.25	76	750	299	93	82	50
6':3x drops	3.41	34.34	0.13	-0.58	64	1000	225	91	82	50
7: 3xdr.	3.26	34.48	0.13	-0.36	64	1060	153	86	78	46
7':3xdr+ann.	3.27	34.50	0.13	-0.47	64	850	78	91	77	31
8:3xdr+H_2O_2	3.47	34.38	0.12	-0.32	69	900	75	89	65	22
9:3x dr.+Ni	3.51	34.38	0.12	-0.32	69	810	33	38	25	13

individual grains grow into conglomerates (figure 7 d) and the surface roughness of this sample decreases.

The spectral dependence of the haze ratio of the ZnO samples in the range of 400 - 1000 nm is presented in figure 9 a. The values of the haze ratio of the electrodeposited ZnO layers are much higher than those of the SnO_2:F coating. Higher values of haze ratio are observed for the samples deposited with higher quantity of the added ZnO suspension. It has to be noted that the layers with higher values of the average roughness determined by AFM (table 4) demonstrate haze ratio in reflection. Figures 9 b show the variation of the haze ratio with the roughness of the deposited films at two wavelengths of 550 nm. It is seen very clearly (from figure 9 b and table 4), that the value of the haze ratio increases with increasing the surface roughness with maximal values of about 93%, 86% and 50% at 400 nm, 550 nm and 1000 nm, respectively. The results demonstrate that higher values of haze ratio in reflection in the spectral range of 400 - 1100 nm have ZnO layers with larger average grain size and higher average surface roughness. ZnO layers with similar properties could be applied as light trapping structures in thin film solar cells.

3.2. Application of ZnO nanostructured array layers in solar cells

3.2.1 Application as ARC on SHJ solar cells

Electrochemically deposited ZnO nanostructured layer with nanometre sized features have been

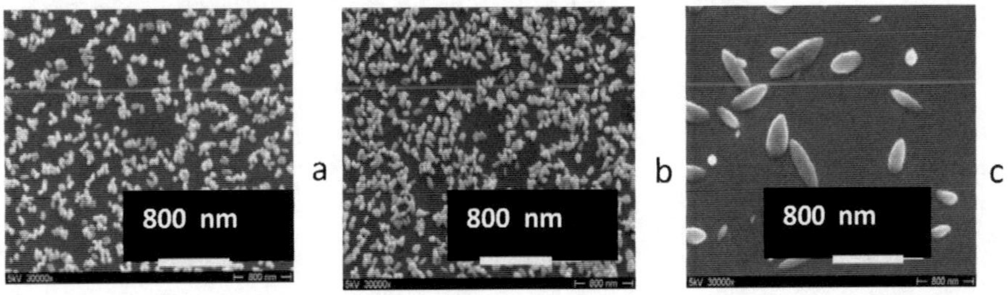

Figure 10. SEM of the ZnO NS array deposited on the samples 1-C6 (a) 1-C4 (b) and 1-C2(c) [45].

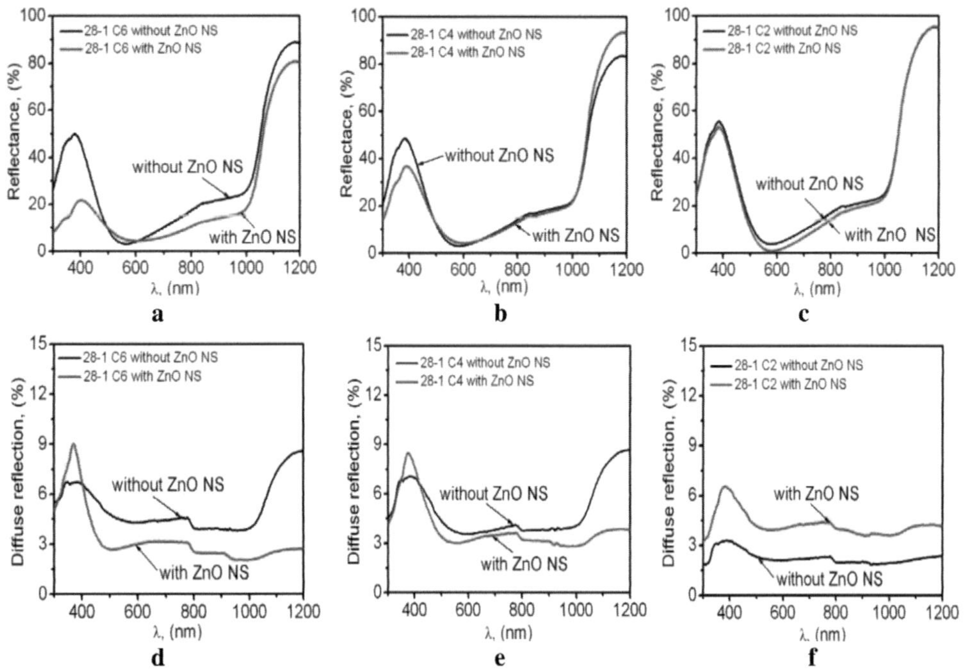

Figure 11. Spectra of specular reflectance (a, b, c) and diffused reflection (d, e, f) of samples 1-C6 (a, d), 1-C4 (b, e) and 1-C2 (e, f) without and with deposited ZnO NS array film [45].

Table 5. The deposition conditions of the ZnO NS array films and comparison of the parameter of the SHJ without and with ZnO NS growth [45].

Solar cells parameters	1-C2	1-C4	1-C6
I_{sc}, (mA), with ZnO NS	30.637	30.689	30.237
I_{sc} ,(mA), without ZnO NS	30.514	29.839	28.979
ΔI_{sc} ,(mA)	0.123	0.850	1.258
ΔI_{sc} ,(%), relative increasing	0.4	2.85	4,3
V_{oc}, (mV),with ZnO NS	695	694	689
V_{oc} ,(mV), without ZnO NS	697	697	691
FF, (%), with ZnO NS	72.74	72.90	71.61
FF, (%), without ZnO NS	74.00	73.50	73.02
ΔFF	1.26	0.60	1.41
η, (%), with ZnO NS	15.492	15.535	14.917
η, (%), without ZnO NS	15.733	15.278	14.521
Δη, (%)	-0.241	0.257	0.396
Δη, (%), relative changes	-1.5	1.68	2.7
t, (min)	1.5	1.5	1.5
$I_{el.-chem.}$, (mA)	3	6	6
$U_{el-chem}$, (mV)	1000	1000	950

applied as antireflective coating (ARC) on the front surface of heterojunction Si solar cells (SHJ) – Ag grid/ITO/p-a-Si:H/(i)a-Si:H/(n)/c-Si/(i)/a-Si:H/(n+)/a-Si:H/Al [44,45]. The ZnO NS were deposited for 1.5 min with applying of different electrochemical potential: in the case of solar cell 1-C6 – at 950 mV, of 1-C4 and 1-C2 - at 1000 mV. The ZnO nanostructures have been deposited between the fingers of the Ag grid electrodes on SHJ solar cells with sheet resistance of ITO of 80Ω/□ and different thickness of ITO (80 nm in the case of 1-C6 and 1-C4 and 150 nm for 1-C2, respectively). In the case of samples 1-C6 and 1-C4 the nano-rods sized about 100-200 nm are distributed in different surface area with different density. The SEM of the sample 1-C2 reviles the presence of largest nano-ZnO features about 800 nm with relatively large distance between them. Differences in the ZnO nanostructures distribution on the surface of different solar cells can be explained by the difference in the surface properties, thickness and resistivity of the ITO layers and by the different conditions applied during electrochemical deposition on solar cells.

The spectra of specular reflectance and diffused reflection of the three samples of the series, before and after the deposition of ZnO nanostructured array films, presented in figure 11 show that the values of reflectance decrease which is more pronounced in the sample 1-C6. The value of diffused reflection decreases in the samples 1-C6 and 1-C4 and increases in the case of sample 1-C2. The differences in the surface morphology and microstructure of the textured ZnO films determine the difference in the optical reflection properties, as reported in [46].

The study demonstrates that the ZnO nanostructured array films grown on the front side of the SHJ

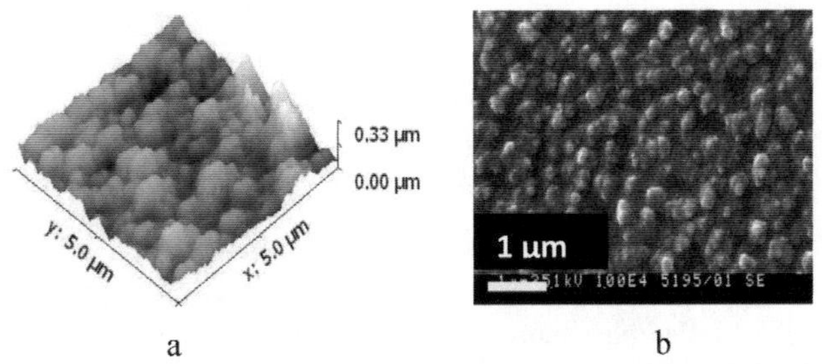

Figure 12. AFM (a) and SEM (b) imges of ZnO NRs array on stainless steel (SS) substrate [47].

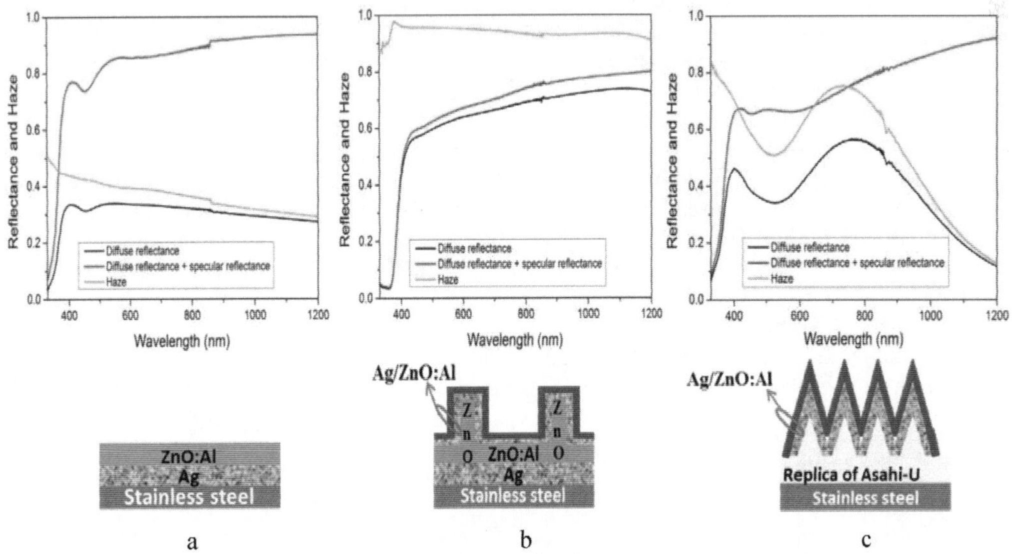

Figure 13. Spectra of diffused reflection, total reflection and haze in reflection of ZnO NRs array deposited on different structured SS substrates [47, 48].

Table 6. Parameters of thin film a-Si:H solar cells with different scattering substrates [47]

cells №	**Substrate structure**	**V_{oc}, mV**	**J_{sc}, mA/cm²**	**FF**	**Eff.**
1	SS/BR/NR	803	13.4	0.47	5.11%
2	SS/BR/NR/BR	850	13.8	0.62	7.28%
3	*SS/Replica of Asahi-U texture/BR*	851	12.4	0.62	6.56%

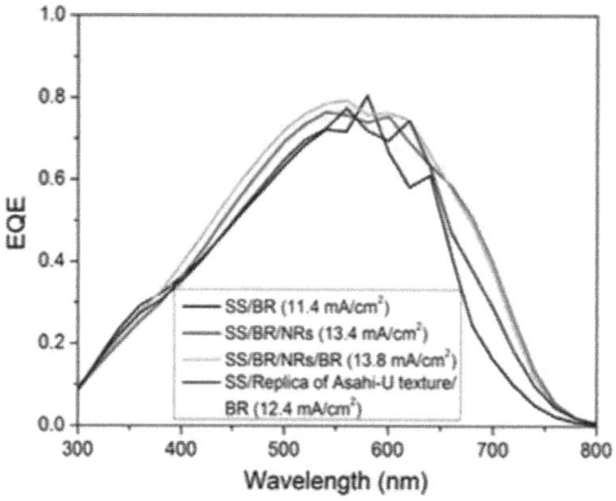

Figure 14. EQE spectra of thin film solar cells on different SS substrates [47].

solar cells act as ARC leading to decreasing of the reflectance, improvement of the short circuit current (by about 4.3 %) and efficiency (by about 2.7%) of the thin film SHJ solar cells (table 5). An improvement in the short circuit current and efficiency of the SHJ solar cells is achieved in the samples where the values of reflectance and diffuse reflection decrease after ZnO NS array films deposition on the front side. The higher increase in *Jsc* and efficiency, η, is observed in the samples where larger decreasing of the values of reflectance and diffuse reflection are obtained as a result of deposition of ZnO NS array films leading to improvement of the light harvesting [44].

3.2.2. ZnO nano-rod arrays as back reflector in thin film solar cells

The ZnO nano-rods (NRs) prepared by an electrochemical process are implemented in the a-Si:H thin films n+-i-p$^+$ solar cells on flexible stainless steel (SS) foil, aiming at increasing the J_{sc} without deterioration the open circuit voltage, V_{oc}, and the fill factor, *FF* [47]. Figures 12 a and b show the AFM and SEM images of ZnO NRs array on SS substrate. The surface morphology consists of nano-rods with diameter size of single NR between 200 and 500 nm and high of about 200-300 nm. The thin film solar cell structure deposited on such kind of flexible substrate is SS/back reflector (BR)/n-type a-Si:H/i-a-Si:H/p-μc-Si:H/ITO/Ag grid. The BR consists of a stack of Ag and AZO, with thicknesses of respectively 300 nm and 80 nm. In addition to the components of the basic BR (figure 13 a) there is the scattering medium consisting of either ZnO NRs coated by a stack of Ag (300 nm) and AZO (80nm) (figure 13 b) or a nano-imprinted replica of Asahi-U texture (figure 13 c).

The spectra of reflectance, diffused reflectance and haze in reflectance of three different back reflectors on SS substrate are shown in figure 13. The highest value of haze ratio in reflectance demonstrates the BR with ZnO NRs covered by two layers structure consisting of Ag (300 nm thick) and ZnO:Al (80 nm thick). The parameters of the solar cells obtained from I-V characteristics demonstrate that when the ZnO NRs are coated with Ag and AZO (cell № 2), the solar cell shows 0.4 mA/cm^2 higher J_{sc} than the reference cell of № 1 and 1.4 mA/cm^2 higher J_{sc} than the cell of № 3 deposited on substrate with the nano-imprinted replica of Asahi-U texture, and highest value of efficiency without deterioration of the V_{oc} and *FF* (table 6). The optical enhancement of the solar cell by ZnO NRs is explicitly demonstrated in the external quantum efficiency (EQE) spectra in the wavelength range from 400 nm to 800 nm (figure 14) [47,48].

3.2.3. Application of NS ZnO layer in gas sensors

The gas sensitivity of the electrochemically-deposited ZnO nanostructured (NS) thin films on QCM was studied by exposure to NO_2. The resulting structure was kept in air until saturation in the frequency value was reached. The temperature in the test chamber was kept constant (25 °C ± 0.5 °C) by a thermostat. The concentration of the NO_2 gas flowing into the thermal stabilized measurement chamber was 500 ppm. The resonance frequency was measured by a frequency counter an recorded on a PC. The experimental setup for measuring the sensing properties of the ZnO-QCM system was described in detail in [49]. The frequency accuracy was ~ 0.1 Hz. The oscillators consisted of thermo-stable polished AT-cut quartz plates with a diameter of 8 mm with gold electrodes with a thickness of 600 nm and a surface area of 0.1256 cm2 in an HC-49 type package. A 16-MHz base resonance frequency was thus obtained [50].

Figure 15 shows a SEM micrograph of a NS ZnO thin film deposited at constant charge density for 60 min. The structure of this thin film consists of planar hexagonal ZnO nano-whiskers (NW) with a thickness of about 100 – 300 nm and different lengths (0.5 – 1μm), stacked with the narrow side to the substrate surface. The thickness of the ZnO film is 0.5 μm. The SEM micrograph reveals the porous surface morphology of the deposited layer.

The ability of the ZnO thin films to sorb NO_2 was evaluated by the QCM method. For this purpose, the resonance frequency shift was measured. The sorption process was investigated in a gas flow of NO_2 with a concentration of 500 ppm by following the change in the frequency, Δf, of the QCM as a function of the mass loading. Figure 16 shows this change as a function of the exposure time. The highest rate of frequency change, S_p = 11.5 Hz/min, was registered in the first two minutes. After that, the rate decreased more slowly, with the frequency reaching a constant value within about five minutes after exposure to NO_2. When the gas flow was switched off, the frequency returned to its initial value within two minutes due to desorption of the gas, the fastest change occurring in the first minute. The rate of change was calculated for the first five minutes after exposure to the NO_2 in order to estimate the sensitivity of the electrodeposited ZnO layer. The average value of the frequency-change rate, S_p, and of the total sorbed mass, Δm, were evaluated on the basis of the time-frequency characteristics measured using the relationship between Δf and Δm for AT-cut quartz plates as given by the Sauerbrey equation (equation (5)) [41]. In our experiments, the frequency shift measured, Δf, was 38 Hz, and the sorbed mass, Δm, was calculated to be 8.39 ng. The S_p value of 7.6 Hz/min was calculated as the ratio of the frequency shift to the time of saturation (five minutes). The electrochemically-deposited ZnO layer with grown nano-whiskers had a highly developed surface interacting with the NO2 molecules. When the concentration of the gas in the chamber is raised, the NO_2 molecules are adsorbed in the ZnO layer; as a result, the resonator's mass increases and its frequency decreases. When the gas flow in the chamber is stopped, the NO_2 molecules start desorbing and the frequency of the quartz resonator returns to its initial value.

The results obtained demonstrated a potential application of the electrochemically-deposited nanostructured ZnO layers as room-temperature gas sensors of NO_2.

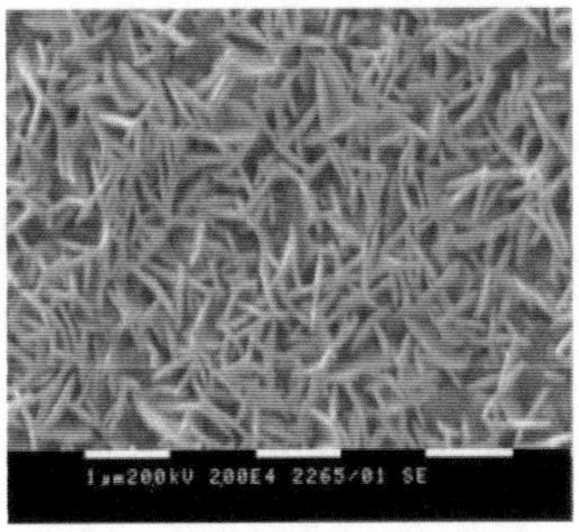

Figure 15. SEM image of a ZnO layer deposited electrochemically for 60 minutes on a quartz resonator [50].

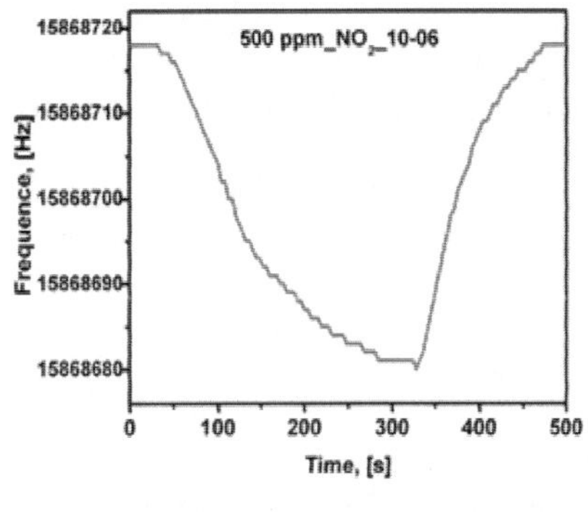

Figure 16. Frequency-time characteristic of a quartz crystal microbalance with an electrodeposited nanostructured ZnO layer under and after exposure to a 500-ppm NO_2 flow [50].

4. Conclusion

The described results in this review article show that ZnO nano-polycrystalline and nanostructures arrays layers offer very interestic and promising properties due to the nanoscale structures and larger developed effective surface area. The results obtained demonstrated a potential application of the magnetron sputtered undoped and doped ZnO thin films and electrochemically-deposited nanostructured ZnO layers as room-temperature gas sensors of NH_3 and NO_2.

The proposed novel solar cell architectures that make use of nanomaterials have demonstrated increasing light harvesting and improvement of the parameters of the solar cells as a result of the effects due to the nano-scaled structures. Future advances in and the rate at which large scale application of the new challenging effects using ZnO in development of new devices can be achieve, will depend entirely on the amount of resources the society chooses to invest and the scientific researches' efforts in the field of the nanotechnology. All this will determinate the next period as period of **Demonstration and large industrial applications of nanostructured ZnO.**

Acknowledgements

The work has been funded by the 7 FP of EU project NanoPV № 24331. The author thanks to assoc. prof. Dr V Georgieva, assist. Dr. K Lovchinov, eng. B Georgieva, for the assistance in the measurements of the materials and devices properties; to assist. M. Kamenova, eng. P Andreev, eng. A Rachkova, eng.. Nichev and eng. M Petrov for collaboration in deposition of the investigated materials; to Dr. O Angelov - for RF magnetron sputtering of ZnO undoped and doped thin films and to Dr M. Ganchev - for the partial technical support in equipment establishments for electrochemical deposition.

Reference

[1] Ellmer K and Klein A ZnO and its application 2008 *Transparent Conductive Zinc Oxide Basics and Applications in Thin Solar Cells,* ed K Ellmer, A Klein A and B Rech (Heidelberg: Springer-Verlag) ISBN 978-3-540-73611-0, p.1-33
[2] Somerville A A 1912 *Phys.Rev.*I **34** 311
[3] Wagner C and Schottky W 1930 *Z. Phys. Chem.* B **11** 163
[4] Fritsh O 1935 *Ann. Phys.* **5** 375
[5] Shalimova K V 1951 *Compt Rendus Russ. Acad Sci.***78** 1127 (in Russian)
[6] Borissov M, Kunev S. 1955 *Z. Phys.Chem.* **205** 56
[7] Heiland G, Mollwo E 1959 *Solid State Phys.* **8** 191
[8] Hutson A R 1960 *Phys. Rev. Lett.* **4** 505
[9] Seiama T, Kato A, Fujiishi K and Nagatani M 1962 *Anal. Chem.* **34** 1502
[10] Matsuoka M 1971 *Jpn.J.Appl.Phys.***10** 736
[11] Hickernell 1976 *Proc.IEEE* **64** 631
[12] Vanheusdalen K, Seager C H, Warren W L Tallant D R and Voigt J A 1996 *Appl. Phys. Lett.* **68** 403
[13] Katayama M 1999 *Thin Solid Film* **341** 140
[14] Hung L S and Chen C H 2002 *Mater. Sci. Eng.* R **39** 143
[15] Rech B and Wagner H 1999 *Appl. Phys. A* **69** 155
[16] Paporte T and Lincot D 1999 *Appl. Phys. Lett.* **75** 3817
[17] Popova R, Lukanov A, Ivanova I, Angelov O, Dimova-Malinovska D, Dushkin C and Mikli V 2010 *Nanoscience & Nanotechnology* **10** 166
[18] Nomura K, Ohta H, Takagi A, Kamiya T, Hirano M, and Hosono H 2004 *Nature* **432** 488
[19] Kopenkampt R, Boedecker K, Lux-Steiner M C, Poschenrieder M, Zenia F, Levy-Clement C 2000 *Appl. Phys. Lett.* **77** 2575
[20] Dimova-Malinovska D, Andreev P, Sendova-Vassileva M, Nichev H and Mikli V 2009 Proc. of 24 PVSECE, (*Hambourg, Germany, 21-25 Sept.2009)* pp 391-394

[21] Dimova-Malinovska D, Tzenov N, Tzolov M and Vassilev L 1998 *Mat. Sci. Engineering* **B52** 59
[22] Dimova-Malinovska D, Nichev H, Angelov O, Grigorov V and Kamenova M 2007 *Supperlattice and Microstructure* **42** 123
[23] Dimova-Malinovska D, Angelov O, Nichev H and Pivin J-C 2007 *JOAM* **1** 248
[24] Dragoman D and Dragoman M 2002 *Optical Characterization of Solids*, (Heidelberg: Springer-Verlag)
[25] Mott N F and. Davis E A 1979 *Electronic Processes in Non Crystalline Materials* (Oxford: Clarendon) p 289
[26] Saito N, *J. Appl. Phys.* 1985 **58** 3504
[27] Dimova-Malinovska D, Nichev H and Angelov O 2008 *Phys.Stat. Sol.* (c) **5** 3353
[28] Warren B E 1990 *X-ray Diffraction* (New York: Dover Publ. Inc.)
[29] Zhu B, Sun X, Guo S, Zhao X, Wu J, Wu R, and J. Liu J 2006 *Jap. J. Appl. Phys.* **45** 7860
[30] Powder Diffraction File 00-036-1451, PDF-2 Database Sets, International Centre for Diffraction Data Newton Square, PA, 1993
[31] Warren B E, 1990 *X-ray Diffraction* (New York: Dove Publications Inc.)
[32] Sze S M 1995 *Semiconductor Sensitivity, Chemical Sensors* (New York: Wiley) p 383
[33] Watson J 1984 *Sensors Actuators* **5** 29
[34] Lim H-J, Lee D Y and Oh Y-J 2006 *Sensors and Actuators* B **125** 405
[35] Ma Y, Wang W L, Liao K J and Kong C Y 2002 *J. Wide Band Mater.* **10** 113
[36] Pan Z W, Dai Z R and Wang Z I 2001 *Science* **291** 1947
[37] Huang M H, Mao S, Feick H, Yan H, Wu Y, Kind H, Weber E, Russo R and Yang P 2001 *Science* **292** 1897
[38] Rao G S and Rao D T 1999 *Sensors and Actuators* B **55** 166
[39] Dimova-Malinovska 1974 *"Deposition and study of the ZnO thin films properties" Ph. D. Thesis* (Moscow Power Engineering Institute, Russia 1974) (in Russian)
[40] Dimova-Malinovska D, Nichev H, Georgieva V, Angelov O, Pivin J-C and Mikli V 2008 *Phys. Status Solidi (a)* **1** DOI 10.1002/pss.200778887
[41] Sauerbrey G 1959 *Z. für Physik* **155** 206
[42] Dimova-Malinovska D, Andreev P, Sendova-Vassileva M, Nichev H and Starbova 2010 *Energy procedia* **2** 55
[43] Lovchinov K, Angelov O, Nichev H, Mikli V and Dimova-Malinovska D 2012 *J. Phys.: Conference Series* **356** 012025
[44] Mews M, Schulze T F, Mingirulli N and Korte L 2013 *Energy Procedia* **38** 855
[45] Petrov M, Lovchinov, Mews M, Lendertz C and Dimova-Malinovska D 2015 *Phys. Status Solidi* A **212** 166
[46] Hu Y H, Chen Y-C, Xu H-J, Gao H, Jiang W-H, Hu F and Wang Y X 2010 *Engineering* **2** 973
[47] Zhang D, Petrov M, Dimova-Malinovska D, Dörenkämper M, Bakker Kl, Schropp R and Soppe W 2014 Proc. 29th EUPVSEC *(Amsterdam, Nederland, 22-26 Sept.2014)* pp1631-1633
[48] Dimova-Malinovska D 2016 *J. of Phys.: Conference series* **682** 012004
[49] Georgieva V, Stefanov P, Spassov L, Raicheva Z, Atanassov M and Tincheva T 2009 *J. Optoelectr. Adv. Mater.* **11** 1363
[50] Lovchinov K, Petrov M, Ganchev M, Georgieva V, Nichev H, Georgieva B and Dimova-Malinovska D 2014 *J. Phys.: Conference Series* **514** 012023

Investigation of ring polymers in confined geometries

Zoryana Usatenko[1]*, Piotr Kuterba[2], Hassan Chamati[3], Joanna Halun[2]

[1]Institute of Physics, Faculty of Physics, Mathematics and Computer Science, Cracow University of Technology, Cracow 30-084, Poland

[2] Institute of Physics, Jagiellonian University, Cracow 30-348, Poland

[3] Institute of Solid State Physics, Bulgarian Academy of Sciences, 72 Tzarigradsko Chaussée, Sofia 1784, Bulgaria

E-mail: zusatenko@pk.edu.pl

Abstract. The investigation of a dilute solution of phantom ideal ring polymers and ring polymers with excluded volume interaction (EVI) in a good solvent confined in a slit geometry of two parallel repulsive walls and in a solution of colloidal particles of big size were performed. Taking into account the correspondence between the field theoretical ϕ^4 $O(n)$-vector model in the limit $n \to 0$ and the behavior of long-flexible polymers, the depletion forces and the forces which exert phantom ideal ring and ring polymer with EVI on the walls were obtained in the framework of the massive field theory approach at fixed space dimensions d=3 up to one-loop order. Additionally, the investigation of a dilute solution of phantom ideal ring polymers in the slit geometry of two inert walls and mixed walls with one repulsive and one inert wall were performed. Taking into account the Derjaguin approximation the depletion forces between big colloidal particle and a wall as well as between two big colloidal particles were calculated. The obtained results give possibility to understand the complexity of physical effects arising from confinement and chain topology and can find practical application in new types of micro- and nano-electromechanical devices.

1. Introduction

The investigation of a dilute solution of ring polymers in confined geometries such as slit geometry of two parallel walls or inserted in a solution of mesoscopic colloidal particles is a task of great interest not only from the experimental and numerical, but also from the theoretical point of view. As it was shown in a series of atomic force microscopy (AFM) experiments [1, 2] biopolymers such as DNA very often present ring topology. Such situation takes place, for example, in the case of Escherichia coli (E.coli) bacteria with a chromosome which is not a linear polymer, but has a ring topology [3]. The biopolymers of DNA of some viruses such as bacteriophages λ that infect bacteria oscillate between linear and ring topology [4, 5]. The linear form of DNA is encountered in mature viruses, however inside of the host cell DNA of phages adopts a ring topology [6]. Besides, the physical effects arising from confinement and chain topology play a significant role in shaping individual chromosomes and in the process of their segregation, especially in the case of elongated bacterial cells [7].

Ring polymers with specified knot type were chemically synthesized a long time ago [8]. Ring topology of polymer chains has influence on statistical mechanical properties of these polymers, for example on scaling properties [9, 10] and shape [2, 11] because it restrains the accessible phase space. Looking back into the history of investigation of the statistical mechanical properties of ring polymers we should mention that it was established [12], that ring polymers with more complex knots are more compact and have a smaller radius of gyration and this decreases their ability to spread out under confinement. A series of papers [13, 14] have been devoted to the investigation of ring polymers compressed or squeezed by a force in a slab. For example, the results of Monte Carlo simulations performed in reference [13] suggest that the knotted ring polymers will exert higher entropic forces on the walls of the confining slit than unknotted or linear polymer chains. It was stated [13] that the knotted ring polymers expanded as the width of the slit increased in contrast to the behaviour of unknotted (or linear) polymer chains whose size showed a plateau after a certain width of slit was reached. The entropic force exerted on the walls arising from confinement to a slit of a knotted ring polymer was calculated using a bead-spring model by Matthews *et al.* in [5]. It was found [5] that in the case of a narrow slit more complex knot types in a ring polymer exert higher forces on the confining walls of the slit in comparison to unknotted polymers of the same length, and for relatively wide slits the opposite situation takes place. Recently advanced Monte Carlo simulation techniques [15] have been used in order to study the effect of nanoslit confinement on topological properties of circular model DNA, which was modeled as a semi-flexible polymer chain.

Thus, at the moment most of the papers dedicated to the investigation of ring polymers compressed in confined geometries like slit or squeezed by a force in a slab of two parallel walls are performed using Monte Carlo simulation techniques, and present analytical results are not complete. The above mentioned arguments stimulate us to apply the massive field theory approach in a fixed space dimensions $d < 4$ for the investigation of ring polymers confined in the slit geometry of two parallel walls or immersed in the solution of big mesoscopic colloidal particles of different size. This method, as it was shown in the case of infinite [16, 17], semi-infinite [18, 19] systems, and confined geometry [20, 21] gives good agreement with the experimental data and with the results of Monte Carlo simulations.

2. The model and the polymer-magnet analogy

We are interested in the investigation of ring polymers confined in the slit geometry of two parallel walls and in the solution of mesoscopic spherical colloidal particles of one sort or two different sorts. We consider the dilute polymer solution, where different polymers do not overlap and the behaviour of such polymer solution can be described by a random walk (RW) of a single polymer for ideal polymer in Θ - solvent [22, 23] or self-avoiding walk (SAW) for real polymer with excluded volume interaction (EVI) in a good solvent at the temperatures above the Θ - temperature [22, 23]. Taking into account the polymer-magnet analogy developed by de Gennes [22, 23], the scaling properties of long-flexible polymers in the limit of an infinite number of steps N may be derived from a formal $n \to 0$ limit of the field theoretical ϕ^4 $O(n)$- vector model at its critical point. In this case the $1/N$ value plays the role of a critical parameter analogous to the reduced critical temperature in magnetic systems. In the case when the polymer solution is in contact with solid substrates, the monomers interact with the surfaces. We assume that the surfaces are impenetrable. It means that the potential $U(z)$ of the interaction between the monomers of a polymer chain and a wall tends to infinity $U(z) \to \infty$ when the distance between the wall and polymer is less than the monomer size l. The deviation from the adsorption threshold ($c \propto (T - T_a)/T_a$) (where T_a is adsorption temperature) changes sign at the transition between the adsorbed (the so-called normal transition, $c < 0$) and the nonadsorbed state (ordinary transition, $c > 0$) [24, 25] and it plays the role of a second critical parameter. The value c corresponds to the adsorption energy divided by $k_B T$ (or the surface enhancement in field theoretical treatment). The adsorption threshold for long-flexible polymer chains takes place, when $1/N \to 0$ and $c \to 0$.

As was mentioned by de Gennes [22, 23], the partition function $Z(\mathbf{x},\mathbf{x}')$ of a single polymer with two ends fixed at $\mathbf{x}$ and $\mathbf{x}'$ is connected with the two-point correlation function $G^{(2)}(\mathbf{x},\mathbf{x}') = \langle\vec{\phi}(\mathbf{x})\vec{\phi}(\mathbf{x}')\rangle$ in ϕ^4 $O(n)$- vector model for n-vector field $\vec{\phi}(\mathbf{x})$ with the components $\phi_i(x)$, $i = 1, ..., n$ (and $\mathbf{x} = (\mathbf{r}, z)$) via the inverse Laplace transform $\mu_0^2 \to L_0$:

$$Z(\mathbf{x},\mathbf{x}';N,v_0) = \mathcal{I}\mathcal{L}_{\mu_0^2 \to L_0}(\langle\vec{\phi}(\mathbf{x})\vec{\phi}(\mathbf{x}')\rangle\,|_{n\to 0}) \tag{1}$$

in the limit, where the number of components of n tends to zero. The conjugate Laplace variable L_0 has the dimension of length squared and is proportional to the total number of monomers N which form the polymer chain.

The effective Ginzburg-Landau-Wilson Hamiltonian describing the system in semi-infinite ($i = 1$) or confined geometry of two parallel walls ($i = 1,2$) is [24]:

$$\mathcal{H}[\vec{\phi},\mu_0] = \int d^dx\Big\{\frac{1}{2}(\nabla\vec{\phi})^2 + \frac{\mu_0{}^2}{2}\vec{\phi}^2 + \frac{v_0}{4!}(\vec{\phi}^2)^2\Big\} + \sum_{i=1}^{2}\frac{c_{i_0}}{2}\int d^{d-1}r\vec{\phi}^2, \tag{2}$$

where the conjugate chemical potential μ_0 is the "bare mass" in field - theoretical treatment, v_0 is the "bare coupling constant" which characterizes the strength of the excluded volume interaction (EVI). In the case of the slit geometry the walls are located at the distance L one from another in z- direction in such way that the surface of the bottom wall is located at $z = 0$ and the surface of the upper wall is at $z = L$. Each of the two system surfaces is characterized by a certain surface enhancement c_{i_0}, where $i = 1,2$.

The fundamental two-point correlation function of the free theory corresponding to the effective Ginzburg-Landau-Wilson Hamiltonian equation (2) in a mixed $(\mathbf{p},z)$ representation is:

$$G^{(2)}_{ij}(\mathbf{p},\mathbf{p}';z,z') = (2\pi)^{d-1}\delta_{ij}\delta(\mathbf{p}+\mathbf{p}')\tilde{G}_{\parallel}(\mathbf{p};z,z';\mu_0,c_{1_0},c_{2_0},L), \tag{3}$$

where the free propagator $\tilde{G}_{\parallel}(\mathbf{p};z,z';\mu_0,c_{1_0},c_{2_0},L)$ of model (2) was obtained in one of our previous papers [20]. In the case when the ends of the polymer chain $\mathbf{x} = (\mathbf{r},z)$ and $\mathbf{x}' = (\mathbf{r}',z')$ in the partition function (1) coincide, such partition function corresponds to the partition function of a phantom ring polymer, i.e. a ring polymer where we perform the summation over all possible knot structures, as it was mentioned by Eisenriegler [25]. Taking into account the above mentioned arguments the free propagator of model (2) describing a ring polymer with two coinciding ends $\mathbf{x} = \mathbf{x}'$ in the slit geometry of two parallel walls can be obtained by analogy to what was done in [20] and has a form:

$$\begin{aligned}\tilde{G}_{\parallel}(\mathbf{p};z,z;\mu_0,c_{1_0},c_{2_0},L) = &\frac{1}{2\kappa_0}\Big\{\big[\kappa_0^2+\kappa_0(c_{1_0}+c_{2_0})+c_{1_0}c_{2_0}\big]e^{\kappa_0 L} - \big[\kappa_0^2-\kappa_0(c_{1_0}+c_{2_0})+c_{1_0}c_{2_0}\big]e^{-\kappa_0 L}\Big\}^{-1}\\ &\times\Big\{\big[\kappa_0^2+\kappa_0(c_{1_0}+c_{2_0})+c_{1_0}c_{2_0}\big]e^{\kappa_0 L} + \big[\kappa_0^2-\kappa_0(c_{1_0}+c_{2_0})+c_{1_0}c_{2_0}\big]e^{-\kappa_0 L}\\ &+\big[\kappa_0^2+\kappa_0(c_{2_0}-c_{1_0})-c_{1_0}c_{2_0}\big]e^{\kappa_0(L-2z)}\\ &+\big[\kappa_0^2-\kappa_0(c_{2_0}-c_{1_0})-c_{1_0}c_{2_0}\big]e^{-\kappa_0(L-2z)}\Big\}\end{aligned} \tag{4}$$

with $\kappa_0 = \sqrt{p^2+\mu_0^2}$, where $\mathbf{p}$ is the value of parallel momentum associated with $d-1$ translationally invariant directions in the system. The interaction between polymer and the walls is implemented by the boundary conditions. In the case of two repulsive walls (where the segment partition function and thus the partition function for the whole polymer tends to zero as any segment approaches the surface of the walls) the Dirichlet-Dirichlet boundary conditions (D-D b.c.)(see also [20, 24, 26]) take place: $\vec{\phi}(\mathbf{r},0) = \vec{\phi}(\mathbf{r},L) = 0$ or $c_1 \to +\infty, c_2 \to +\infty$. In the case of two inert walls Neumann-Neumann boundary conditions (N-N b.c.) take place: $\frac{\partial\vec{\phi}(\mathbf{r},z)}{\partial z}|_{z=0} = \frac{\partial\vec{\phi}(\mathbf{r},z)}{\partial z}|_{z=L} = 0$ or $c_1 = 0, c_2 = 0$, and for the mixed case of one repulsive and one inert wall Dirichlet-Neumann boundary conditions (D-N b.c.) are: $\vec{\phi}(\mathbf{r},0) = 0, \frac{\partial\vec{\phi}(\mathbf{r},z)}{\partial z}|_{z=L} = 0$ or $c_1 \to +\infty, c_2 = 0$. The present calculations are valid for the case of wide slit limit $y \gtrsim 1$ (where $y = \frac{L}{R_x}$) and are not able to describe the case of dimensional crossover from d to $d-1$ dimensional system which arises for $y \ll 1$. The $d-1$ dimensional system is characterized by another critical temperature and a new critical fixed point. Nevertheless, some assumptions allowed us to describe the region of narrow slit for the case of grand canonical ensemble, as it was proposed in one of our previous papers (see reference [20]). Besides, in the case of infinitely large wall separations, the slit system decomposes into two half-space systems.

It should be mentioned that the most common parameter in polymer physics used for denoting polymers size which is observable in experiments is the radius of gyration R_g (see references [25, 27]): $R_g^2 = \chi_d^2\frac{R_x^2}{2}$, where χ_d is a universal numerical prefactor depending on the dimension d of the system and $R_x^2 = \frac{l^2N^{2\nu}}{3}$ (where ν is 0.5 for ideal polymers and 0.588 for real polymers with EVI). For ideal polymers the relation: $\chi_d^2 = \frac{d}{3}$ takes place and for real polymers at $d = 3$ the following takes place [27]: $\chi_3^2 = 0.958$.

3. Thermodynamic description

We consider the dilute solution of phantom ideal ring polymers and ring polymers with EVI immersed in the slit geometry of two parallel walls and allow the exchange of polymer coils between the slit and the reservoir. The polymer solution in the slit is in equilibrium contact with an equivalent solution in the reservoir outside of the slit. We follow the thermodynamic description of the problem as it was given in [20, 26] and perform the calculation in the framework of the grand canonical ensemble where the chemical potential μ is fixed. As it was shown in [26], the free energy of the interaction between the walls in the grand canonical ensemble is defined as the difference of the free energy of an ensemble where the separation of the walls is fixed at a finite distance L and where the walls are separated infinitely far from each other:

$$\delta F^R = -k_B T \mathscr{N} \ln\left(\frac{\mathcal{Z}_{\parallel}^R(L)}{\mathcal{Z}_{\parallel}^R(L\to\infty)}\right) = -k_B T \mathscr{N}\left\{\ln\left(\frac{\mathcal{Z}_{\parallel}^R(L)}{\mathcal{Z}}\right) - \ln\left(\frac{\mathcal{Z}_{\parallel}^R(L\to\infty)}{\mathcal{Z}}\right)\right\}, \tag{5}$$

where $\mathscr{N}$ is the total amount of polymer coils in the solution and T is the temperature. The $\mathcal{Z}_{\parallel}^R(L)$ value is the partition function of one ring polymer located in a volume V containing two walls at a distance L apart. For the sake of convenience we normalized the partition functions $\mathcal{Z}_{\parallel}^R(L)$ and $\mathcal{Z}_{\parallel}^R(L\to\infty)$ to $\mathcal{Z} = V\hat{\mathcal{Z}}_b$, where $\hat{\mathcal{Z}}_b = \mathscr{I}\mathscr{L}_{\mu_0^2 \to \frac{R_x^2}{2}}[\frac{1}{2\mu_0}]$.

Following the thermodynamic description proposed in [20,26] the corresponding *reduced free energy of interaction* δf per unit area, $A = 1$, for the case of ring polymer confined in the slit geometry of two parallel walls after performing Fourier transform in the direction parallel to the surfaces and integration over $d^{d-1}r$ may be written in the form:

$$\delta f^R = \frac{\delta F^R}{n_p k_B T} = L - \int_0^L dz \frac{\hat{\mathcal{Z}}_I^R(z)}{\hat{\mathcal{Z}}_b} + \int_0^\infty dz\left(\frac{\hat{\mathcal{Z}}_{HS_1}^R(z)}{\hat{\mathcal{Z}}_b} - 1\right) + \int_0^\infty dz\left(\frac{\hat{\mathcal{Z}}_{HS_2}^R(z)}{\hat{\mathcal{Z}}_b} - 1\right). \tag{6}$$

Here $n_p = \mathscr{N}/V$ is the polymer density in the bulk solution, the functions $\hat{\mathcal{Z}}_I^R(z)$ and $\hat{\mathcal{Z}}_{HS_i}^R(z)$ are equal to: $\hat{\mathcal{Z}}_I^R(z) = \mathscr{I}\mathscr{L}_{\mu_0^2\to L_0} G^{(2)}(\mathbf{p} = 0; z, z)|_{n\to 0}$, and $\hat{\mathcal{Z}}_{HS_i}^R(z) = \mathscr{I}\mathscr{L}_{\mu_0^2\to L_0} G_{HS_i}^{(2)}(\mathbf{p} = 0; z, z)|_{n\to 0}$, where $G^{(2)}(\mathbf{p} = 0; z, z)$ and $G_{HS_i}^{(2)}(\mathbf{p} = 0; z, z)$ with $i = 1,2$ are correlation functions of model (2) describing the system in the slit geometry of two parallel walls and in semi-infinite geometry, respectively. The *reduced free energy of interaction* δf^R, according to equation (6), is a function of the dimension of length and dividing it by another relevant length scale, for example, the size of polymer in bulk, e.g. R_x yields a universal dimensionless scaling function for the *depletion interaction potential*:

$$\Theta^R(y) = \frac{\delta f^R}{R_x}, \tag{7}$$

where $y = L/R_x$ is a dimensionless scaling variable. The resulting scaling function for the *depletion force* between two walls induced by the polymer solution is denoted as:

$$\Gamma^R(y) = -\frac{d(\delta f^R)}{dL} = -\frac{d\Theta^R(y)}{dy}. \tag{8}$$

As it is known [25], the force exerted on the surfaces of a confining slit by polymer is equal to the *depletion force* with opposite sign: $K^R(y) = \frac{d(\delta f^R)}{dL}$.

4. Phantom ideal ring polymer in the slit geometry of two parallel walls

Let's consider at the beginning the case of phantom ideal ring polymer under Θ-solvent condition trapped in the slit geometry of two parallel walls situated at the distance L one from another. According to equations (7) and (8) in the wide slit region $L/R_x \gtrsim 1$ (or $y \gtrsim 1$) for the dimensionless scaling functions for the *depletion interaction potential* $\Theta^R(y)$ and the *depletion force* $\Gamma^R(y)$ between two repulsive (or inert) walls induced by the dilute solution of ideal ring polymers we obtain:

$$\Theta_{DD}^{R,id}(y) = -2ye^{-2y^2}, \qquad \Gamma_{DD}^{R,id}(y) = 2e^{-2y^2} - 8y^2e^{-2y^2}. \tag{9}$$

The universal dimensionless scaling functions for the *depletion interaction potential* and the *depletion force* in the case of one repulsive and one inert wall (D-N b.c.) in the limit $y \gtrsim 1$ has a form:

$$\Theta_{DN}^{R,id}(y) = 2ye^{-2y^2}, \qquad \Gamma_{DN}^{R,id}(y) = -2e^{-2y^2} + 8y^2e^{-2y^2}. \tag{10}$$

We obtained that phantom ideal ring polymers due to the complexity of chain topology and from entropical reasons prefer to escape from the space not only between two repulsive walls but also in the case of two inert walls. As a result, the attractive *depletion force* arises between the walls. As it is possible to see from equation (10), the scaling function for the *depletion force* in the case of mixed boundary conditions becomes repulsive in the region $y \gtrsim 1$.

The region of narrow slit corresponds to the situation of $L/R_x \ll 1$ (or $y \ll 1$). We can state that in the case of very narrow slit with two repulsive walls the ring polymers would pay a very high entropy to stay in the slit or even enter it. The contribution from the partition function of polymer chain in the slit $\frac{1}{L}\int_0^L dz \frac{\hat{Z}_I^R(z)}{\hat{Z}_b}$ in equation (6) vanishes strongly as $y \to 0$. Taking into account in the case of narrow slit of two repulsive walls only the corresponding bulk and the surface contributions in equation (6) gives possibility to obtain the asymptotic solutions for the scaling functions $\Theta_{DD}^{R,id}(y)$ and $\Gamma_{DD}^{R,id}$ in the form: $\Theta_{DD,narr}^{R,id}(y) \approx y - \frac{\sqrt{2\pi}}{2}$, and $\Gamma_{DD,narr}^{R,id}(y) \approx -1$. The quantities Θ and Γ are normalized to the overall polymer density n_p. So, the above results simply indicate that the force is entirely induced by free chains surrounding the slit, or, in other words, by the full bulk osmotic pressure from the outside of the slit. It means that in the case of narrow slit with two repulsive walls no chain has remained inside the slit. The advantage of the proposed procedure is that no expansion is necessary in this case of narrow slit region and it should be equally valid for polymer chains with EVI. In figures 1(a) and 1(b) we present the results for the force K^R (as functions of L for different values of the radius of gyration R_g) which exerts phantom

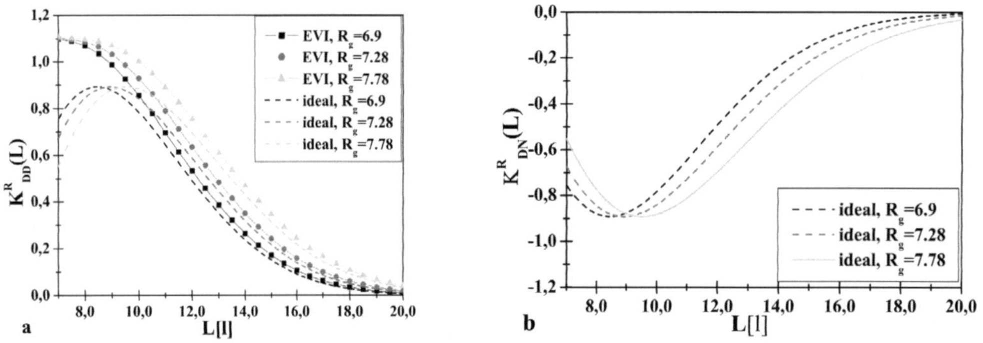

Figure 1. The force $K^R(L)$ for phantom ideal ring polymer and ring polymer with EVI in a good solvent immersed between: a) two repulsive walls; b) one repulsive and other one inert wall for different values of the radius of gyration R_g.

ideal ring polymer on the confining two repulsive (or two inert) walls and two mixed walls, respectively. It should be mentioned that calculations, presented in figure 1(a) and figure 1(b), were performed for different values of the radius of gyration [5]: $R_g(12_1) = 6.9 \pm 0.01[l]$, $R_g(9_1) = 7.28 \pm 0.01[l]$, and $R_g(6_1) = 7.78 \pm 0.01[l]$ which correspond to the ring polymers with different knot types: $12_1; 9_1; 6_1$, where C_p is a standard notation [29], C denotes the minimum number of crossings in any projection on a plane and p is used in order to distinguish knot types with the same C. It should be mentioned that region of validity of the obtained results for the entropically induced forces K^R for phantom ideal ring polymers [see figure 1(a), figure 1(b)] is defined by the value of $y = L/R_x \gtrsim 1$ with $R_x = \sqrt{2}R_g$.

5. Dilute solution of ring polymers with EVI in the slit of two repulsive walls

As a next step let's consider the dilute solution of ring polymers with EVI in a good solvent immersed in the slit geometry of two parallel repulsive walls. As it is known [22, 23], in a good solvent the effects of the EVI between monomers play a crucial role so that the polymer coils occupy large space and are less compact than in the case of ideal polymers. The influence of the EVI on the depletion functions and corresponding partition functions can be obtained via using perturbation treatment in the framework of the massive field theory approach in a fixed space dimensions $d = 3$ up to one-loop order approximation in the presence of confining slit geometry of two parallel repulsive walls. The calculation of the partition functions $\hat{\mathcal{Z}}_i^R(z)$ and $\hat{\mathcal{Z}}_{HS_j}^R(z)$ which allows to obtain the reduced free energy of interaction δf^R in equation (6) are connected with the calculation of the correlation functions $G^{(2)}(\mathbf{p} = 0; z, z)$ and $G^{(2)}_{HS_i}(\mathbf{p} = 0; z, z)$ with $i = 1, 2$, where the terms describing the EVI effects are taken into account via using perturbation treatment. The above mentioned correlation functions were renormalized in accordance with the renormalization scheme proposed in references [18, 20]. In order to make the theory UV finite in RG sense directly in $d = 3$ space dimensions we perform the standard mass renormalization $\mu_0^2 = \mu^2 - \delta\mu_0^2$ with $\delta\mu_0^2 = \frac{(n+2)}{6}\frac{v_0}{(2\pi)^{d-1}}\int\frac{d^{d-1}q}{2\kappa_q}$ and the coupling constant renormalization $v_0 = \mu v$ of the corresponding correlation functions by analogy to what was proposed by Parisi [17]. Besides, the surface enhancement renormalization $c_{j_0} = c_j + \delta c_j$ of the correlation functions in the case of D-D b.c. reduces to an additive renormalization as it took place in the case of semi-infinite geometry [18] and slit geometry (see reference [20]). In order to be concise, we do not present here the results of the complicated calculations and just discuss the limiting cases of wide and narrow slit regimes. The corresponding expression for the universal dimensionless scaling function for the *depletion interaction potential* can be written in the form:

$$\Theta_{DD}^{R,EVI}(y) = -2ye^{-2y^2} + \frac{\tilde{v}}{2}\left(\sqrt{2\pi}i\mathrm{erfc}[\sqrt{2}y]D + e^{-2y^2}B(y)\right), \tag{11}$$

where we introduced $D = \frac{1}{4}\left[10 - 2i\ln 2 - 2\psi^{(0)}\left(-\frac{1}{2}\right) + \sqrt{\pi}\psi^{(0)}\left(\frac{1}{2}\right) - 2\gamma_E\right]$, $B(y) = 2y\sqrt{\pi}\psi^{(0)}\left(-\frac{1}{2}\right) - \frac{1}{y}$ and $v = b_n\tilde{v}$, with $b_n = \frac{6}{n+8}\frac{(4\pi)^{3/2}}{\Gamma(1/2)}$ at space dimensionality $d = 3$ and the limit $n \to 0$ was assumed. It should be mentioned that $\gamma_E = 0.577$ is the Euler's constant, $\psi^{(0)}(z)$ - is the digamma function, and $\mathrm{erfc}[z] = 1 - \mathrm{erf}[z]$ is the complementary error function. The calculations were performed at the corresponding stable fixed point $\tilde{v}^* = 1$ obtained from ressummed beta functions of the underlying bulk field theory in the framework of one-loop approximation scheme. Taking into account equations (8) and (11), the result for the *depletion force* between two repulsive walls is:

$$\Gamma_{DD}^{R,EVI}(y) = 2e^{-2y^2} - 8y^2e^{-2y^2} - \frac{\tilde{v}}{2}e^{-2y^2}\left[\frac{1}{y^2} + 2\sqrt{\pi}\psi^{(0)}\left(-\frac{1}{2}\right) - D - 4yB(y)\right]. \tag{12}$$

The obtained results for the scaling functions $\Gamma_{DD}^R(y)$ between two repulsive walls in the dilute solution of phantom ideal ring polymers and ring polymers with EVI are presented in table 1.

In the region of very narrow slit $y \ll 1$ the *depletion potential* becomes:

$$\Theta_{DD,narr}^{R,real}(y) \approx y - \frac{\sqrt{2\pi}}{2} + \frac{\tilde{v}}{8}\left(ln2 - \frac{1}{2}\right)\sqrt{2\pi}$$

and the *depletion force* is: $\Gamma^{R,real}_{DD,narr}(y) \approx -1$.

Table 1. The dimensionless scaling function for the *depletion force* Γ^{R}_{DD} between two repulsive walls as function of $y = L/R_x$

y	1.3	1.5	1.7	1.9	2.1
ideal ring	-0.3922	-0.1777	-0.0652	-0.0197	-0.0049
ring with EVI	-0.3912	-0.1750	-0.0638	-0.0192	-0.0048

In figure 1(a) we present the results for the entropically induced force K^R (as a function of L for different values of the radius of gyration R_g) which exerts ring polymer with EVI on the confining two repulsive walls. We observed that in the wide slit region ring polymers with less complex knot types (with bigger radius of gyration) in a ring topology exert higher forces on the confining walls. Our results for entropically induced force K^R are in agreement with previous results obtained by Matthews *et al.* [5] using a bead-spring model.

6. Ring polymers in a solution of mesoscopic spherical colloidal particles

Let's consider the situation of the dilute solution of ring polymers immersed in the solution of mesoscopic colloidal particles. We focus our attention on the investigation of the depletion interaction for two cases: (a) between a big spherical colloidal particle of radius R and the wall; (b) between two colloidal particles of big sizes with different radii R_1 and R_2 and different adsorbing or repelling properties with respect to polymers in the solution. The interaction of the dilute solution of ring polymers with particles and walls is implemented by the corresponding boundary condition, as it was mentioned above. The difference between the forces with and without the particle (or particles) yields the depletion interaction of the particle with the wall (or between two particles). Taking into account the Derjaguin approximation [30], which describes the sphere of the big colloidal particle of the radius R (with $R \gg L$ and $R \gg R_x$) by a superposition of immersed plates with local distance $h(\rho) = a + R - \sqrt{R^2 - \rho^2}$ from the wall, where a is the nearest distance from the particle to the wall and ρ is the width of the fringe itself we performed calculations of the *depletion force*: $\frac{-d(\Phi^R(\tilde{y})/n_p k_B T)}{d\tilde{y}}$ between colloidal particle and the wall in a dilute solution of ring polymers. In the case of two big spherical colloidal particles with different radii $R_1 \neq R_2$ when $R_i \gg L$ and $R_i \gg R_x$, $i = 1,2$ we also can use the Derjaguin approximation for calculation of the *depletion force* with some modification and taking into account that $h(\rho) = a + R_1 - \sqrt{R_1^2 - \rho^2} + R_2 - \sqrt{R_2^2 - \rho^2}$, where a, in this case, is the nearest distance from the particle to other particle. It should be mentioned, that the *depletion interaction potential* $\Phi^R(\tilde{y})/n_p k_B T$ is equal to: $2\pi \tilde{R} R_x^2 \int_{\tilde{y}}^{\infty} dy \Theta^R(y)$ with $\tilde{y} = a/R_x$ where $\tilde{R} = R$ for the case of the big particle of radius R near the wall and $\tilde{R} = R_1 R_2/(R_1 + R_2)$ for the case of two particles with different radii $R_1 \neq R_2$.

Taking into account the results for the scaling function $\Theta^R(y)$ between two parallel repulsive (D-D b.c.) and inert (N-N b.c.) walls [see equation (9)] or for the mixed case of D-N b.c. [see equation (10)] induced by the dilute solution of ideal ring polymers in accordance with [30] for the *depletion interaction potential* $\Phi^R(\tilde{y})/n_p k_B T$ between the big spherical colloidal particle and the wall (or between two big spherical colloidal particles) we obtain:

$$\frac{\Phi^{id}(\tilde{y})}{n_p k_B T} = \pm \pi R_x^2 \tilde{R} \eta^{id}(\tilde{y}), \tag{13}$$

where $\eta^{id}(\tilde{y}) = -e^{-2\tilde{y}^2}$. The sign "+" corresponds to the case of D-D b.c. or N-N b.c. and the sign "−" to the case of D-N b.c. Based on the *depletion interaction potential* we can calculate the *depletion force* for certain b.c.'s. The *depletion force* between the big spherical colloidal particle and the wall (or between two big spherical colloidal particles) for the case of D-D b.c. or N-N b.c. and for the mixed case of D-N b.c. has the following form:

$$-\frac{d}{d\tilde{y}}\frac{\Phi^{id}(\tilde{y})}{n_p k_B T} = \pm 2\pi R_x^2 \tilde{R}\vartheta^{id}(\tilde{y}), \tag{14}$$

where $\vartheta^{id}(\tilde{y}) = -2\tilde{y}e^{-2\tilde{y}^2}$ and the sign "+" corresponds to D-D b.c. or N-N b.c. and the sign "−" to the case of D-N b.c. In a similar way taking into account equation (11) we performed calculations for the *depletion force* between the big colloidal particle and the wall (or between two big spherical colloidal particles of different radii) induced by the dilute solution of ring polymers with EVI in a good solvent. The obtained results for the corresponding scaling function $\vartheta^{EVI}(\tilde{y})$ for the *depletion force* in the case of D-D b.c. induced by the dilute solution of ideal ring polymers and ring polymers with EVI in a good solvent are presented in table 2.

Table 2. The dimensionless scaling function $\vartheta(\tilde{y})$ for the *depletion force* which arises between the big spherical colloidal particle and the wall (or between two big spherical colloidal particles) in the case of D-D b.c. as function of $\tilde{y} = a/R_x$.

$\tilde{y}$	1.3	1.5	1.7	1.9	2.1
ideal ring	-0.0885	-0.0333	-0.0105	-0.0028	-0.0006
ring with EVI	-0.0873	-0.0326	-0.0103	-0.0027	-0.0006

7. Conclusions

The present investigations show that the scaling functions for the *depletion interaction potential* and the *depletion force* between two repulsive or two inert walls [see equations (9),(11),(12) and table 1] or in a mixed case of one repulsive and one inert walls [see equation (10)] induced by the dilute solution of phantom ideal ring polymers and real ring polymers with EVI are characterized by completely different behavior than in the case of the dilute solution of linear polymer chains immersed in the slit geometry of two parallel walls [see [20]]. We would like to recollect that in the case of linear ideal polymer chains in confined geometry of two inert walls, in reference [20] we obtained that the scaling functions for the *depletion interaction potential* and the *depletion force* were equal to zero. The obtained in the present paper results indicate that ring polymers due to the complexity of chain topology and from entropical reasons prefer to escape from the space not only between two repulsive walls but also in the case of two inert walls. As a result, the attractive *depletion force*, that arises, decreases as the width of slit increases. Besides, the *depletion force* between one repulsive and one inert walls induced by the dilute solution of ideal ring polymers becomes repulsive in the region $y \gtrsim 1$. Similar behaviour was observed for binary liquid mixture confined in the slit geometry of two parallel walls with mixed boundary conditions as it was shown in [28]. Such behaviour of the dilute solution of phantom ideal ring polymers in confined geometries of two parallel walls with mixed boundary conditions can be used for the production of new types of micro- and nano- electromechanical devices, because it allows to reduce the static friction in such systems. As it is possible to see from table 1, the absolute value of the scaling function for the *depletion force* between two repulsive walls for the dilute solution of ring polymers with EVI is smaller than for the dilute solution of ideal ring polymers in the region of validity of the obtained results, e.g. for $y \gtrsim 1$.

In order to compare our results with the results obtained in the framework of other methods we performed calculations of the entropically induced forces K^R exerted by ring polymer on the surfaces of the confining slit. The results for the forces K^R exerted by ring polymers on two repulsive, two inert and for the mixed case of one repulsive and one inert walls as functions of L for different values of

R_g are presented in figure 1(a) and figure 1(b), respectively. We observed that in the wide slit region ring polymers with less complex knot types (with bigger radius of gyration) in a ring topology exert higher forces on the confining repulsive walls [see figure 1(a)] at the same length L. As it is possible to see from figure 1(b) in the mixed case of one repulsive and one inert walls in the wide slit region the opposite is true. Figure 2 presents the ratio $K^R_{DD}(L)/K^{lin}_{DD}(L)$ of the force for ring polymers K^R_{DD} and the corresponding force for linear polymers K^{lin}_{DD} in the slit geometry of two parallel repulsive walls as function of the distance L (in l units) between the walls for different values of the radius of gyration. As it is possible to see from figure 2 our results for the ratio of the resulting entropic forces $K^R_{DD}(L)/K^{lin}_{DD}(L)$ are in good qualitative agreement with the previous results obtained by Matthews *et al.* (see figure 2 in [5]) in the framework of bead-spring model. The difference between our results and the results obtained by Matthews *et al.* arises due to the fact that in [5] calculations were performed for relatively short polymers with polymer length of order $N \sim 300$ units for different values of the radius of gyration. The force $K^{lin}(L)$ was obtained from the results presented in one of our previous papers [20]. Besides, taking into account the Derjaguin approximation [30] we obtained the results for the *depletion*

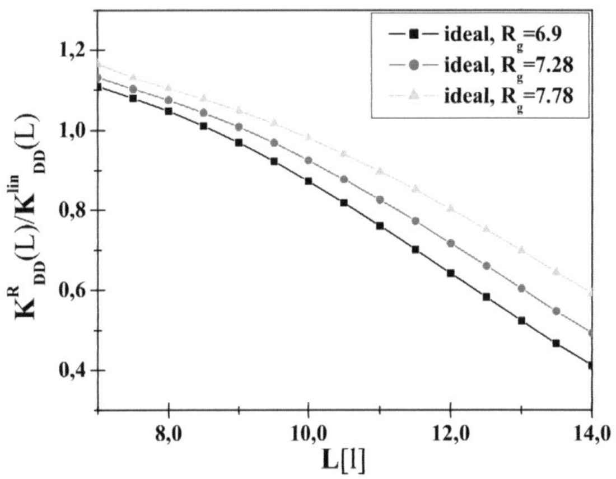

Figure 2. The ratio $K^R(L)/K^{lin}(L)$ of the force for ring polymers and the force for linear polymers in the slit geometry of two parallel repulsive walls as function of the distance L (in l units) between the walls for different values of the radius of gyration R_g.

interaction potential and the *depletion force* for the dilute solution of ring polymers immersed in the solution of mesoscopic spherical colloidal particles. We performed investigations of the depletion effect for mesoscopic colloidal particles with different radii and different adsorbing or repelling properties with respect to ring polymers in the solution. As it is possible to see from the obtained in equation (14) and table 2 results the absolute value of the depletion force between the spherical colloidal particle and the wall is bigger than for the case of two spherical colloidal particles. This feature is universal and does not depend on the boundary condition type. In the case when two colloidal particles have the same radius ($R_1 = R_2$), the corresponding *depletion force* is twice smaller than for the case of particle near the wall. Further analytical and experimental investigations of the dilute and semidilute solutions of ring polymers with EVI immersed in the slit geometry of two inert walls or mixed walls is a task of great interest and is currently under consideration.

Acknowledgments
Z.U. and H.C. would like to thank for financial support under a bilateral agreement between the Polish Academy of Sciences and the Bulgarian Academy of Sciences.

References

[1] Marek J et al. 2005 *Cytometry Part* A **63A** 87
[2] Witz G, Rechendorff K, Adamcik J and Dietler G 2011 *Phys. Rev. Lett.* **106** 248301
[3] Berg J, Tymoczko J and Stryer L 2002 *Biochemistry* 5th ed. (New York: W H Freeman and Co.)
[4] Arsuaga J, Vazquez M, Trigueros S, Sumners D W and Roca J 2002 *Proc. Natl. Acad. Sci. U.S.A.* **99** 5373
[5] Matthews R, Louis A A and Yeomans J M 2011 *Molecular Physics* **109** 1289
[6] Miller R V 1998 *Sci. Am.* **278** 66
[7] Jun S and Mulder B 2006 *Proc. Natl. Acad. Sci. U.S.A.* **103** 12388
[8] Dietrich-Buchecker C O and Sauvage J P 1989 *Angew. Chem. Int. Edn. Engl.* **28** 189
[9] Dobay A et al. 2003 *Proc. Natl. Acad. Sci. U.S.A.* **100** 5611
[10] Ercolini E et al. 2007 *Phys. Rev. Lett.* **98** 058102
[11] Alim K and Frey E 2007 *Phys. Rev. Lett.* **99** 198102
[12] Quake S 1994 *Phys. Rev. Lett.* **73** 3317
[13] Janse van Rensburg E J 2007 *J. Stat. Mech.: Teor. and Exp.* P03001
[14] Janse van Rensburg E J, Orlandini E, Tesi M C and Whittington S G 2008 *J. Phys. A: Math. Theor.* **41** 015003
[15] Micheletti C and Orlandini E 2012 *Macromolecules* **45** 2113
[16] Parisi G 1980 *J.Stat.Phys.* **23** 49
[17] Parisi G 1988 *Statistical Field Theory* (Redwood City:Addison-Wesley)
[18] Diehl H W and Shpot M 1998 *Nucl. Phys.* B **528** 595
[19] Usatenko Z 2014 *J. Stat. Mech.: Teor. and Exp.* P09015
[20] Romeis D and Usatenko Z 2009 *Phys.Rev.* E **80** 041802
[21] Usatenko Z 2011 *J. Chem. Phys.* **134** 024119
[22] de Gennes P G 1972 *Phys.Lett.* A **38** 339
[23] de Gennes P G 1979 *Scaling Concepts in Polymer Physics* (New York:Ithaca/Cornell University Press)
[24] Diehl H W 1986 *Phase Transitions and Critical Phenomena* vol 10, ed C Domb and J L Lebowitz (London: Academic Press) pp 75-267
[25] Eisenriegler E 1993 *Polymers Near Surfaces* (Singapore:World Scientific Publishing Co.Pte.Ltd.)
[26] Schlesener F, Hanke A, Klimpel R and Dietrich S 2001 *Phys.Rev.* E **63** 041803
[27] des Cloizeaux J and Jannink G 1990 *Polymers in Solution* (Oxford:Clarendon Press)
[28] Hertlein C, Helden L, Gambassi A, Dietrich S and Bechinger C 2008 *Nature* **451** 172
[29] Orlandini E and Whittington S 2007 *Rev. Mod. Phys.* **79** 611
[30] Derjaguin B V 1934 *Kolloid-Z.* **69** 155

Structural, morphological, and optical study of titania-based nanopowders suitable for photocatalytic applications

M Šćepanović[1], M Grujić-Brojčin[1], B Abramović[2] and A Golubović[1]

[1]Center for Solid State Physics and New Materials, Institute of Physics, University of Belgrade, Pregrevica 118, 11080 Belgrade, Serbia
[2] Department of Chemistry, Biochemistry and Environmental Protection, Faculty of Sciences, University of Novi Sad, Trg D. Obradovića 3, 21000 Novi Sad, Serbia

E-mail: maja@ipb.ac.rs

Abstract. Systematic investigation of the relationship between structural, morphological, optical and photocatalytic properties of the titania-based nanopowders is presented. A series of pure and doped titania catalysts with various (anatase and brookite) phase compositions have been prepared by sol–gel or hydrothermal route. The crystal structure and composition of the synthesized samples have been extensively characterised by XRD and Raman scattering measurements. The nanopowder morphology has been studied using microscopic methods (SEM, AFM, and STM), whereas the porous structure has been revealed by the analysis of nitrogen sorption data. The optical and electronic properties have been studied by spectroscopic ellipsometry. All investigated properties have been correlated to photocatalytic activity, tested in degradation of the pharmaceutically active substances (such as metoprolol and alprazolam) induced by UVA or visible radiation. Based on this correlation, the physical properties which contribute most to the increase in photocatalytic activity of synthesized nanopowders have been determined, in order to optimize the synthesis conditions which could lead to the maximal efficiency in degradation of particular pollutant.

1. Introduction

Titanium dioxide (titania, TiO_2) is known as a polymorphic material with three naturally occurring crystalline modifications (anatase, rutile, and brookite), composed of TiO_6 octahedra sharing the edges and corners in a different manners [1]. As environmentally-friendly material, titania with various content of these polymorphs has been recognized as the most promising photocatalyst suitable for decomposition of various organic pollutants in the water and air. The possibility to improve photocatalytic performance of porous TiO_2 nanocrystalline powders by changing the synthesis conditions, as well as doping with different ions (usually the transition metals and rare earth elements) attracted extensive attention in recent decades. More than 16,000 publications indexed in the Web of Science database have been devoted to the photocatalysts based on TiO_2 (and/or comparison of the photocatalytic efficiency of some other catalysts to the efficiency of TiO_2) in the last 15 years, with continuous growth in the number of publications per year (figure 1). In spite of such a large number of publications, targeted modification of the TiO_2 properties (such as content, crystallinity and stoichiometry of different TiO_2 phases; crystallite size and micro strain; specific surface area, pore size distribution, shape and complexity; energy of band gap; the type and amount of surface O–H groups

[2]) which may be crucial for improving its photocatalytic efficiency in degradation of some specific pollutants still requires further research.

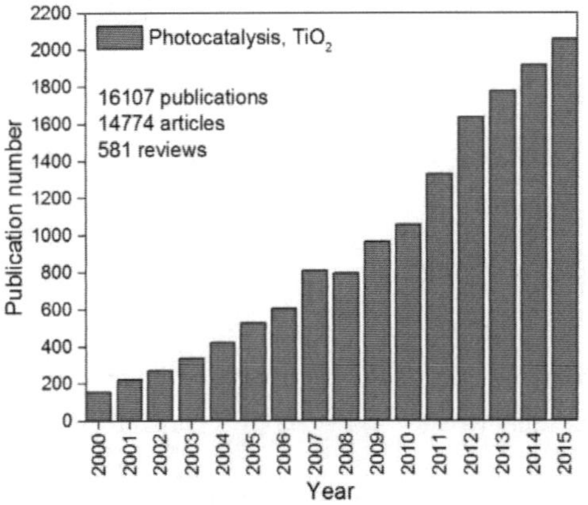

Figure 1. Number of publications per year indexed in the Web of Science database retrieved using the words photocatalysis and TiO_2 as topics.

Among a large number of pollutants, the influence of pharmaceuticals is considered as an emerging environmental problem, due to long term usage worldwide. Namely, continuous input and persistence of human and veterinary medicinal compounds to the aqueous system, even at low concentrations, may result in a potential risk for aquatic and terrestrial organisms. For all of these reasons, our recent research has been focused on the modification of the structural, compositional, morphological, surface, electronic and optical properties of TiO_2 nanopowders by changing synthesis and doping conditions, tailored to improve photocatalytic efficiency of TiO_2 in degradation of two pharmaceuticals, metoprolol and alprazolam. According to the literature, β-blocker metoprolol (MET) tartrate salt, a cardiovascular drug known among most frequently detected pharmaceuticals in wastewaters, is listed in concentration of 10–390 ng L^{-1} [3]. Alprazolam is one of the 5th generation benzodiazepines, a group of psychiatric substances acting on the central nervous system [4], occurring in the environment in the range of ng L^{-1}–μg L^{-1} [5]. Although in trace concentrations, the presence of those pharmaceuticals in the environment is considered potentially dangerous, as they are persistent against biological degradation, retaining their chemical structure long enough [6]. To obtain photocatalyst efficient enough in comparison to commercial titania Degussa P25 (with anatase to rutile phase ratio of about 80:20) in MET degradation under UVA light irradiation we have synthesized anatase TiO_2 nanopowders by sol-gel route with several synthesis parameters varied [2, 7]. The La-doped anatase TiO_2 nanopowders have also been synthesized via sol–gel route [8], to study the influence of doping on TiO_2 efficiency in MET degradation. The brookite type TiO_2 nanopowders have been prepared by combination of sol–gel and hydrothermal methods and their activity in photocatalytic degradation of alprazolam has been investigated and compared to the activity of Degussa P25 [9]. Several characterization techniques (including the X-ray diffraction (XRD), scanning electron microscopy (SEM), non-contact atomic force microscopy (nc-AFM), scanning tunnelling microscopy and spectroscopy (STM/STS), energy dispersive X-ray spectroscopy (EDS), nitrogen physisorption measurements, Fourier transform infrared (FTIR) and Raman spectroscopy, spectroscopic ellipsometry (SE)) have been employed to correlate structural, morphological, surface, electronic and optical properties of synthesized TiO_2 nanopowders with their photocatalytic activity under ultraviolet (UVA) irradiation. The experimental details related to the applied techniques and photocatalytic measurements, can be found in our previous papers [2, 7-9].

2. Improving the efficiency of sol–gel synthesized TiO_2 in photocatalytic degradation of MET

In order to produce highly photocatalytically active nanocrystalline TiO_2 powders by sol–gel route described in details in our previous papers [2, 7], several synthesis parameters, including the type of precursor, alcohol and basic solution, as well as temperature and duration of calcination, have been varied. The first series of the samples were calcined for 7 h, whereas the list of varied synthesis parameters is given in table 1.

Table 1 The list of TiO_2 samples and corresponding sol–gel synthesis parameters [2].

Sample name	Precursor	Alcohol	Basic solution	Calcination temperature [°C]
$T(BE)_{550}$	Tetrabutyltitanate (B)	Ethanol (E)	/	550
$T(CEA)_{550}$	Titanium tetrachloride (C)		NH_4OH (A)	
$T(CPA)_{550}$		2-propanol (P)		
$T(CAA)_{550}$	Titanium tetrachloride (C)	Amyl alcohol (A)	NH_4OH (A)	550
$T(CMA)_{550}$		Methanol (M)		
$T(CGA)_{550}$		Glycerol (G)		
$T(CAN)_{400}$				400
$T(CAN)_{550}$	Titanium tetrachloride (C)	Amyl alcohol (A)	NaOH (N)	550
$T(CAN)_{650}$				650

The XRD patterns of TiO_2 nanopowders (figure 2) have shown the presence of anatase diffraction peaks (JCPDS card 21-1272, space group $I4_1/amd$) with slightly shifted positions [2], indicating a change in lattice parameters in comparison to the bulk values for anatase. Further analysis of the XRD data enabled the assessment of the anatase nanocrystallites size to about 20 nm [2]. Also, the peaks corresponding to the rutile TiO_2 phase (JCPDS card 21-1276, space group $P4_2/mnm$) have been observed in the diffractogram of the sample $T(CAN)_{650}$ and the amount of rutile phase in this sample is estimated to be ~30%. However, as diffraction peaks of brookite TiO_2 phase were not seen in XRD patterns of TiO_2 nanopowders, the Raman spectroscopy method had to be used for the analysis of brookite content.

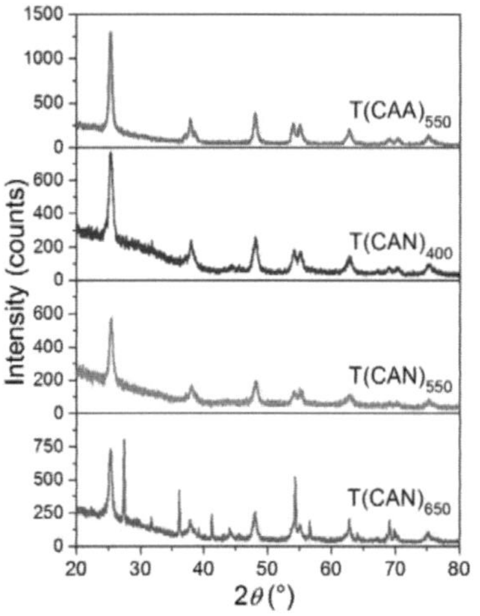

Figure 2. XRD diffractograms of some synthesized TiO_2 samples [2].

Typical Raman spectrum of synthesized TiO_2 nanopowder is shown in figure 3a in the range of 50–1000 cm^{-1}. This spectrum, as well as the spectra of all other synthesized nanopowders, is dominated by characteristic anatase Raman modes [2]. Detailed analysis of the most intensive anatase E_g Raman mode has shown nanocrystalline anatase structure with variable stoichiometry and microstrain in all spectra, whereas average anatase crystallite size has been estimated between 17 and 25 nm. Besides, low-intensity Raman modes, ascribed to brookite phase [2], can be also observed in the range from 210 to 380 cm^{-1} of all spectra (inset of figure 3b). To estimate the amount of brookite phase, the sum of the integrated intensities of Lorentzian peaks, originating from the brookite modes (ΣI_B), was compared to the intensity of Lorentzian peak related to the B_{1g} mode of anatase phase ($I_A(B_{1g})$). Although the estimation of brookite modes intensities is unreliable to some extent, due to mode overlapping, a variation in brookite content with synthesis conditions may be roughly estimated from the intensity ratio of brookite to anatase modes, $\Sigma I_B/I_A(B_{1g})$, shown in figure 3b.

The IR reflectivity spectra of representative TiO_2 nanopowders shown in figure 4 are dominated by the features in the range of 450−1000 cm^{-1}, which can be related to the surface vibration of Ti−O bonds in anatase or mixed anatase/rutile phase ($T(CAN)_{650}$ and Degussa P25). The main spectral features in the region of 1200−4000 cm^{-1} (shown magnified in the inset of figure 4) appear around 1260 cm^{-1} (assigned to carbonate or nitrate species at TiO_2 surface [2]), between 1400 and 1800 cm^{-1} (probably originated from carbonates or hydrogen carbonates groups [10]), and in the ranges of 2800−3000 and

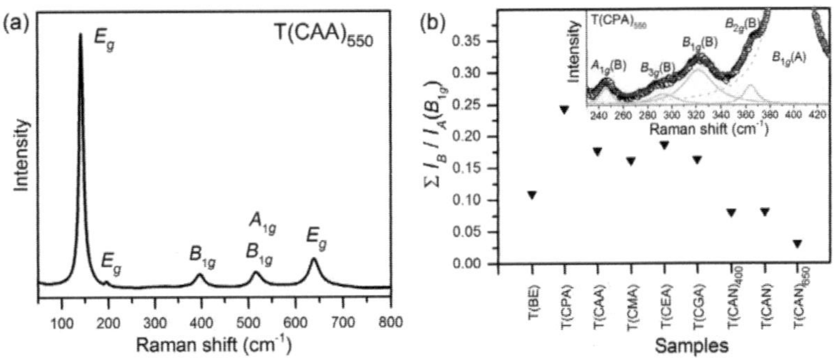

Figure 3. Raman spectrum of T(CAA)$_{550}$ nanopowder with characteristic anatase modes; (b) The intensity ratio of brookite modes, ΣI_B, to the anatase B_{1g} mode, $I_A(B_{1g})$ in TiO_2 nanopowders. The inset: Lorentzians of brookite modes in the experimental Raman spectra of sample T(CPA)$_{550}$ [2].

3200–3800 cm^{-1}. The bands in the range of 2800–3000 cm^{-1} can be assigned to the symmetric and asymmetric stretching mode of the C–H bond, originating from residues which remain in TiO_2 samples even after calcination [11]. However, the most of attention here is devoted to characterizing O–H groups present at the surface of TiO_2, knowing that in many cases the nature and distribution of these groups are determining factor of catalyst behaviour. So, the vibration band between 3000 and 3150 cm^{-1} has been associated to the O–H stretching mode of water molecules [12], whereas the one at about 1630 cm^{-1} has been assigned to O–H bending modes [13]. The bands in frequency range of 3150–3500 cm^{-1} are usually attributed to H-bounded O–H groups, and those in the 3400–3600 cm^{-1} region to mixed O–H groups [12]. The bands in the region from 3600 to 3800 cm^{-1} originate from stretching modes of different types of free hydroxyl groups adsorbed on the crystalline surfaces of titania or from defects located at the surface [13]. It should be noted that IR reflectivity spectra of TiO_2 nanopowders have enabled identification of specified surface groups, but did not allow the relation between the amounts of different kinds of these groups at the surface of TiO_2 nanoparticles to be determined.

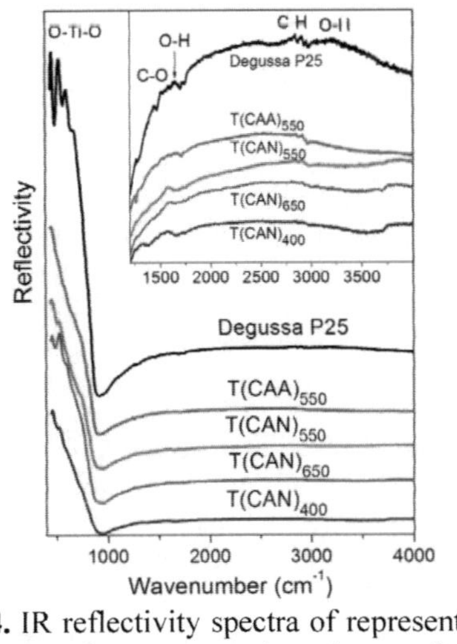

Figure 4. IR reflectivity spectra of representa-tive TiO_2 nanopowders. Inset: enlarged part of the IR spectra in the region of 1200–4000 cm^{-1}. Spectra of some powders are upshifted for clarity [2].

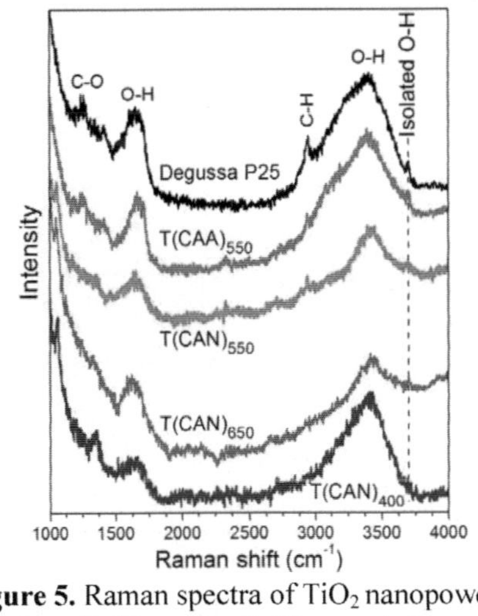

Figure 5. Raman spectra of TiO_2 nanopowders in the C–H and O–H spectral region (1000–4000 cm^{-1}). The spectra of some powders are upshifted for clarity [2].

More information on the amount of O–H groups in our powders we have obtained from Raman spectra in the C–H and O–H spectral region shown in figure 5. This region is dominated by the broad feature between 2800 and 3800 cm^{-1}, which can be assigned to O–H stretching vibration of water molecules adsorbed on the surface of the particles and condensed in the pores of TiO_2 nanopowders [14], as well as the feature located between 1625 and 1670 cm^{-1} which can be related to the bending modes of adsorbed water [10]. In addition, the peaks in the frequency range of 1200–1500 cm^{-1} can be associated with carbonate groups [13], whereas the peak at 2940 cm^{-1} can be assigned to the stretching mode characteristic for CH-species [15]. However, low intensity peak located at ~ 3700 cm^{-1} is the most interesting one in this analysis. Namely, this feature, assigned to the O–H stretching mode of isolated (free) hydroxyls [10], can be related to the ability of the anatase samples to form the surface O–H groups, with the important influence on their catalytic activity [16]. According to several authors, the hydroxyl groups with vibration frequency at ~ 3700 cm^{-1} are the most photocatalytically active among all hydroxyl groups at the surface of TiO_2 nanoparticles [13].

The influence of the different synthesis parameters on the photocatalytic efficiency of TiO_2 nanopowders in degradation of MET under UVA light irradiation are presented in figure 6, together with the results obtained using commercially available Degussa P25 under the same conditions. From figure 6a, which illustrates the influence of the type of precursor on TiO_2 nanopowders photocatalytic activity, it can be seen that the activity of TiO_2 nanopowders synthesized with titanium tetrachloride as precursor is much higher, but still somewhat lower than of Degussa P25. The influence of the type of alcogel on photocatalytic activity of TiO_2 nanopowders (figure 6b) was examined by using different alcohols during the synthesis. As can be seen, the highest photocatalytic activity showed the catalysts obtained using methanol and amyl alcohol. The figure 6c shows the influence of calcination temperature on the photocatalytic activity of TiO_2 nanopowders on MET photodegradation. It can be seen that the TiO_2 nanopowders calcined at ≥ 550 °C showed the highest photocatalytic activity, even slightly higher than Degussa P25.

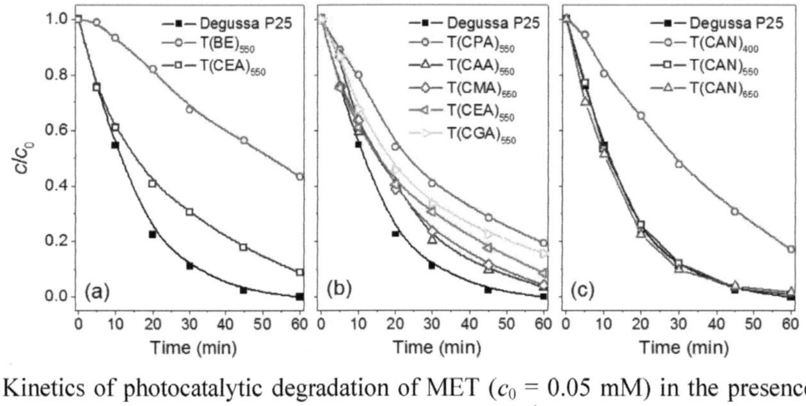

Figure 6. Kinetics of photocatalytic degradation of MET (c_0 = 0.05 mM) in the presence of sol–gel synthesized TiO_2 nanopowders and Degussa P25 (1.0 mg mL^{-1}). The influence of different synthesis parameters: (a) precursor, (b) alcohol and (c) calcination temperature [2].

Structural analysis of synthesized nanopowders by XRD and Raman spectroscopy, briefly presented here, has shown that different mixtures of TiO_2 crystal modifications (anatase, rutile, and brookite) were obtained with proper choice of synthesis parameters. It has been noticed that the content of brookite decreased photocatalytic activity in synthesized nanopowders with other properties comparable, whereas the amount of rutile did not have critical impact on the photocatalytic activity. So, it could be concluded that structural properties of nanopowders had some impact on their photocatalytic activity, but not decisive. On the other side, the results obtained from BET (Brunauer-Emmett-Teller) analysis for the specific surface area, and especially pore size distribution [2], have led to unambiguous conclusions that larger surface area and greater number of pores with dimensions in mesoporous region (> 2 nm), produced higher photocalytical degradation of MET, as a material with relatively large molecules (with molecular size of 0.610 nm × 1.347 nm [7]). Also, isolated hydroxyl groups on the surface of TiO_2, which can trap holes to generate the photocatalytically important active species, i.e. $^{\bullet}OH$ radicals [17], have appeared as one of decisive factors in high photocatalytic efficiency of TiO_2 in degradation of MET. Namely, as can be seen from figure 5, the Raman mode at ~3700 cm^{-1}, originating from isolated O−H groups, appears in the spectra of all samples with relatively high photocatalytic activity (including Degussa P25), and it is hardly detected in $T(CAN)_{400}$ with relatively poor activity.

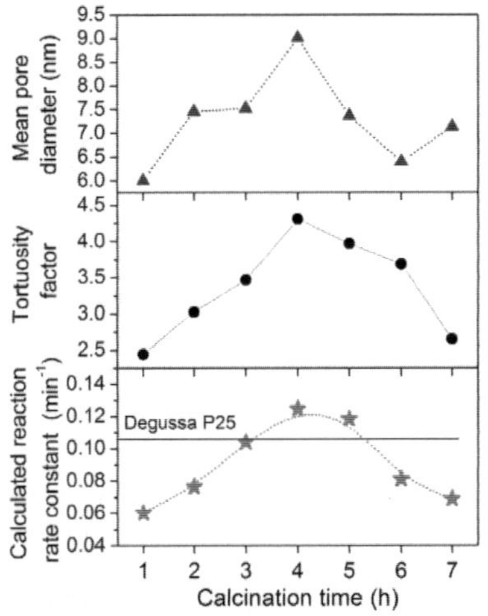

Figure 7. The dependence of (a) mean pore diameter, (b) predicted tortuosity factor, and (c) MET pseudo-first order rate constant value determined for the first 20 min of irradiation on calcination time [7].

In the second series of anatase nanopowders prepared by sol−gel method, the structural and morphological properties have been modified by varying the duration of calcination at 550 °C from 1 to 7 h, with the other synthesis parameters same as for the CAN samples from the first series. The XRD analysis has shown that extending of calcination time caused slight growth of crystallites in synthesized samples (from 13 to 17.5 nm), with the other structural properties similar [7]. Raman scattering data confirmed the anatase as dominant TiO_2 phase, with the presence of small amount of brookite in some samples [7]. The nonstoichiometry, revealed from the additional broadening and blueshift of the most intensive Raman E_g mode, could not be clearly correlated to the calcination time [7]. The analysis of nitrogen sorption data has shown that the specific surface area was greater in the samples with shorter calcination time (up to 4 h), whereas the greatest mean pore diameter was registered in the sample calcined for 4 h (figure 7a). Additionally, the corrugated pore structure model was employed to evaluate tortuosity factor which provides the information on the connectivity among the pores, essential to describe transport dynamics in porous media [18]. This model has shown that tortuosity factor increased with increasing calcination time up to 4 h, and then gradually decreased for a longer time of calcination (figure 7b). The study of photocatalytic efficiency in degradation of MET has demonstrated that samples calcined for 4 and 5 hours have displayed higher photocatalytic performance than Degussa P25, whereas the sample calcined for 3 hours has shown comparable activity, as can be seen from the dependence of reaction rate constant on calcination time shown in figure 7c. Since the analysis of XRD and Raman spectroscopy data have shown that there is no great variation in structure and phase composition of synthesized powders, the main reason for different efficiency in photocatalytic degradation of MET should have been seeking in porous properties of these powders. However, the powder calcined for 1 h, with the greatest specific surface area S_{BET} =182 g cm^{-3}, has the poorest photocatalytic efficiency of all examined samples. At the same time, some nanopowders with smaller S_{BET} value have relatively good photocatalytic activity [7]. On the other hand, maximal photocatalytic efficiency was reached in the sample calcined for 4 h, in which mean pore diameter has the greatest value, as well as tortuosity factor. Moreover, calculated reaction rate constant in degradation of MET (figure 7c) behaves similarly as mean pore diameter with calcination time varied. The same tendency is especially expressed in dependence of tortuosity factor on calcination time (figure 7). These relationships confirm the crucial role of the pore size distribution and tortuosity factor as a measure of the pore structure complexity in efficient photocatalytic degradation of relatively large pollutant molecules such as MET.

3. Surface modification of sol–gel synthesized TiO_2 nanoparticles induced by La-doping

The influence of La-doping in the range of 0.5−6.0 mol% on structural, morphological, surface and electronic properties of TiO_2 photocatalysts, prepared by sol−gel method, have been extensively characterized in order to correlate the change in these properties with TiO_2 photocatalytic efficiency in degradation of MET in aqueous solution under UVA light irradiation. The XRPD and Raman

IOP Conf. Series: Journal of Physics: Conf. Series **794** (2017) 012003 doi:10.1088/1742-6596/794/1/012003

measurements have revealed the anatase phase as dominant in all synthesized nanopowders, with crystallite size decreasing from ~15 nm in pure TiO_2 to ~12 nm in La-doped samples [8]. The analysis of nitrogen sorption data has shown that all samples are fully mesoporous, with a mean pore diameters in the range of ~6–8 nm, as well as the increase of specific surface area and pore structure complexity with doping [8]. The SE has apparently shown that direct bandgap energy has been gradually increased with the increase of La content (figure 8). The AFM, STM, and STS techniques have been used successfully to evaluate the surface morphology and electronic properties of La-doped nanopowders. It can be seen from nc-AFM images of pure and 6 mol% La-doped TiO_2 samples (Figs. 9a and 9b, respectively) that investigated nanopowders consist of very small nanocrystals (up to 20 nm) and greater agglomerated particles. In the close-up image of the sample with 6 mol% La (figure 9c) one could observe a granular structure (up to 2 nm) covering the surface of some nanoparticles. Since TiO_2 nanoparticles with both smooth and rough surface in that sample have also been distinguished by the STM, STS measurements have been used to obtain information on surface bandgap in the nanoparticles with different surface morphologies. The figure 10(a) shows a typical dI/dV versus V characteristic of isolated TiO_2 particles with smooth surface and its STM image in the inset. From presented STS data, the surface bandgap of those particles has been estimated as ~3.2 eV, a value close to bandgap of anatase phase. However, in the same sample, for isolated TiO_2 particles covered with rough granular structure (figure 10(b)), the higher value of surface bandgap (~4 eV) has been found, implied that granular structure was most probably consisted of La_2O_3. These results have supported the findings of the SE that gradual increase of direct bandgap energy with La-doping could be attributed to the surface modification of TiO_2 nanoparticles due to the presence of La_2O_3.

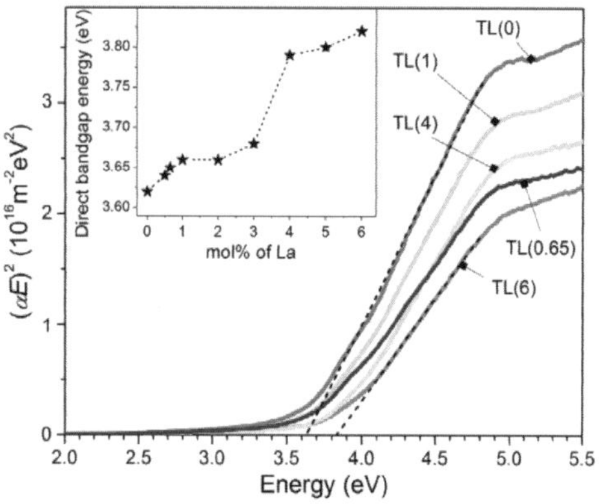

Figure 8. The plot of $(\alpha E)^2$ versus E for direct bandgap transition in pure and La-doped TiO_2 nanopowders samples with characteristic tangent lines. Inset: the dependence of estimated direct bandgap energy on mol% of La [8].

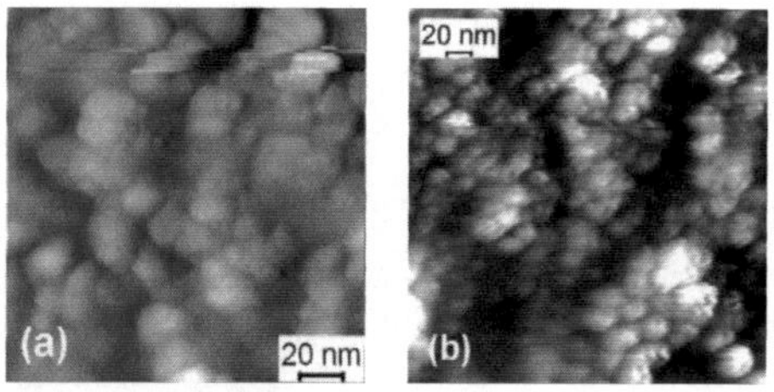

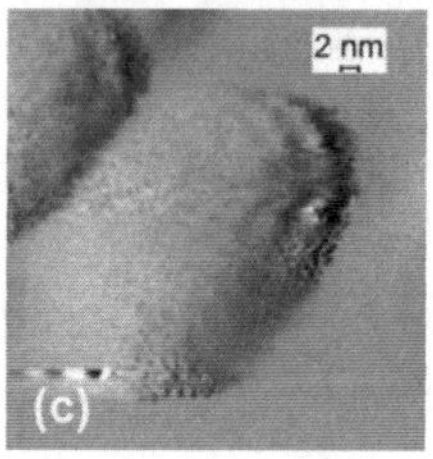

Figure 9. The nc-AFM images of TiO_2 samples: (a) pure, (b) doped with 6 mol% La, and (c) close-up image of the second one [8].

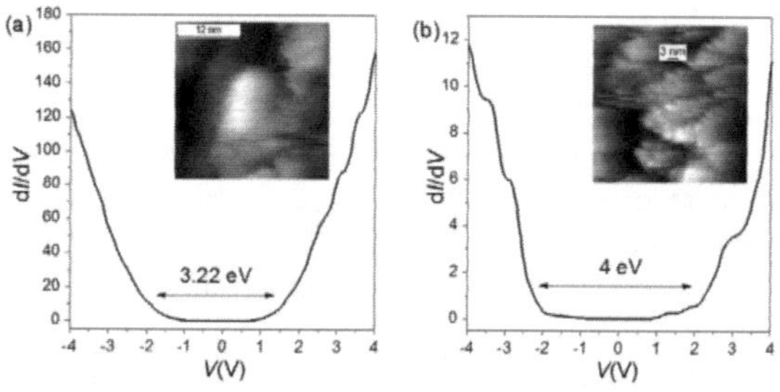

Figure 10. Differential dI/dV versus voltage, V, characteristics of isolated TiO_2 nanoparticles of the sample doped with 6 mol% La, with (a) smooth and (b) rough surface obtained by STS, together with corresponding STM images in the insets [8].

The study of photocatalytic activity has revealed that all doped samples showed increased photocatalytic activity compared to pure TiO_2, among which the 0.65 mol% La-doped sample appeared to be the most efficient (figure 11). The improvement in photocatalytic properties of La-doped samples can be ascribed to the formation of La_2O_3 at the grain boundaries of anatase nanoparticles. This effect inhibits the growth of crystal grains, decreasing of anatase crystallite size and produce more complex pore structure [19], which is confirmed by the results of XRPD, Raman and physisorption measurements. It should be also noted that the presence of isolated hydroxyl groups on the surface of TiO_2 nanoparticles, which could represent decisive factor for high photocatalytic activity in degradation of MET [2], has been registered only in the nanopowders with low level of La-doping - 0.65 and 1 mol% [8]. This implies that reduction of photocatalytic activity of TiO_2 nanopowders with further increase of La-doping amount may be related to the blocking of the active centres due to the accumulation of excess La_2O_3 on the surface of anatase grains, registered by both AFM and STM measurements.

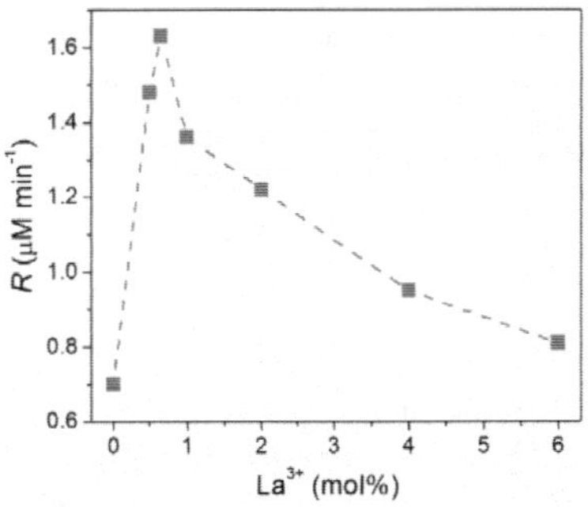

Figure 11. The influence of La doping content in TiO_2 nanopowders on MET (0.05 mM) photocatalytic degradation rate determined for the first 30 min of irradiation [8].

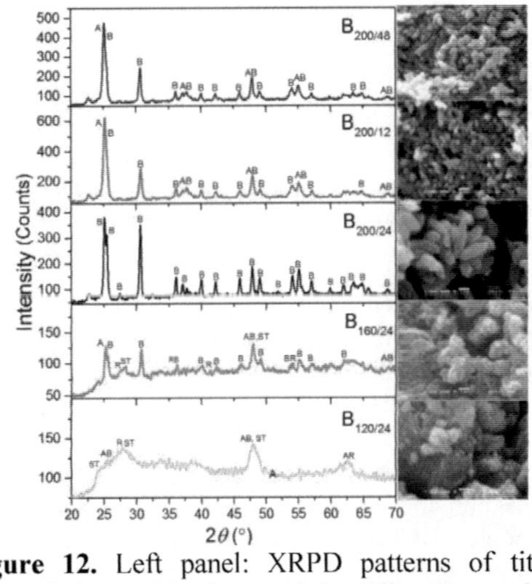

Figure 12. Left panel: XRPD patterns of titania nanopowders, with characteristic diffraction peaks of brookite (B), anatase (A), rutile (R) and sodium titanate (ST) phase. Right panel: SEM micrographs of synthesized nanopowders with the scale bars corresponding to 0.1 μm.

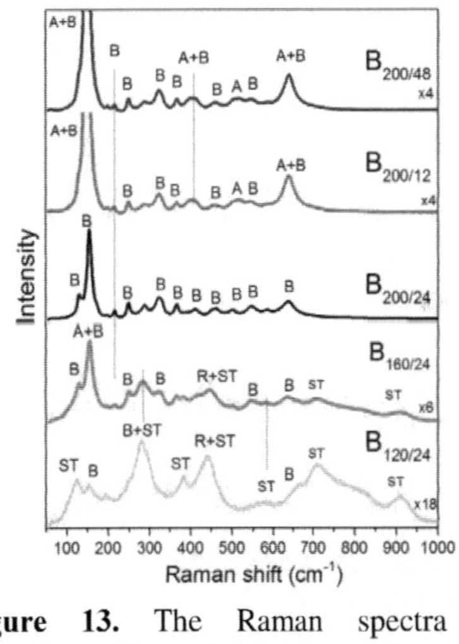

Figure 13. The Raman spectra of hydrothermally synthesized powders, with the most intensive Raman modes assigned to brookite (B), anatase (A), rutile (R), and sodium titanate (ST). The spectra of some powders are upshifted for clarity [9].

4. Brookite type TiO_2 nanopowders prepared using hydrothermal route in photocatalytic degradation of alprazolam

Photocatalytic degradation of alprazolam has been investigated in aqueous solutions containing nanocrystalline brookite-type powders, synthesized by combination of sol–gel and hydrothermal method, with titanium tetrachloride ($TiCl_4$) as a precursor, and hydrothermal temperature and reaction time varied in the range of 120–200 °C and 12–48 h, respectively. The XRPD patterns (figure 12, left panel) have shown that pure brookite phase (JCPDS card 29-1360, space group *Pbca*) has been obtained only in the sample $B_{200/24}$, synthesized at 200 °C, after 24 h of hydrothermal process. In all other samples anatase phase also appears, whereas rutile and sodium titanate phases have been noticed in the samples synthesized at lower temperatures. The presence of different titania phases has also been confirmed by Raman spectra shown in figure 13.

The SEM micrographs (figure 12, right) have shown spindle-like particles in brookite-rich samples synthesized at 200 °C, whereas BET measurements have detected mesoporous structure in these samples [9]. The dependence of $(\alpha E)^{1/2}$ plots on the photon energy for an indirect transition, as well as the $(\alpha E)^2$ plots for a direct transition, obtained from SE data, are shown in Figs. 14(a) and 14(b), respectively. The energies of indirect and direct transition in pure brookite sample ($B_{200/24}$), estimated as ~ 2.5 eV and ~3.85 eV, respectively, seemed to have higher values compared to the other samples synthesized at 200 °C ($B_{200/12}$ and $B_{200/48}$), consisted from anatase and brookite.

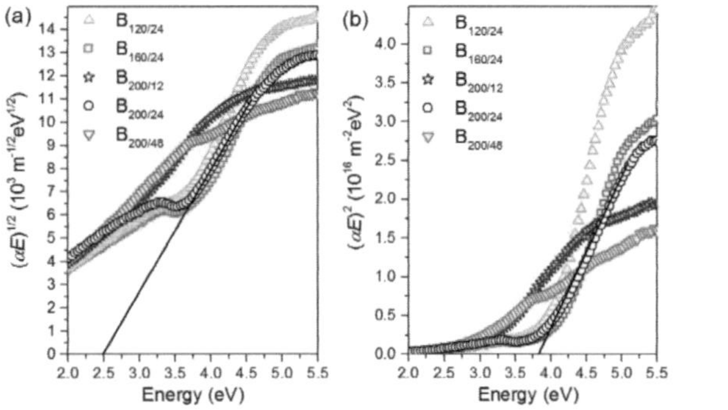

Figure 14. The plot of (a) $(\alpha E)^{1/2}$ and (b) $(\alpha E)^2$ versus E for indirect and direct bandgap transition in synthesized powders, respectively, with characteristic tangent lines for pure brookite sample [9].

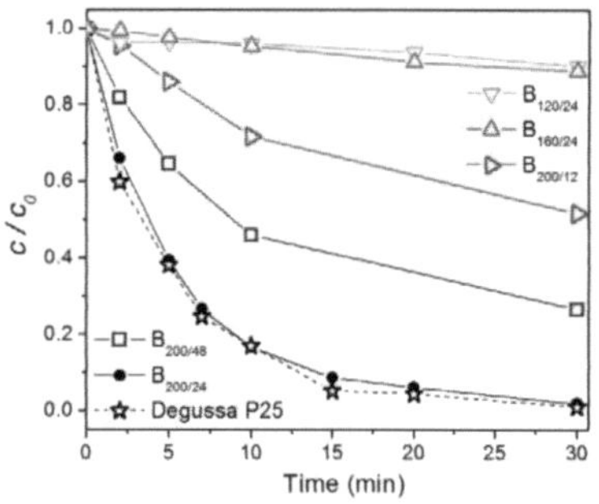

Figure 15. Photocatalytic efficiency of synthesized brookite type powders and Degussa P25 in the removal of alprazolam (c_0 = 0.03 mM) [9].

The kinetics curve of alprazolam removal using synthesized brookite type nanopowders have revealed that the sample consisted of pure brookite has shown the highest efficiency in the photodegradation of alprazolam, removing more than 98% of alprazolam in the first 30 minutes of photocatalytic reaction. The other brookite-rich powders, synthesized at 200 °C, have also shown high photocatalytic efficiency, whereas the powders consisted dominantly of sodium-titanate, produced at lower temperatures (120 and 160°C), have been almost photocatalytically inactive in degradation of alprazolam. Improved photocatalytic efficiency of the pure brookite sample in photodegradation of alprazolam has been mainly ascribed to increased amount of brookite phase in the hydrothermally synthesized TiO_2 samples, rather than brookite crystallite size and specific surface area of nanopowders. Such improvement may be also related to higher energy of direct and indirect transitions in pure brookite sample, as well as a slight increase of the most frequent pore diameter with the increase of brookite content in brookite-rich samples.

5. Conclusion

This study has shown that different mixtures of TiO_2 crystal modifications (anatase, rutile, and brookite) could be obtained with proper choice of synthesis methods and parameters. It has been also demonstrated that any changes in sol–gel and/or hydrothermal synthesis conditions, however small, could have significant effects on the synthesised TiO_2 nanoparticles and consequently on their photocatalytic properties. According to this, careful control of the overall synthesis conditions and the intentional variation of some individual parameters allowed us to synthesize pure and La-doped anatase nanopowders suitable for photocatalytic degradation of MET, as well as brookite-type TiO_2 nanopowders for degradation of alprazolam, with photocatalytic performance comparable or even better than commercially available Degussa P25 TiO_2 nanopowder. At the same time it has been shown that decisive factors for enhanced photocatalytic efficiency in degradation of different pollutants, such as MET and alprazolam, did not have to be the same. For instance, the sample consisted of pure brookite has shown the highest efficiency in the photodegradation of alprazolam, whereas a higher content of brookite (although its highly disordered phase) was not favourable for photocatalytic degradation of MET. The high efficiency of TiO_2 catalysts in photodegradation of MET, as a material with relatively large molecules, has been more correlated with large specific surface area of nanopowders and great number of well-connected pores with dimensions in mesoporous region (> 2 nm). Also, the presence of isolated hydroxyl groups on the surface of TiO_2 which can trap holes to generate $^{\bullet}OH$ radicals, as the photocatalytically important active species, has appeared as one of crucial factors in high photocatalytic efficiency of TiO_2 in degradation of MET. These findings not only have pointed to different degradation mechanisms of the investigated pharmaceutical compounds in the presence of different catalysts, but also have launched further research to deepen understanding of those mechanisms.

Acknowledgements
This work was financially supported by the Ministry of Education, Science and Technological Development (Republic of Serbia), under the Projects No. III45018 and ON172042.

References
[1] Zhu T and Gao S-P 2014 *J. Phys. Chem. C* **118** 11385
[2] Šćepanović M, Abramović B, Golubović A, Kler S, Grujić-Brojčin M, Dohčević-Mitrović Z, Babić B, Matović B and Popović Z V 2012 *J. Sol-Gel Sci. Technol.* **61** 390
[3] Klavarioti M, Mantzavinos D and Kassinos D 2009 *Environ. Int.* **35** 402
[4] Cabrera C G, de Waisbaum R G and Nudelman N S 2005 *J. Phys. Org. Chem.* **18** 156
[5] Bound J P, Kitsou K and Voulvoulis N 2006 *Environ. Toxicol. Phar.* **21** 301
[6] Ikehata K, Naghashkar N J and El-Din M G 2006 *Ozone Sci. Eng.* **28** 353
[7] Golubović A, Abramović B, Šćepanović M, Grujić-Brojčin M, Armaković S, Veljković I, Babić B, Dohčević-Mitrović Z and Popović Z V 2013 *Mater. Res. Bull.* **48** 1363
[8] Grujić-Brojčin M, Armaković S, Tomić N, Abramović B, Golubović A, Stojadinović B, Kremenović A, Babić B, Dohčević-Mitrović Z and Šćepanović M 2014 *Mater. Charact.* **88** 30
[9] Tomić N, Grujić-Brojčin M, Finčur N, Abramović B, Simović B, Krstić J, Matović B and Šćepanović M 2015 *Mater. Chem. Phys.* **163** 518
[10] Soria J, Sanz J, Sobrados I, Coronado J M, Maira A J, Hernandez-Alonso M D and Fresno F 2007 *J. Phys. Chem. C* **111** 10590
[11] Goncalves R H, Schreiner W H and Leite E R 2010 *Langmuir* **26** 11657
[12] Araña J, Doña-Rodrıguez J M, González-Díaz O, Tello Rendón E, Herrera Melián J A, Colón G, Navío J A and Pérez Peña J 2004 *J. Mol. Catal. A-Chem.* **215** 153
[13] Begin-Colin S *et al.* 2009 *J Phys. Chem. C* **113** 16589
[14] Watson S, Beydoun D, Scott J and Amal R 2004 *J. Nanoparticle Res.* **6** 193
[15] Brownson J R S, Tejedor–Tejedor M I and Andersonbrownson M A 2005 *Chem. Mater.* **17** 6304
[16] Yang D, Liu H, Zheng Z, Yuan Y, Zhao J, Waclawik E R, Ke X and Zhu H 2009 *J. Am. Chem. Soc.* **131** 17885
[17] Liu G, Sun C, Cheng L, Jin Y, Lu H, Wang L, Smith S C, Lu G Q and Cheng H-M 2009 *J. Phys. Chem. C* **113** 12317
[18] Androutsopoulos G P and Salmas C E 2000 *Ind. Eng. Chem. Res.* **39** 3747
[19] Cao G, Li Y, Zhang Q and Wang H 2010 *J. Hazard. Mater.* **178** 440

IOP Conf. Series: Journal of Physics: Conf. Series **794** (2017) 012004 doi:10.1088/1742-6596/794/1/012004

Nanophotonics of biomaterials and inorganic nanostructures

P Petrik[1], E Agocs[1], B Kalas[1,2], B Fodor[1,2], T Lohner[1], J Nador[1], A Saftics[1,3], S Kurunczi[1], T Novotny[1], E Perez-Feró[1], R Nagy[1], A Hamori[1], R Horvath[1], Z Hózer[1] and M Fried[1]

[1] Centre for Energy Research, Hungarian Academy of Sciences, Konkoly Thege Miklós út 29-33, H-1121 Budapest, Hungary

[2] Doctoral School of Physics, University of Pécs, 7624 Pécs, Ifjúság útja 6, Hungary

[3] George Olah Doctoral School, Faculty of Chemical Technology and Biotechnology, Budapest University of Technology and Economics, Műegyetem rkp. 3, H-1111 Budapest, Hungary

E-mail: petrik.peter@energia.mta.hu

Abstract.
Optical methods have been used for the sensitive characterization of surfaces and thin films for more than a century. The first ellipsometric measurement was conducted on metal surfaces by Paul Drude in 1889. The word 'ellipsometer' was first used by Rothen in a study of antigen-antibody interactions on polished metal surfaces in 1945. The 'bible' of ellipsometry has been published in the second half of the '70s. The publications in the topic of ellipsometry started to increase rapidly by the end of the '80s, together with concepts like surface plasmon resonance, later new topics like photonic crystals emerged. These techniques find applications in many fields, including sensorics or photovoltaics. In optical sensorics, the highest sensitivities were achieved by waveguide interferometry and plasmon resonance configurations. The instrumentation of ellipsometry is also being developed intensively towards higher sensitivity and performance by combinations with plasmonics, scatterometry, imaging or waveguide methods, utilizing the high sensitivity, high speed, non-destructive nature and mapping capabilities. Not only the instrumentation but also the methods of evaluation show a significant development, which leads to the characterization of structures with increasing complexity, including photonic, porous or metal surfaces. This article discusses a selection of interesting applications of photonics in the Centre for Energy Research of the Hungarian Academy of Sciences.

1. Introduction

Both destructive (e.g. Secondary Ion Mass Spectrometry [SIMS], Sputtered Neutrals Mass Spectrometry, X-Ray Photoelectron Spectrometry, Auger Electron Spectrometry, Glow-Discharge Mass Spectrometry, Raman Depth Profiling) and non-destructive (e.g. Rutherford Backscattering Spectrometry, Elastic Recoil Detection, Angle-Dependent Soft X-Ray Emission Spectroscopy, Grazing Incidence X-Ray Diffraction) depth profiling techniques for thin films are typically limited in measurement speed (minutes to hours) and lateral resolution (millimeters), and favorable in terms of depth resolution (nanometers or tens of nanometers) and sensitivity (detection limits from 1 at% down to 10^{-7} at%) [1]. A third group of techniques based on cross-sectioning (e.g. Transmission Electron Microscopy [TEM], Scanning Electron Microscopy

[SEM], Scanning Auger Electron Microscopy, Raman Mapping, etc.) provide high resolution in both lateral and vertical directions, but require extensive sample preparations. Spectroscopic ellipsometry belongs to the group with limited lateral resolution and high sensitivity (for layer thickness, composition and vertical resolution), but its most important feature to our topic is the high speed combined with the non-destructive nature. This allows either the mapping of surfaces (instrumentation for imaging and mapping have long been existing [2, 3]) and the real time monitoring of solid state [4] or bio processes [5]. This article provides a short review of both bio and inorganic applications of ellipsometry. Because the field is large, we mainly focus on the results of our group from the recent years.

2. Characterization of biomaterials

2.1. Past developments

Already in the first significant book on ellipsometry published in 1977 by R. M. A. Azzam and N. M. Bashara, a 10-pages chapter titled 'Applications of ellipsometry in biology and medicine' was devoted to bioellipsometry [6]. This book was followed by other excellent ones on ellipsomery [7–10], many of them dealing with biological applications of ellipsometry as well (e.g. Chapter 12 of Ref. [8] or Chapter 7.4.2 of Ref. [9]). The first applications mentioned in Ref. [6] are by Poste et al. [11], Vroman et al. [12] and Rothen [13]. The word 'ellipsometer' was first used by Rothen in an article from the year 1945 [14], already motivated by the measurement of film thicknesses with 1 Å accuracy on polished metallic surfaces for antigen antibody investigations. These studies were based on single-wavelength ellipsometry, taking advantage of the high sensitivity and the *in situ* capability of monitoring the interface during immunological reactions and protein adsorption processes. The first spectroscopic ellipsometer was constructed in the middle of the '70s [15], and spectroscopic ellipsometry has been applied to biological materials already in the '80s [16]. There have been many direction of the development, including imaging [17, 18] or plasmonic [19] ellipsometry.

2.2. Recent developments

Depending on the optical properties of the surrounding media, ellipsometry has a typical sensitivity of $\approx 10^{-4}$ in refractive index units (which corresponds to ≈ 1 ng/mm^2 for surface mass density changes in a silicon substrate), when using a standard flow cell configuration, in which the sample surface is measured through the liquid [20]. A significant increase of the sensitivity can be reached by surface plasmon enhancement [19, 21]. In this configuration the interface is reached from the substrate, and measured by the evanescent field at the proximity of the interface. Using this concept, illuminating the interface through a semi-cylinder (figure 1), a wide range of incident angles can be used with a standard spectroscopic ellipsometer [22].

The performance of this setup has been demonstrated for different plasmonic gold layer thicknesses [23]. It can also be shown that using the ellipsometric approach with phase measurement capability, a sensitive region other than that for the optimum plasmonic configuration can also be found on the thickness-incident angle plane close to small (less than 10 nm) gold layer thicknesses (see figure 2). The best sensitivity values are $\approx 10^{-6}$ in refractive index units that have a potential to get close to advanced waveguide biosensors [24, 25] after further development and optimization.

Also, designing special lenses, the light can be focused on the plasmon layer through the semi-cylinder (figure 1) reaching a spot diameter below 0.5 mm. The focusing allows the use of a small volume (10 μL) flow cell with a channel length of 8 mm. When scanning the spot along the long axis of the channel that coincides with the axis of the cylinder, multi-surface (multi-channel) measurements are possible within one experiment [22]. The internal reflection, plasmon-enhanced measurement can also be combined with imaging ellipsometry [26]. A different method

Figure 1. Schematic drawing of a 10-μL Kretschmann-Raether flow cell (left, ϵ_a, ϵ_m and ϵ_0 denote the dielectric fucntions of the BK7 glass semi-cylinder, the gold layer and the water in the flow cell, respectively; θ_a is the angle of incidence, d stands for the thickness of the gold layer), as well as the realized construction that can readily be put on the mapping stage of the ellipsometer (middle, see also Ref. [22]). To the right, the schematic view of the dual-channel configuration is shown. By preparing a sample surface divided into a processed (in this case covered by nanoparticles) and an unprocessed reference area, both of them can be measured in the same process by stepping the measurement spot along the axis of the cylinder that coincides with the long axis of the flow cell.

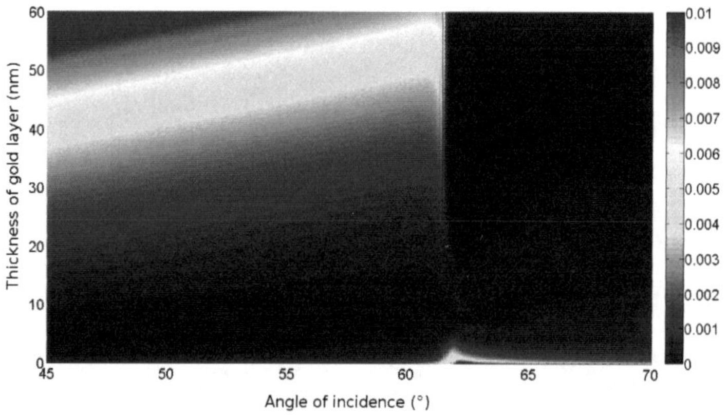

Figure 2. Calculated map of the absolute value of the complex reflection coefficient, $|\rho| = |r_p/r_s|$, as a function of the angle of incidence and layer thickness. r_p and r_s denote the reflection coefficients of light polarized parallel and perpendicular to the plane of incidence, respectively. The deep blue areas with the smallest values represent the most sentivite regions. Besides the usual plasmon resonance area in the top right part of the graph, there is another sensitive region for smaller angles of incidence (below 60°) and small (below 10 nm) thicknesses (bottom left part of the image).

of the surface field enhancement is the design of suitable multi-layer structures to generate Bloch surface waves (figure 3 from Ref. [27]).

2.3. Combination of techniques

The combination of characterization methods offers a range of benefits, such as the simultaneous measurement of the same process in space and time, as well as the complementary nature of sensitivities in terms of material properties or distance from the surface. The combined

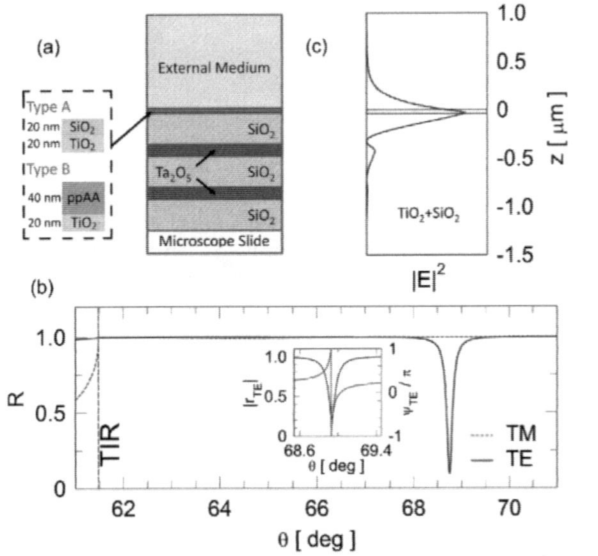

Figure 3. Schematic drawing of a one-dimensional photonic crystal to generate Bloch surface waves (reprinted from Ref. [27]).

application of quartz crystal microbalance (QCM) and spectroscopic ellipsometry unifies the mass sensitivity of QCM and the thickness sensitivity of spectroscopic ellipsometry [28–30]. The optical density can also be related to the mass. However, as usual, it is an important and complex task to turn the high sensitivity into a high accuracy, i.e. into a validated quantity verified by reference methods and calibration samples.

Using internal reflection ellipsometry [31], the increased sensitivity can only be exploited close to the interface, and the sensitivity exponentially decreases with increasing distance from the surface as the evanescent field vanishes. To overcome this issue, surface sensitive waveguide interferometry has been combined with standard flow cell ellipsometry to measure from both sides of the interface *in situ* and simultaneously (figure 4). In this combination, the high surface sensitivity of waveguide interferometry (typically better than 1 pg/mm^2) is extended with a spectroscopic and complex modeling capability of ellipsometry (although at a one or two orders of magnitude lower sensitivity), also at larger distances from the surface – because this measurement is performed from the other side, through the liquid. This technique has been demonstrated for the deposition of basic protein molecules, such as fibrinogen [32].

Further possibilities for the development of the internal reflection plasmonic design is the extension of the wavelengths towards the ultra-violet spectral region to utilize the absorption of protein molecules, using special optical components, as well as the application of structured substrates, for example to create Bloch surface waves for an increased sensitivity ($\approx$2 pg/mm^2 was demonstrated in Ref. [27], however, only for single-wavelength ellipsometry, see figure 3). The enhanced electric field at the sensor surface increases the sensitivity of detecting the adsorption at the interface of the external medium and the Bragg structure. A key target of future developments is the increase of sensitivity combined with spectroscopic and complex modeling capabilities.

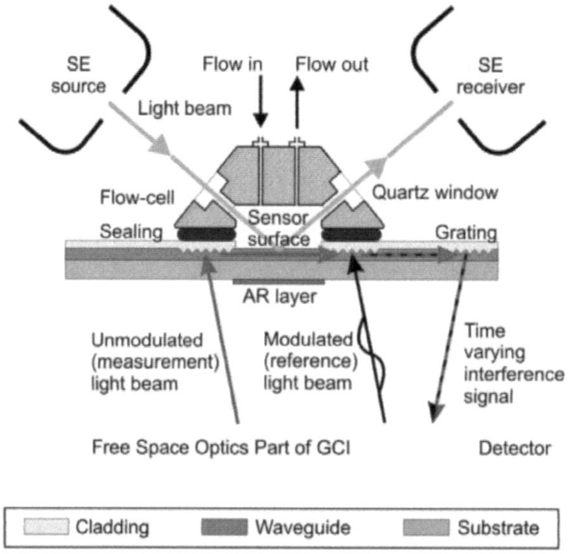

Figure 4. Schematic drawing on the combination of spectroscopic ellipsometry and waveguide interferometry [32].

3. Characterization of inorganic structures

Although the ellipsometric method is more than hundred years old [33], and it has been demonstrated already in 1945 for biological applications [14] as shown above, the technique has become widely used only from the second half of the '80s (see figure 1 in Ref. [34]), because computers were needed for both the control of the instrument and for the evaluation of the measured data. Also, the first major application of ellipsometry was the measurement of the surface oxide on single-crystalline silicon wafers and other thin films in the microelectronics [35]. The reason is that being an extremely sensitive method, the samples had to be perfect on the nanometer scale, so that simple optical models can be used. The trend is today to measure less perfect samples like e.g. biological layers, however, the evaluation of these structures requires complex models with numerous parameters. To fit these models we need strong computers, sophisticated software, and a lot of measurement data, typically taken in a broad wavelength range and at many angles of incidence.

A recent example is shown in figure 5 demonstrating a multiple-angle of incidence ellipsometry measurement on a multi-layer structure with thicknesses in the nanometer range. It has been revealed that the minute changes in the amorphization of the substrate during RF-plasma deposition of Nb_2O_5 layers can sensitively be followed by ellipsometry. It has also been shown that a possible correlation between the thickness and the refractive index of these very thin layers can be circumvented by calculating the product of these two quantities defined as the amount of damage [36], similar to de Feijter's method for calculating the adsorbed amount of proteins from the product of the refractive index and the layer thickness [37].

The development of ellipsometric modeling is not only directed toward complex multi-layer structures, but also nanostructures such as plasmonic, dielectric and other nanoparticles, solid state or biological nanostructures or periodic photonic structures. Figure 6 shows the ellipsometric modeling of Si nanowires with different lengths. It is shown that the wavelength range used for the fit is not only required to be adapted to the length of the nanowires, but the characteristic feature sizes can also be determined based on the used wavelength range [38, 39].

Surface roughness 2.25 ± 0.04 nm
Nb-oxide (TL model) 359.2 ± 0.2 nm
SiO_2 2.7 ± 0.1 nm
EMA, 87% c-Si + 14% ia-Si 6.8 ± 0.3 nm
c-Si substrate

Wall potential [kV]	Duration of rf sputtering deposition [sec]	Thickness of ion-implantation amorphized Si layer [nm]
2.0	100	1.35
1.5	150	1.02
1.0	300	0.84

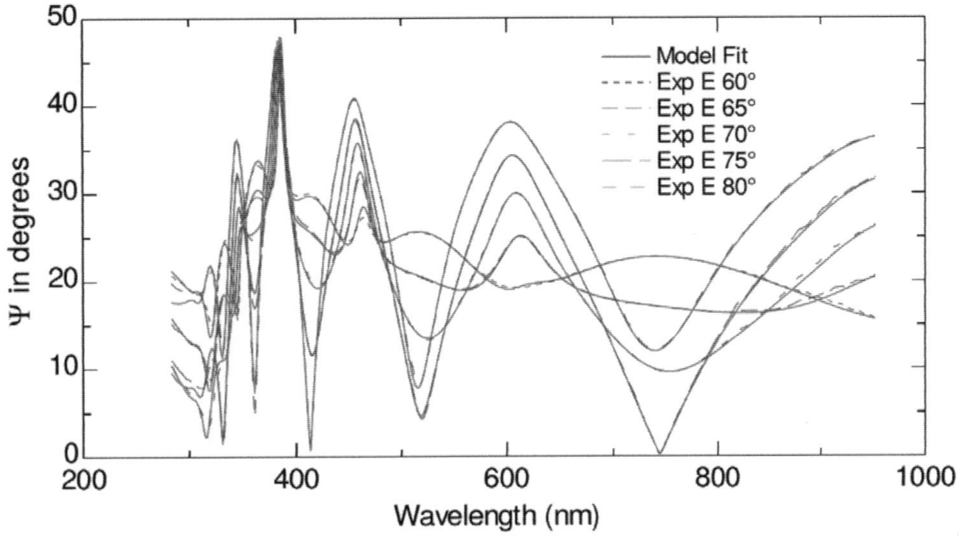

Figure 5. Measured and fitted ellipsometric angles of Ψ at different angles of incidence (shown in the legend) as a function of wavelength for Nb_2O_5 layers created by 120 minutes of RF-sputtering at a wall potential of 2.0 kV. The top left table shows the optical model (the abbreviations of TL and EMA stand for Tauc-Lorentz and Effective Medium Approximation, respectively). The thickness of the ion-implantation amorphized Si layer determined from ellipsometry is shown in the top right table for different wall potentials and sputter durations.

This work also shows how different kinds of anisotropies can be distinguised based on the fit results using different models and assumptions.

Another example of the rapid increase of the research on nanostructures is plasmonics as shown in figure 7, and its application in photovoltaics, typically following by a delay of 10-15 years (see also Ref. [40]).

As pointed out in the introduction, besides the high sensitivity, the major advantage of optical techniques is the high speed and the non-destructive nature. For example, even if it would be technically possible to measure on a large surface point-by-point using SIMS or SEM, it would take so much time, and it would require so much effort, that it's use wouldn't be a benefit. The most perspective capability of the optical techniques like ellipsometry is not the characterization of a completely unknown sample or structure, but the high throughput and high sensitivity testing and monitoring of processes, as in the above examples for bio processes, and also for solid state processes e.g. in photovoltaics. Figure 8 shows a prototype of a divergent

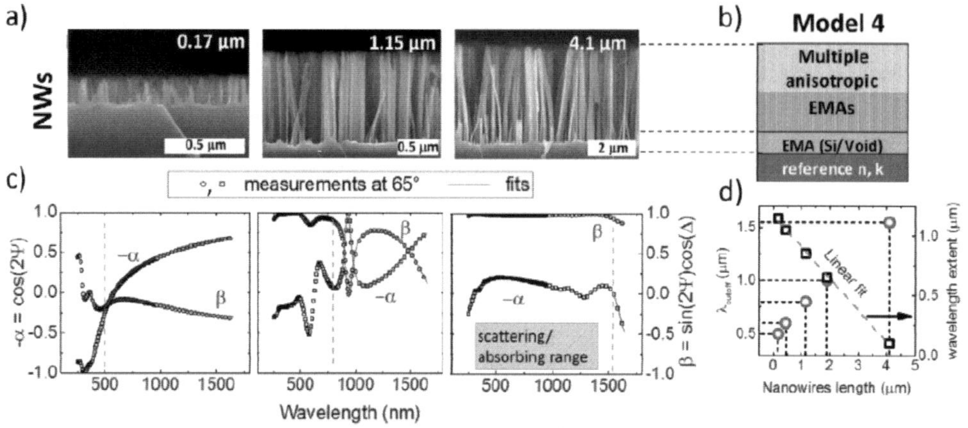

Figure 6. (a) Cross-sectional electron microscopy images of Si nanowires of different lengths; (b) effective medium approximation-based optical model used for the ellipsometric evaluations; (c) measured and fitted ellipsometric values (α and β) at an angle of incidence of 65°; (d) cut-off wavelengths and the width of the wavelength range (denoted as 'wavelength extent') as a function of the length of the nanowires.

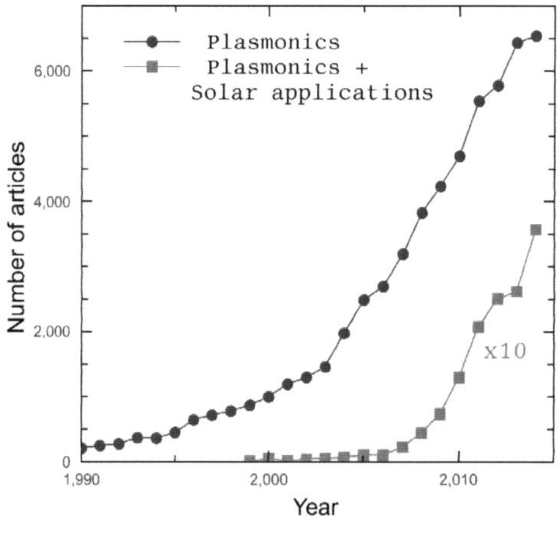

Figure 7. Number of publications over the years having the keyword 'Plasmon*' (denoted as 'Plasmonics') or 'Plasmon AND (Solar OR Photovoltaic)' (denoted as Plasmonics + Solar applications, with the number multipled by 10) in the title, keywords or abstracts.

light source ellipsometer that has been developed in our group for more than a decade [3]. The latest versions are capable of mapping a 30-by-30 point area within 2 minutes, even on surfaces as large as 90-by-90 cm [41]. The most important feature is that these maps can be recorded spectroscopically, which means that each point of the 30-by-30 area includes a whole spectrum in a range depending on the given configuration, which can always be adapted to the layer structures and materials to be measured. Using this approach, maps can not only be taken on the layer thicknesses even in multi-layer structures, but also on certain material properties like crystallinity, resistivity, band gap, and a lot more, depending on the possibilities to relate the

IOP Conf. Series: Journal of Physics: Conf. Series **794** (2017) 012004 doi:10.1088/1742-6596/794/1/012004

dispersion of the material to its physical properties. These properties can be tested real time, during processing on solar panels, displays, sensors, etc., on large areas, non-destructively.

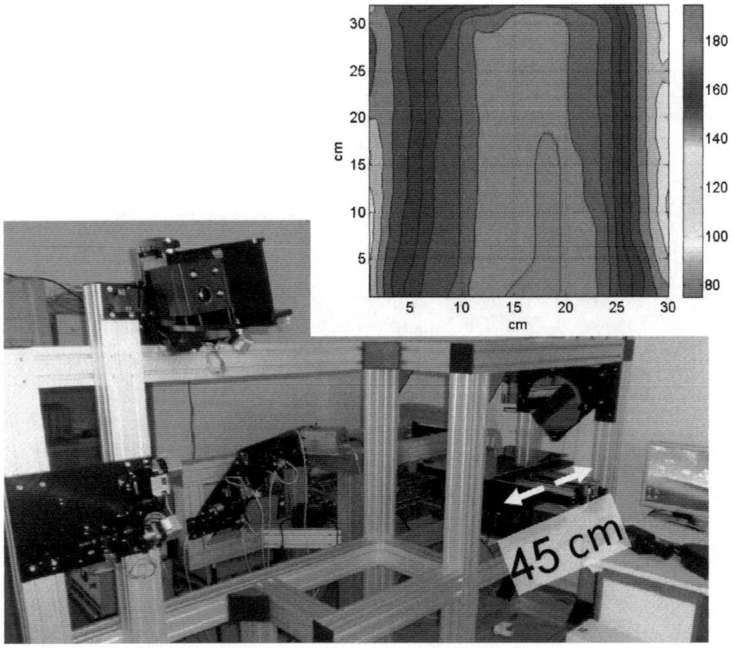

Figure 8. Prototype of a divergent light source mapping ellipsometer for sample sizes of 300-450-600-900 mm. The graph shows the thickness map of a ZnO layer deposited on a Mo-covered glass sheet on a 30 cm by 30 cm area that can be measured within 2 minutes using our mapping device. Using a commercial tool with point-by-point measurement, a map with the same resolution requires approximately 20 minutes.

Another example of the development in non-conventional directions toward new configurations and new materials is the investigation of Zr alloy tubes used for nuclear fuel cladding [42]. It was shown that the surface properties of such tubes can be measured by proper focusing directly on the tubes with diameters as small as 9.1 mm. Figure 9 shows the measurement setup, in which the light of the ellipsometer is focused on the outer surface of the Zr tube, allowing the determination of the thickness and the optical properties of the surface oxide. From a detailed investigation it is revealed that the oxide properties gradually change in depth. Consequently, improved optical models involving grading, multi-sample analysis and B-spline parameterization of the dielectric function are required [43]. As mentioned above, the power of the optical characterization is not to discover an unknown sample and structure. Those properties that can be measured by electron microscopy should not be measured by ellipsometry. However, ellipsometry can be measured on a large series of samples for qualifying, checking and testing the properties, it can measure samples in line, non-destructively, and a further possibility is to perform investigations during processing, like oxidation, which can lead to the understanding of the dynamics and kinetics of those processes. Finally, there are a range of material properties (refractive index, band gap, critical points, joint density of states and band structure related properties) that can most sensitively be determined by optical techniques.

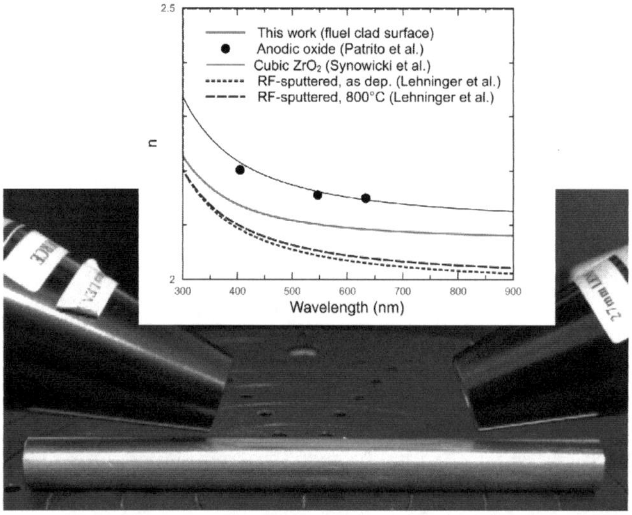

Figure 9. Setup for the measurement of a Zr tube for nuclear fuel cladding. The inset shows the refractive index of the ZrO_2 layer determined using the multi-sample method, together with other references from the literature.

4. Conclusions

In this summary it has been demonstrated that there are large opportunities in the development of optical characterizations for both biological and inorganic nanostructures and thin films. There is a significant progress in the instrumentation (plasmonics, combined methods) as well as in the evaluations (grading, anisotropy, scatterometry, multi-sample method and special parameterizations). The two straightforward major motivations are the improvement of the sensitivity and the extension of the method towards new materials and structures. Being an indirect method, the verification is also of primary importance using complementary techniques. Once verified, the optical techniques provide a way for the quick, high-sensitivity and non-destructive testing of the sample properties.

Acknowledgments

Support from OTKA K115852, M-ERA-NET 117847, ENIAC E450EDL, SEA4KET, TÉT-12-DE-1-2013-0002, TÉT-14-FR-1-2015-0041, Momentum, ERC_HU and EK (T123) internal research projects are gratefully acknowledged.

References

[1] Abou-Ras D, Caballero R, Fischer C H, Kaufmann C A, Lauermann I, Mainz R, Mönig H, Schöpke A, Stephan C, Streeck C, Schorr S, Eicke A, Döbeli M, Gade B, Hinrichs J, Nunney T, Dijkstra H, Hoffmann V, Klemm D, Efimova V, Bergmaier A, Dollinger G, Wirth T, Unger W, Rockett A A, Perez-Rodriguez A, Alvarez-Garcia J, Izquierdo-Roca V, Schmid T, Choi P P, Müller M, Bertram F, Christen J, Khatri H, Collins R W, Marsillac S and Kötschau I 2011 *Microsc. Microanal.* **17** 728

[2] Vaupel M, Yunfeng S and Zhimin Y 2008 *phys. stat. solidi a* **205** 772–778

[3] Fried M, Juhasz G, Major C, Petrik P, Polgar O, Horvath Z and Nutsch A 2011 *Thin Solid Films* **519** 2730–2736

[4] Collins R W, An I, Nguyen H V and Lu Y 1993 *Thin Solid Films* **233** 244

[5] Arwin H 2011 *Thin Solid Films* **519** 2589

[6] Azzam R M A and Bashara N M 1977 *Ellipsometry and Polarized Light* (Elsevier)

[7] Riedling K 1987 *Ellipsometry for Industrial Applications* (Wien and New York: Springer Verlag)
[8] Irene E A 2005 *Handbook of ellipsometry* ed Irene E G and Tomkins H G (Norwich, NY: William Andrew)
[9] Fujiwara H 2007 *Spectroscopic Ellipsometry: Principles and Applications* (New York: Wiley)
[10] Losurdo M and Hingerl K 2012 *Ellipsometry at the Nanoscale* (Heidelberg: Springer-Verlag)
[11] Poste G and Moss C 1972 *Progress in Surface Science* ed Davison S G (Amsterdam: North-Holland)
[12] Vroman L and Adams A C 1969 *Proceedings of the Symposium on Recent Developments in Ellipsometry* ed MBashara N, Buckman A B and Hall A C (Amsterdam: North-Holland)
[13] Rothen A 1974 *Progress in Surface and Membrane Science* ed Cadonhead D A, Danielli E F and Rosenberg M D (New York: Academic Press)
[14] Rothen A 1945 *Rev. Sci. Instr.* **16** 26
[15] Aspnes D E and Studna A A 1975 *Appl. Optics* **14** 220
[16] Arwin H 1986 *Appl. Spectroscopy* **40** 313
[17] Jin G, Jansson R and Arwin H 1996 *Rev. Sci. Instruments* **67** 2930
[18] Bae Y M, Oh B, Lee W, Lee W H and Choi J 2004 *Biosensors and Bioelectronics* **20** 895–902
[19] Poksinski M and Arwin H 2004 *Thin Solid Films* **455-456** 716–721
[20] Kurunczi S, Nemeth A, Hulber T, Kozma P, Petrik P, Jankovics H, Sebestyen A, Vonderviszt F, Fried M and Barsony I 2010 *Appl. Sur. Sci.* **257** 319–324
[21] Oates T W H, Wormeester H and Arwin H 2011 *Progress in Surface Science* **86** 328–376
[22] Nador J, Kalas B, Saftics A, Agocs E, Kozma P, Korosi L, Szekacs I, Fried M, Horvath R and Petrik P 2016 *Opt. Express* **24** 4812–4823
[23] Petrik P, Agocs E, Kalas B, Kozma P, Fodor B, Nador J, Major C and Fried M *SPIE Proceedings* 95290W
[24] Kozma P, Hámori A, Kurunczi S, Cottier K and Horvath R 2011 *Sensors and Actuators B* **155** 446–450
[25] Patko D, Cottier K, Hamori A and Horvath R 2012 *Opt. Express* **20** 23162
[26] Liu L, Viallat A and Jin G 2014 *Sensors and Actuators B* **190** 221–226
[27] Sinibaldi A, Anopchenko A, Rizzo R, Danz N, Munzert P, Rivolo P, Frascella F, Ricciardi S and Michelotti F 2015 *Anal. and Bioanal. Chemistry* **407** 3965–3974
[28] Höök F, Vörös J, Rodahl M, Kurrat R, Böni P, Ramsden J, Textor M, Spencer N, Tengvall P, Gold J and Kasemo B 2002 *Colloids and Surfaces B* **24** 155–170
[29] Iturri Ramos J J, Stahl S, Richter R P and Moya S E 2010 *Macromolecules* **43** 9063–9070
[30] May R A, Flaherty D W, Mullins C B and Stevenson K J 2010 *J. Phys. Chem. Lett.* **1** 1264–1268
[31] Bortchagovsky E G 1997 *Proceedings of SPIE - The International Society for Optical Engineering* **3094** 239–249
[32] Agocs E, Kozma P, Nador J, Hamori A, Janosov M, Kalas B, Kurunczi S, Fodor B, Ehrentreich-Förster E, Fried M, Horvath R and Petrik P 2016 *accepted for publication in Appl. Surf. Sci.*
[33] Drude P 1889 *Ann. Phys. Chem.* **36** 532
[34] Aspnes D E 2014 *Thin Solid Films* **571** 334–344
[35] Irene E A 1993 *Thin Solid Films* **233** 96
[36] Fried M, Juhász G, Major C, Petrik P and Battistig G 2009 *17th IEEE Conference on Advanced Thermal Processing of Semiconductors, RTP 2009* 8–11
[37] De Feijter J a, Benjamins J and Veer F A 1978 *Biopolymers* **17** 1759–1772
[38] Kozma P, Fodor B, Deak A and Petrik P 2010 *Langmuir* **26** 16122–16128
[39] Petrik P, Agocs E, Volk J, Lukacs I, Fodor B, Kozma P, Lohner T, Oh S, Wakayama Y, Nagata T and Fried M 2014 *Thin Solid Films* **571** 579
[40] Petrik P 2016 *Characterization of Photovoltaic Materials and Devices by Spectroscopic Ellipsometry* ed Fujiwara H and Collins R W (Springer)
[41] Fried M 2014 *Thin Solid Films* **571** 345–355
[42] Hózer Z, Győri C, Matus L and Horváth M 2008 *J. Nucl. Materials* **373** 415
[43] Petrik P, Sulyok A, Novotny T, Perez-Feró E, Kalas B, Agocs E, Lohner T, Lehninger D, Khomenkova L, Nagy R, Heitmann J, Menyhard M and Hózer Z 2016 *accepted for publication Appl. Surf. Sci.*

Network structure of Mo-oxide glasses

M Fabian[1,2], E Svab[2], M Milanova[3] and K Krezhov[4]

E-mail: fabian.margit@energia.mta.hu

[1]Centre for Energy Research, 1121 Budapest, Konkoly-Thege st. 29-33, Hungary
[2]Wigner Research Centre for Physics, 1121 Budapest, Konkoly-Thege st. 29-33, Hungary
[3]Institute of General and Inorganic Chemistry, 1113 Sofia "Acad. G. Bonchev" str., bld. 11, Bulgaria
[4]Institute for Nuclear Research and Nuclear Energy, 1784 Sofia, 72 Tzarigradsko Chaussee, Bulgaria

Abstract. The structure of molybdate glasses have been investigated by neutron and high-energy X-ray diffraction coupled with Reverse Monte Carlo (RMC) simulation technique. From the modelling the partial atomic correlation functions $g_{ij}(r)$, the coordination number distributions CN_{ij} and bond angle distributions have been revealed. For binary $90MoO_3$-$10Nd_2O_3$ glass composition the fraction of MoO_4/MoO_6 was 0.55/0.25. Three type of ternary system have been studied, where the most important structural units was authenticated. For MoO_3-Nd_2O_3-B_2O_3 sample mixed MoO_4-BO_4 and MoO_4-BO_3 linkages form pronounced intermediate-range order. In case of MoO_3-ZnO-B_2O_3 series the BO_3 and BO_4 units are linked to MoO_4 and/or ZnO_4, forming mixed MoO_4-$BO_4(BO_3)$, MoO_4-ZnO_4 and ZnO_4-$BO_4(BO_3)$ bond-linkages.

1. Introduction

Rare-earth molybdates display a variety of crystal and amorphous structures and possess useful electro-physical, optical, high ion and electron conductivity properties derived from the parent oxide $La_2Mo_2O_9$ [1,2]. The molybdates $LiLn(MoO_4)_2$, Ln=Nd, Pr, Ho are excellent LN-doped laser hosts [3].

Zinc boromolybdate materials are known as low melting dielectric materials and they have high potential in several applications, due to their stable and unique structure with favourable optical properties. In contrast to the most molybdates with relatively well-characterised crystalline and magnetic structures, e.g. [4-7], the preparation and structural information on amorphous molybdate systems is not ample. Glass formation in molybdate systems, containing different kind of oxides, has been examined by our team as a part of long term academic projects, concerning non-traditional glass forming systems [8-10]. The phase diagram for several oxides, MoO_3-$Nd(La)_2O_3$-B_2O_3 was set up [9] and the basic structural and optical characteristics were investigated by XRD, DTA, SEM, UV-VS spectra and IR spectroscopy [10].

The purpose of this work is to verify the competition for the glass network formation between non-traditional network formers - MoO_3, ZnO - and to get a deeper insight into the structural network characteristics of new series of molybdates containing modifiers Nd_2O_3, MgO and classical glass former B_2O_3. For this study three new amorphous sample series of the glassy MoO_3-Nd_2O_3-MgO, MoO_3-Nd_2O_3-B_2O_3 and MoO_3-ZnO-B_2O_3 system have been prepared by rapid quench technique.

In this paper, the structure of molybdate glasses is investigated using neutron and high-energy X-ray diffraction experiments, and for structure modelling the Reverse Monte Carlo (RMC) computer simulation technique [11] was applied. We expect to obtain new structural data, especially on the characteristics of the partial atomic pair correlation functions, the first neighbour atomic distances, coordination numbers and network former units and their linkages.

2. Experimental details and methods

2.1. Samples

The samples with the nominal composition $90MoO_3$-$10Nd_2O_3$ (hereafter referred to as Mo90Nd10), ternary $80MoO_3$-$15Nd_2O_3$-5MgO (Mo80Nd15Mg5), $75MoO_3$-$12.5Nd_2O_3$-12.5MgO (Mo75Nd12.5Mg12.5); $50MoO_3$-$25Nd_2O_3$-$25B_2O_3$ (Mo50Nd25B25), $40MoO_3$-$30Nd_2O_3$-$30B_2O_3$ (Mo40Nd30B30) and $20MoO_3$-$30Nd_2O_3$-$50B_2O_3$ (Mo20Nd30B50); $30MoO_3$-50ZnO-$20B_2O_3$ (Mo30Zn50B20), $20MoO_3$-50ZnO-$30B_2O_3$ (Mo20Zn50B30) and $10MoO_3$-50ZnO-$40B_2O_3$ (Mo10Zn50B40) (mol%) have been prepared from reagent grade powders (B_2O_3, MoO_3, Nd_2O_3, ZnO, MgO) melted in a platinum cruciable at 1200-1300ºC for 30 min. B_2O_3 enriched with ^{11}B isotope (99.6%) has been used in order to avoid the high neutron absorption cross section of the ^{10}B isotope. The melting temperature was limited to 1300°C taking into account the volatility of the components. The glasses were obtained by pouring the melts onto an iron plate and by pressing with another iron plate.

2.2. Neutron and X-ray diffraction experiments

Neutron diffraction (ND) data have been obtained on the two-axis 'PSD' diffractometer at the 10 MW Budapest research reactor (λ_0=1.068Å) [12], on the time-of-flight 'HIPD' instrument [13] at the LANSCE pulsed neutron source (for Mo-Nd-Mg-O samples) and on the '7C2' diffractometer at the Orphée reactor of the LLB-CEA-Saclay (λ_0=0.726 Å) [14] (for the Mo-Nd-B-O and Mo-Zn-B-O samples series). Glass powders were contained in a cylindrical vanadium sample holder of 8 mm (PSD, 7C2) and 10 mm (HIPD) diameter for the three types of experiments. The structure factors, $S(Q)$s, were evaluated from the raw experimental date using the programme packages available at the facilities. The $S(Q)$s were obtained in the Q=0.45-9.8Å^{-1} (PSD), Q=0.2-15.5Å^{-1} (7C2) and Q=1.5-35Å^{-1} (HIPD) experiments. The agreement between the corresponding $S(Q)$ data was within 1% in the overlapping Q range. The statistics of the PSD data is excellent at relatively low Q values, while the statistics and the information content of the 7C2 and HIPD data is unique above 8Å^{-1}. Therefore the $S(Q)$ patterns were combined by normalizing the PSD data to the 7C2 and HIPD data sets in the 4-8Å^{-1} interval by the least-squares method, and the average values of the two spectra were used for further data treatment, while for Q<4Å^{-1} the PSD data and for Q>8Å^{-1} the 7C2 and HIPD data were used.

The high-energy X-ray diffraction (XRD) measurements were carried out at the BW5 experimental station [15] at HASYLAB, DESY for Mo-Nd-O and Mo-Nd-Mg-O samples. The powdered samples were filled into quartz capillary of 2 mm in diameter. The energy of the radiation was 109.5 keV. Raw data were corrected for detector dead time, background, absorption, and variations in detector solid angle. The synchrotron X-ray radiation makes it possible to reach diffraction data up to high-Q values. For the Mo-Nd-B-O samples X-ray diffraction studies were performed at the beam line ID22 at ESRF. The fine powdered samples were filled into kapton capillary tubes of 1 mm in diameter and measured at room temperature. The energy of the radiation was 70.8 keV. The X-ray $S(Q)$ data were obtained with a good signal-to-noise ratio up to Q_{max}=25Å^{-1} at BW5 station and up to Q_{max}=22 Å^{-1} at ID22, for higher Q-values the experimental data proved to be noisy.

The overall run of the curves for the two radiations is fairly different, because of the differences in the weighting factors, w_{ij} of the partial structure factors, $S_{ij}(Q)$:

$$S(Q) = \sum_{i,j}^{k} w_{ij} S_{ij}(Q) \tag{1}$$

$$w_{ij}^{ND} = \frac{c_i c_j b_i b_j}{\left[\sum_{i,j}^{k} c_i b_j\right]^2} \tag{2}$$

$$w_{ij}^{XD}(Q) = \frac{c_i c_j f_i(Q) f_j(Q)}{\left[\sum_{i,j}^{k} c_i f_j(Q)\right]^2} \tag{3}$$

where c_i, c_j are the molar fractions of the components, b_i, b_j the coherent neutron and $f_i(Q)$, $f_j(Q)$ the X-ray scattering amplitudes, and k is the number of elements in the sample. Note, that the neutron scattering amplitude is constant [16], while the X-ray scattering amplitude is Q-dependent [17].

3. Reverse Monte Carlo simulations

Reverse Monte Carlo modelling have been performed using the RMC^{++} code [11], starting with a disordered atomic configuration. The RMC minimizes the squared difference between the experimental $S(Q)$ and the calculated one from a 3D atomic configuration. The RMC algorithm calculates the one-dimensional partial atomic pair correlation functions $g_{ij}(r)$, and they are inverse Fourier transformed to calculate the partial structure factors, $S_{ij}(Q)$.

For the RMC starting model disordered atomic configuration was built up with a simulation box containing 10.000 atoms. The density values were 0.0733, 0.0668, 0.0670 atoms·Å^{-3} for Mo90Nd10, Mo80Nd15Mg5, Mo75Nd12.5Mg12.5 samples, respectively, and the corresponding RMC half-box lengths were 25.739, 26.548 and 26.522 Å. The density values were 0.0683, 0.0684 and 0.0692 atoms·Å^{-3} were used for the Mo50-Nd25-B25, Mo40-Nd30-$B_2O_3$30 and Mo20-Nd30-B50 samples, respectively, and the corresponding RMC half-box lengths were 26.353, 26.340 and 26.238 Å. The density data 0.087, 0.081 and 0.076 atoms·Å^{-3} and half-box values r_{max}=24.30 Å, 24.90 Å and 25.43 Å for the Mo10-ZnO50-B40, Mo20-Zn50-B30 and Mo30-Zn50-B20 samples, respectively.

In order to generate a realistic structural model, minimum atom-atom distances were chosen to avoid an unphysical overlapping of the atoms and B atoms were imposed to be trigonally and tetrahedrally coordinated. It is reasonable to suppose that boron atoms are surrounded both by three-coordinated and four-coordinated oxygen atoms, forming trigonal BO_3 and tetrahedral BO_4 units as reported for binary neodymium borate glasses [20], for ternary rare-earth boromolybdate MoO_3-$Nd(La)_2O_3$-B_2O_3 [9,10], for B_2O_3-Bi_2O_3-MoO_3 glass [21], similarly, as for binary alkali borate glasses [22 and references therein]. For the starting cut-off distances we have used the characteristic values for B-O 1.2 Å, Mo-O 1.6 Å, Nd-O 2.0 Å, Zn-O 1.9 Å and O-O 2.3 Å. Several RMC runs have been completed by modifying the cut-off distances in the way, that the results of each run have been carefully checked to obtain reliable data for each $g_{ij}(r)$ and coordination number distributions, $CN_{ij}(n)$. The convergence of the RMC calculations was good and the final $S(Q)$s matched very well the experimental ones. (see figure 1).

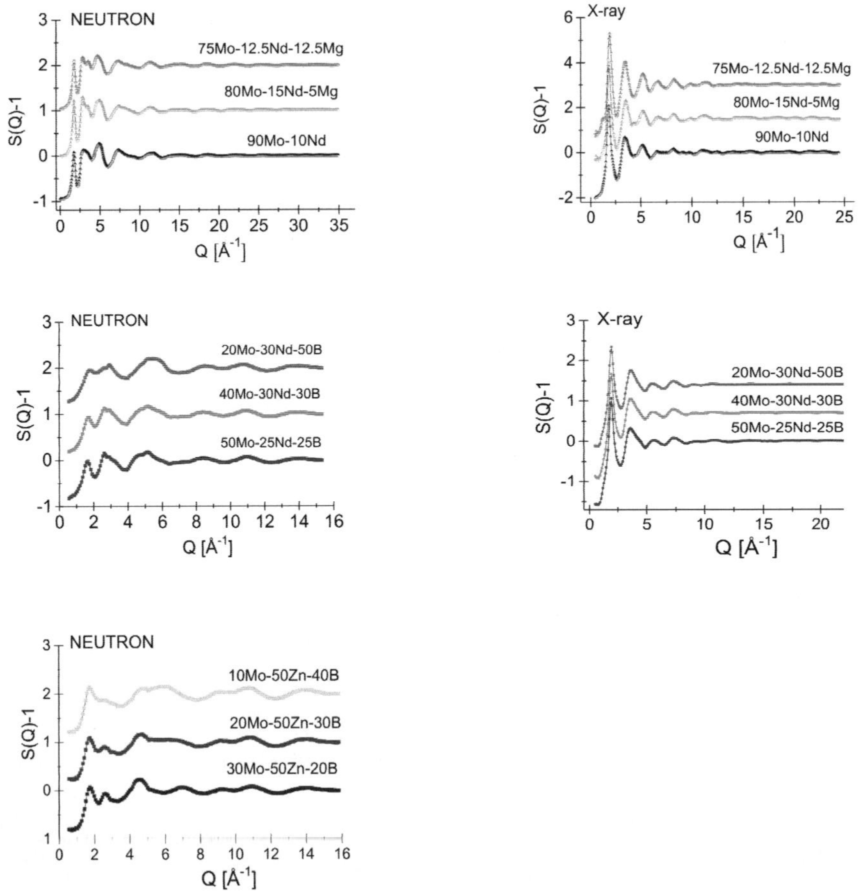

Figure 1. Experimental data (colour) and RMC simulation (solid line): Neutron and X-ray diffraction for Mo-Nd-Mg-O, MO-Nd-B-O and Mo-Zn-B-O series.

The partial atomic pair correlation functions, $g_{ij}(r)$ have been revealed from RMC simulation with a good reproducibility and acceptable statistics. The most important oxygen-linked $g_{i\text{-}O}(r)$ and coordination number distributions are shown in figure 2.

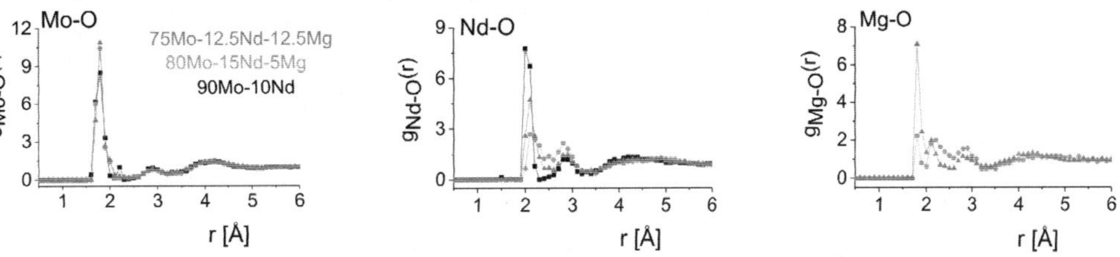

Figure 2a. O-linked partial atomic pair correlation functions obtained from the RMC modelling for the: Mo90Nd10 (black square), Mo80Nd15Mg5 (orange circle) and Mo75Nd12.5Mg12.5 (red triangle).

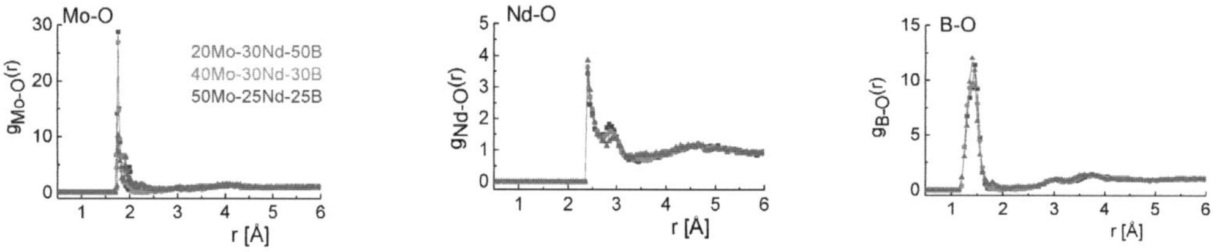

Figure 2b. O-linked partial atomic pair correlation functions obtained from the RMC modelling for the: Mo50Nd25B25 (blue square), Mo40Nd30B30 (magenta circle) and Mo20Nd30B50 (green triangle).

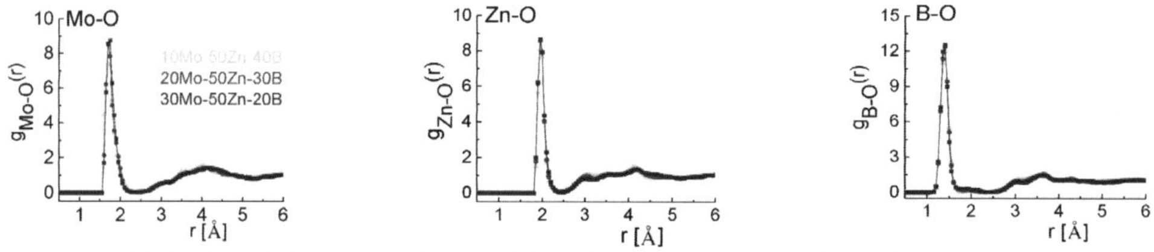

Figure 2c. O-linked partial atomic pair correlation functions obtained from the RMC modelling for the: Mo30Zn50B20 (blue square), Mo20Zn50B30 (red circle) and Mo10Zn50B40 (green triangle).

Table 1. Oxygen-linked atomic distances, r_{ij} (Å).

Sample	r_{ij} (Å)					
	Mo-O (±0.02)	**Nd-O (±0.05)**	**Zn-O (±0.01)**	**B-O (±0.02)**	**Mg-O (±0.1)**	**O-O (±0.1)**
Mo90Nd10	1.80/2.0	2.0/2.8	-	-	-	2.3/2.8
Mo80Nd15Mg5	1.80	2.1/2.8	-	-	1.8	2.45/2.8
Mo75Nd12.5Mg12.5	1.80	2.1/2.9	-	-	1.8/2.1	2.4/2.8
Mo50Nd25B25	1.80/2.0	2.4/2.8	-	1.45	-	2.35/2.8
Mo40Nd30B30	1.75/1.95	2.4/2.9	-	1.42	-	2.3/2.8
Mo20Nd30B50	1.75	2.4/2.9	-	1.40	-	2.3/3.0
Mo30Zn50B20	1.74	-	1.97	1.39	-	2.35/2.80
Mo20Zn50B30	1.72	-	1.97	1.39	-	2.35/2.90
Mo10Zn50B40	1.72	-	1.97	1.38	-	2.35/2.95

Table 2. Average partial coordination numbers, CN_{ij} (atom).

Sample	CN_{ij}				
	Mo-O (±0.2)	**Nd-O (±0.1)**	**Zn-O (±0.5)**	**B-O (±0.5)**	**O-O (±0.5)**
Mo90Nd10	4.9	4.3	-	-	8.0
Mo80Nd15Mg5	4.2	3.1	-	-	6.9
Mo75Nd12.5Mg12.5	4.0	3.1	-	-	6.2
Mo50Nd25B25	4.2	5.1	-	3.35	5.5
Mo40Nd30B30	4.0	5.0	-	3.2	5.2
Mo20Nd30B50	3.5	4.9	-	3.1	4.4
Mo30Zn50B20	3.9	-	4.0	3.1	5.4
Mo20Zn50B30	3.9	-	3.9	3.2	5.1
Mo10Zn50B40	3.9	-	3.8	3.3	5.0

The corresponding first neighbour distances are presented in table 1 and the average $CN_{i\text{-}O}$ in table 2. In the studied system the results obtained from RMC modelling have shown significant correlations between several second neighbour atom pairs, as it is shown in figure 3. This is important, as far as the analysis of these results may give useful information on the linkage of the basic structural units forming the network structure. The second neighbour partial distributions are collected in figure 3.

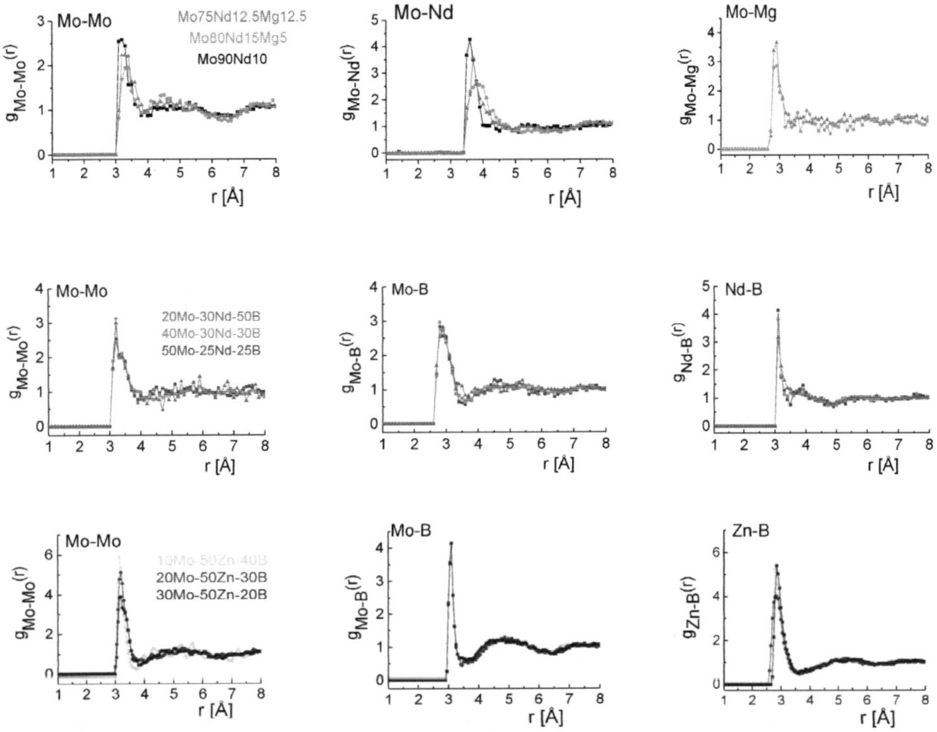

Figure 3. Second neighbour partial atomic pair correlation functions obtained from the RMC modelling for the molybdate glasses.

4. Discussion

In the present study for the Mo-O atom pair correlations a well-defined first neighbour distance is obtained at 1.80±0.05 Å, and the course of the curves is practically the same for all investigated samples. The observed Mo-O distance is somewhat greater than the 1.75(5)Å distance reported for glassy $90MO_3$-$10Nd_2O_3$ from EXAFS measurements [23], and shows similarity to the Mo-O 1.73-1.82Å distances in crystalline $LaBMoO_6$ [24] and in Nd_2O_3-MoO_3 [5]. For the Mo-Nd-B-O samples the Mo-O correlation functions show a sharp peak at an average distance of 1.75±0.05 Å, with a half-width value of ~0.04 Å, while a significantly smaller "sub"-peak appears at an average distance of 1.95±0.05 Å. The position of both the main peak and the sub-peak doesn't show any concentration dependence within the limit of error. In case of Mo-Zn-B-O samples the Mo-O distance shows a slight decrease from 1.74 Å to 1.72 Å with decrease of molybdenum content. This is an indication on the formation of stable Mo-O network structure.

The Nd-modifier shows characteristic Nd-O first neighbour distance at 2.1Å, and a next one at 2.8 Å for the binary Mo90Nd10, while for the other compositions the first neighbour distance shifts to slightly higher distances 2.4-2.8(9) Å. The intensity decreases and the peak is less characteristic inspite of the increased Nd-concentration in the ternary samples. The reason for this result may be in the more complicated network due to the increased number of constituent components. The small peak at 2.85 Å seems to be reproducible, however, the peak position agrees with the second O-O distance, thus its origin may be artificial. The Nd-O distances reported in the literature are somewhat greater than our results, i.e. 2.51-2.73 Å in glassy MoO_3-Nd_2O_3 [8], 2.33-2.5 Å in crystalline Nd_2O_3-MoO_3 [5], 2.24-2.32 Å in $Y_{2-x}Nd_xMo_4O_{15}$ [25]. The Zn-O distance appears at significantly higher value, at 1.97 Å which is in good agreement with the results obtained for other glasses with ZnO (r_{Zn-O}=1.93-1.95 Å [26], 1.83-1.91 Å [27], 1.94-1.97 Å [28]). The B-O distribution function shows a relatively broad first neighbour distance at 1.40±0.05 Å with a half-width value of 0.25 Å, is usually the case for B-O network [18, 22]. In the Mo-Zn-B-O samples the B-O first neighbour distance shows slight concentration dependence, it changes from 1.38 Å to 1.39 Å with increasing boron content, which is in good agreement with previous result obtained for Mo-Nd-B-O glass.

In the high Mo-containing binary Mo90Nd10 sample the average number of oxygen atoms around Mo is 4.9±0.2 atoms. We have investigated the possibility of a mixed network, containing both MoO_4 and MoO_6 units, by RMC simulation. If we suppose that MoO_4 units similarly to the crystalline $LaBMoO_6$ [24], Nd_2O_3-MoO_3 [5] and usually this is the case for molybdate crystals. For glassy molybdate network the presence of MoO_4^{-2} units were reported [9, 10, 21], while the formation of MoO_6 units was also observed, they are present in the network. In case of binary system it would mean that 55% of the Mo atoms are 4-fold oxygen coordinated, while 25% are 6-fold coordinated, while the rest 20% are in 3- and 5-fold coordinated states, with an error about 5%. Taking into consideration the $g_{Mo-O}(r)$ distribution, we can establish that the shorter distance at 1.8 Å can be attributed to the MoO_4 units, while in addition two further oxygen neighbours appear at a longer 2.0 Å distance forming MoO_6 units. According to table 2, with decrease of Mo content the number of oxygen neighbours slightly decreases to 4.0 atoms. For the Mo-Nd-Mg-O, Mo-Nd-B-O and also for the Mo-Zn-B-O ternary samples the average coordination number, as calculated from the RMC modeling, is close to 4. In case of ternary systems distortion of the tetrahedral MoO_4 units can not be excluded.

The average number of oxygen neighbours of the modifier Nd is 4.3 atoms for both the binary Mo90Nd10 and for the boron-containing ternary sample, while it is significantly smaller for the Mg containing samples, 3.1 atoms. The reason may be that Mg is also a modifier, which plays a similar role like Nd, increasing the number of non-bridging oxygen (NBO) atoms.

The B-O coordination number distributions contain both 3- and 4-fold oxygen coordinated boron atoms. The average coordination number slightly decreases from 3.3 to 3.1 atoms with increasing boron content in case of Mo-Nd-B-O system (see table 2). The fraction of BO_3 and BO_4 to the total number of $\{BO_3+BO_4\}$ is an important structural characteristics. It was found that the relative number of 3- and 4-fold coordination depends on the boron concentration. With increasing boron content the relative number of BO_3 units is increasing, while the number of BO_4 is decreasing. For 50MoO_3-25Nd_2O_3-25B_2O_3 sample ~70% of B-atoms are 3-coordinated, while ~30% are 4-coordinated, the average coordination number is 3.3 atoms; for 40MoO_3-30Nd_2O_3-30B_2O_3 glass ~77% of B-atoms are 3-coordinated, while ~23% are 4-coordinated, the average coordination number is 3.2 atoms; for 20MoO_3-30Nd_2O_3-50B_2O_3 sample ~87% of B-atoms are 3-coordinated, while ~13% are 4-coordinated, the average coordination number is 3.1 atoms. With increasing B_2O_3 and Nd_2O_3 content, the number of BO_3 trigonal units increases, while the number of BO_4 units decreases, this is an indication, that conversion of BO_4 to BO_3 takes place.

Interesting B-O coordination number distributions have been revealed for the Mo-Zn-B-O series. The ratio of BO_3 and BO_4 units shows concentration dependence on B_2O_3 content. For the low boron content sample the relative number of 3-fold coordinated boron atoms is significantly higher than that for the higher boron content samples. In consequence, the average CN_{B-O} slightly increases with increasing B_2O_3. We discussed above the relative number of BO_3 and BO_4 units, but we have to keep in mind, that complex borate glasses cannot be viewed as a simple network built only from BO_3 triangles and BO_4 tetrahedrals, but usually they consist of relatively large structural units, such as boroxol, pentaborate, triborate and diborate groups [29,30]. However, a possible analysis of such kind of superunits is out of the scope the present study.

The O-O average coordination number is 8 atoms for the 2-component Mo90Nd10 sample, which is the highest among the investigated samples. The average O-O coordination number reduces to 6 or 5 atoms for the samples with increasing number of contributing components, which may be the consequence of mixed structural units.

The linkage of the network former units is an important point. The RMC modelling also provides information on the second coordination sphere. The basic question is, whether the linkage of the basic structural units show any correlation or not? The question may be raised, whether the Mo-based network is separated from the B-based one, or are there any separated MoO_4 units or not? We have calculated the second neighbour correlations by RMC, and significant correlations have been revealed, as it is shown in figure 3.

The analysis of the metal-metal and boron-metal pair correlation functions show pronounced correlations, the corresponding atomic pair distances are shown in figure 3. We can establish as main points, that the pronounced Mo-Mo correlation can be interpreted as linkage between the MoO_4 units for all samples. The shortest second neighbour distances are obtained for Mo-B 2.85 Å, Zn-B 2.85 Å and Nd-B 3.1 Å, which indicate appearance of mixed linkages. These results indicate the existence of a pronounced intermediate-range ordering. Medium range order was established formed by mixed MoO_4-BO_4 and MoO_4-BO_3 linkages besides the pronounced Mo-Mo second neighbour correlations.

We have calculated the three-particle bond-angle distributions using the final atomic configuration of the RMC algorithm, plotted both as the function of *cos*(Θ (scale below) and Θ(upper scale), where Θ represents the actual bond angle. Figure 4 shows the distributions for the network former atoms (for Mo-Zn-B-O samples): Mo-O-B, B-O-Zn and Mo-O-Zn.

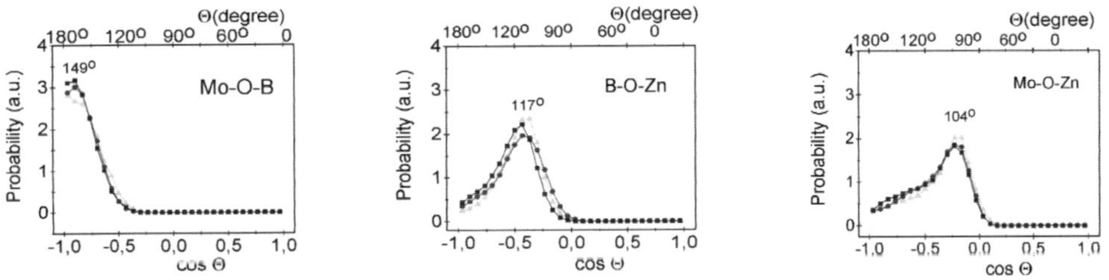

Figure 4. Three-particle bond-angle distributions obtained from RMC simulation for Mo30Zn50B20 (blue square), Mo20Zn50B30 (red circle) and Mo10Zn50B40 (green triangle).

For the Mo-O-B and Mo-O-Zn the peak positions are at 149±3° and 104±5°, respectively, which are very close values to the ideal tetrahedral configuration. B-O-Zn distribution also show a close link between the boron and zinc units. The addition of a modifier initially converts BO_3 triangles into BO_4 tetrahedra, increasing coordination number and strengthens the network. We suppose that the network is formed by BO_3 and BO_4 organized into superstructure units, and linked to MoO_4 or ZnO_4, forming in this manner mixed bond-linkages.

In the these systems the oxygen centred coordination number distributions show that most of Mo atoms favour to be connected through an oxygen to Zn and B atoms, and the majority of the Zn also prefer to be linked to B or Mo forming in this way $^{[4]}$Mo-O-$^{[3],[4]}$B, $^{[4]}$Mo-O-$^{[4]}$Zn, $^{[3],[4]}$B-O-$^{[4]}$Zn linkages.

5. Conclusions

Molybdate glasses have been prepared by melt quenching technique, and the network structure was investigated by neutron and high-energy X-ray diffraction. For data evaluation the RMC simulation technique was applied to obtain a possible 3-dimensional network configuration, which is consistent with the experimental data. From the RMC modelling the partial atomic correlation functions $g_{ij}(r)$ and the coordination number distributions CN_{ij} have been revealed. Formations of MoO_4 units were estabilished with 1.75/1.80 Å Mo-O distances. In the binary glass MoO_4 (55%) and MoO_6 (25%) structural units were revealed. In ternary systems mainly MoO_4 units are present, and with decreasing MoO_3 concentration, the ratio of MoO_6 units roughly decreases. For the B-containing ternary glasses B-O first neighbour distance was obtained at 1.40 Å, the B-O network is formed by BO_3 and BO_4 groups. ZnO proved to be a network former, not a modifier as it is often reported in the literature for similar systems. From the analyses of the obtained structural parameters we have concluded that the glassy network is formed by trigonal BO_3 and tetrahedral BO_4, MoO_4, ZnO_4 groups. Concentration dependence was found for the BO_4/BO_3 fraction, it increases with increasing B_2O_3 content. We have concluded that the network is formed by organization of BO_3 and BO_4 groups. The BO_3 and BO_4 units are linked to MoO_4 or ZnO_4, forming mixed $^{[4]}$Mo-O-$^{[3],[4]}$B, $^{[4]}$Mo-O-$^{[4]}$Zn, $^{[3],[4]}$B-O-$^{[4]}$Zn bond-linkages. Significant medium-range order exists up to ~7 Å.

Acknowledgement
The authors are grateful to L. Alexandrov and Prof. Y. Dimitriev for their helpful contribution in this work. Thanks to the financial support of Hungarian OTKA-PD 109384 and to the bilateral cooperation of Hungarian-Bulgarian Academy of Sciences SNK-63/2013. The research was supported by the EU-FP7-NMI3 No.283883. We are grateful to the Laboratoire Léon Brillouin, Saclay, France, for beam time allocation at 7C2 diffractometer and to its staff for providing assistance.

References
[1] Lacorre P, Goutenoire F, Bohnke O, Retoux R and Laligant Y 2000 *Nature* **404** 856
[2] Georges S, Goutenoire F, Bohnke O, Steil M C , Skinner S J, Wiemhöfer H D and Lacorre P 2004 *J. New Mat. Electrochem. Systems* **7** 51
[3] Kato A, Uchitomi N, Oishi S, Shishido T and Iida S 2006 *Phys. Stat. Sol. (c)* **3(8)** 2709
[4] Naruke H and Yamase T 2003 *J. Solid State Chem.* **173** 407
[5] Barker R S and Evans I R 2006 *J. Solid State Chem.* **179** 1918
[6] Zhao D, Cheng W D, Zhang H and Hang Fang S P 2008 *Dalton Trans.* **28** 3709
[7] Atuchin V V, Chimitova O D, Gavrilova T A, Molokeev M S, Kim S J, Surovtsev N V and Bazarov B G 2011 *J. Cryst. Growth* **318** 683
[8] Alexandrov L, Iordanova R, Dimitriev Y, Hamda K, Ide J and Milanova M 2008 *Adv. Mat. Res.* **39-40** 37
[9] Aleksandrov L, Iordanova R and Dimitriev Y 2009 *J. Non-Cryst. Solids* **355** 2023
[10] Aleksandrov L, Komatsu T, Iordanova R and Dimitriev Y 2011 *J. Phys. Chem. Solids* **72** 263
[11] R McGreevy R L and Pusztai L 1988 *Mol. Simul.* **1** 359
[12] Sváb E, Mészáros Gy and Deák F 1996 *Mat. Science Forum* **228** 247
[13] Proffen Th, Billinge S J L, Egami T and Louca D 2003 *Zeit. Krist.* **218** 132
[14] Ambroise J P and Bellissent R 1984 *Rev. Phys. Appl.* **19** 731
[15] Poulsen H, Neuefeind J, Neumann H B, Schneider J R and Zeidler M D 1995 *J. Non-Cryst. Solids* **188** 63
[16] Hannon A C, ISIS Disordered Materials Database 2006
[17] Waasmaier D and Kirfel A 1995 *Acta Crystallogr. A* **51** 416
[18] Fábián M, Sváb E, Proffen Th and Veress E 2008 *J. Non-Cryst. Solids* **354** 3299
[19] Fabian M, Proffen Th, Ruett U, Veress E and Svab E 2010 *J. Phys.: Condens. Matter* **22** 404206
[20] Burns A E, Winslow D W, Clarida W J, Affatigato M, Feller S A and Brow R K 2006 *J. Non-Cryst. Solids* **352** 2364
[21] Iordanova R, Aleksandrov L, Bachvarova-Nedelcheva A, Ataala M and Dimitriev Y 2011 *J. Non-Cryst. Solids* **357** 2663
[22] Fábián M, Sváb E, Proffen Th and Veress E 2010 *J. Non-Cryst. Solids* **356** 441
[23] Alexandrov L, Iordanova R and Dimitriev Y 2007 *Phys. Chem. Glasses: European J. Glass Sci. Technol. B.* **48(4)** 242
[24] Wang Z, Zhang L, Wang G, Song M, Huang Y and Wang G 2009 *Opt. Mat.* **31** 849

[25] Wang H, Peng J, Chen D F and Hu Z 2006 *Solid State Sci.* **8** 1144
[26] McKeown D A, Muller I S, Buechele A C and Pegg I L 2000 *J. Non-Cryst. Solids* **261** 155
[27] Rada S, Pascuta P, Rus L, Rada M and Culea E 2012 *J. Non-Cryst. Solids* **358** 30
[28] Suzuya K, Itoh K, Kajinami A and Loong C K 2004 *J. Non-Cryst. Solids* **345-346** 80
[29] El Batal F H and Ashour A H 2002 *Mater. Chem. Phys.* **77** 677
[30] Uchino T and Yoko T 1997 *Borate Glasses, Crystals and Melts*, Society of Glass Technology Sheffield 417

IOP Conf. Series: Journal of Physics: Conf. Series **794** (2017) 012006 doi:10.1088/1742-6596/794/1/012006

3D nano surface profilometry by combining the photonic nanojet with interferometry

P C Montgomery, S Lecler, A Leong-Hoï and P Pfeiffer

Laboratoire des Sciences de l'Ingénieur, de l'Informatique et de l'Imagerie (ICube), Unistra-CNRS, UMR 7357, 23 rue du Loess, 67037 Strasbourg cedex, France.

E-mail: paul.montgomery@unistra.fr

Abstract. Interference microscopy has become a standard technique for measuring microscopic surface roughness, being rapid, non-contact and having a high nanometric axial resolution. Nonetheless, a limitation is the lateral resolution of $\lambda/2$ due to diffraction. Amongst the many recently developed techniques for improving the lateral resolution in unlabelled far field imaging, the use of microspheres placed on the sample has recently been extended from 2D imaging to 3D measurement by combining it with interferometry. In the present work, results are shown of combining the high lateral resolution of the photonic nanojet produced by the microsphere with interference microscopy to achieve sub-diffraction limited lateral resolution with nanometric axial resolution. This is achieved by imaging through a microsphere placed on top of the sample in front of the interference objective in a Linnik setup. Results of the 3D measurements of narrow gratings through the microsphere are presented, that are not observable with the same objective in air since they are below the resolution limit. Simulations of the interaction between the photonic jet and the grating leads to a basic understanding of the image formation. This new technique opens new possibilities for high resolution characterization in nanomaterials and the biological sciences.

1. Introduction

Interference microscopy has become a standard technique for measuring microscopic surface roughness and 3D surface structures [1]. While it has the advantages of having a very high nanometric axial resolution and being rapid since it operates using classical imaging in the far field, it suffers from the limits of diffraction, resulting in a lateral resolution of $\lambda/2$, which is over two orders of magnitude worse than the axial resolution. This can be calculated from the criterion of $R_c = K\lambda/NA$, where R_c is the distance between two point sources, NA is the numerical aperture, with $NA = n.\sin\theta$ where θ is the half angle of the light cone that can enter the lens (objective) from a point source, n is the refractive index of the object-space and K is 0.61 or 0.47 depending on whether the resolution definition comes from the Rayleigh criterion or Sparrow criterion respectively [2]. By decreasing the wavelength and increasing the NA, at best the lateral resolution can be improved to a value of 200 nm [3]. Near field techniques have clearly overcome the lateral resolution problem, but remain limited by the need for point scanning [4].

In far field optical microscopy, several techniques have been proposed to improve the lateral resolution (see classification scheme in [5] for example). Notably, stimulated emission depletion (STED) microscopy [6], photoactivated localization microscopy (PALM) [7] and stochastic optical reconstruction microscopy (STORM) [8] have brought the field of optical nanoscopy to the forefront with the Nobel Prize for Chemistry of 2014 [9]. While these techniques result in a lateral resolution of several tens of nm, they nonetheless require labelling with fluorescent dyes or high light intensities, both of which can be destructive to living matter. In addition, markers cannot easily be used in materials characterization.

Some of the main principles of label-free far field nanoscopy for materials characterization were studied at the beginning of the 1990's [10, 11]. A review of the following surge in optical nanoscopy techniques can be found in [5, 12, 13] where we differentiated the so-called super-resolution techniques from those that use nanodetection. Nanodetection techniques allow the study of nanostructures without necessarily resolving them. They include many techniques classed according to the methods used, such as contrast, phase and deconvolution.

Super-resolution techniques achieve real improved lateral resolution [5], such as I^5M (combination of I^2M, interference illumination microscopy and I^3M, incoherent interference illumination microscopy) [14], the use of 4π illumination [15], SL (scattering lens) microscopy [16], FSL (far field superlens), the Pendry hyperlens [17] and the use of microspheres such as in SMON (submerged microsphere optical nanoscopy) [18, 19]. All these techniques are limited to 2D imaging.

Those that have been developed specifically for 3D measurement include SIM (structured illumination microscopy) [20] and TDM (tomographic diffraction microscopy) the tomographic mode of digital holographic microscopy (DHM) [21, 22]. In the latter technique, the improvement in lateral resolution is achieved using a synthetic numerical aperture by varying the illumination angle. In the reflection mode, 130 nm patterns on a Blu-ray® disc have been profiled in 3D at a wavelength of 475 nm and with an axial sensitivity of 5 nm. This latter technique is particularly interesting since it uses the phase of the light to give the extremely high axial resolution.

Other interference based microscopy techniques are the well-known phase shifting microscopy (PSM) and coherence scanning interferometry (CSI, or white light scanning interferometry, WLSI) that are now mature non-contact techniques for measuring surface roughness and microscopic 3D surface structure. They provide quantitative information over large areas with sub-nanometer axial resolution [2] using high sensitivity phase measurement of the wavefront by comparing it with that from light reflected from a reference mirror in a dedicated interference objective. The technique has been very successful in the characterization of very rough surfaces such as biomaterials [23] and colloids [24]. But the lateral resolution is limited by diffraction, as mentioned, to a value of about 200 nm [3]. Recently, we have succeeded in combining the 2D super-resolution qualities of the microsphere technique with the high axial resolution PSM technique of interference microscopy, using a Linnik configuration. At the same time the technique has also been used with a Mirau objective [25], resulting in a 2 to 4 fold improvement in lateral resolution compared with classical imaging, and with the technique of white light fringe envelope detection using a Linnik configuration [26]. 3D super-resolution using microspheres has also been achieved in digital holography [27]. In practice, achieving successful results with this technique is challenging, due to the presence of the microsphere in the interferometer arm. The principle of operation is also little understood.

In the present work, we present our own first nano-3D results using the Linnik configuration in combination with glass microspheres developed independently during the preparation of the aforementioned papers. We demonstrate that the photonic nanojet produced by the glass microsphere plays an important role in the improvement in lateral resolution. High resolution 3D imaging of square profile grating structures is achieved by imaging with white-light illumination through 24 μm diameter glass microspheres placed on top of the sample in front of the interference objective. Results are shown of 3D measurements in air of a 0.6 μm period grating that is not visible with the same objective without the microsphere and of a 1.2 μm period grating whose 3D measurements have been improved through the microsphere. Some first results of simulations of the interaction between the photonic jet and the grating lead to some basic understanding of the image formation.

2. 3D microscope system

The system used is a modified Leitz-Linnik interference microscope [28] using identical x50 (NA = 0.85) objectives and a PIFOC piezoelectric nanopositioner controlled in a closed loop with a capacitive position sensor for the Z-scanning of the fringes over the depth of the sample (figure 1). Images are acquired with a monochrome CMOS camera (PhotonFocus MV1-D2048-96-G2) with a Giga Ethernet connection. The measurement system is controlled by a PC equipped with an Intel® Xeon® CPU processor (2.40 GHz, 8 Go RAM) under a Windows 7 (64 bits) operating system. The control and analysis software was developed in-house under LabVIEW (version 2014, 64 bits, from National Instruments) combined with the IMAQ Vision module.

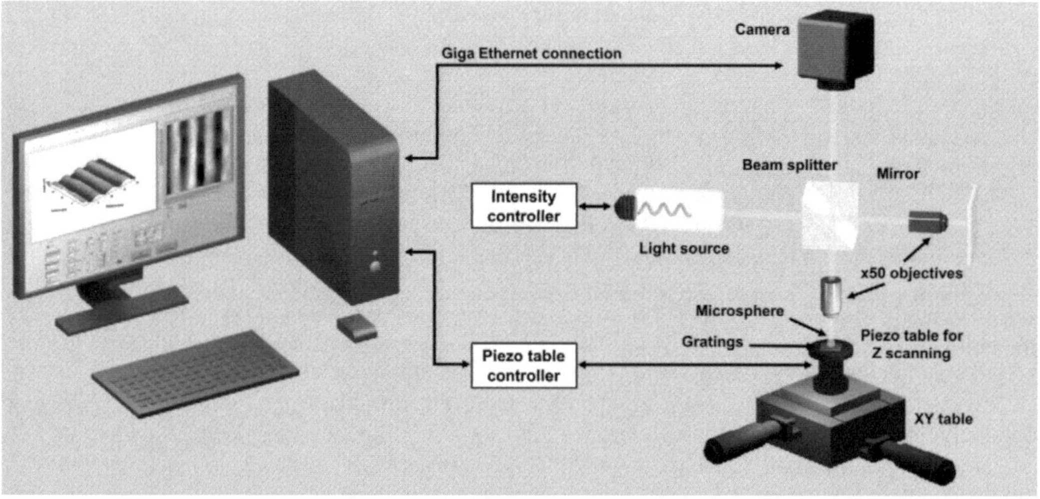

Figure 1. Schematic layout of the Linnik interference system used for nano-3D measurements.

The microspheres are carefully placed on the sample (figure 2(a)) so that they are positioned between the objective and the area of the sample to be observed (figure 2(b)). The Linnik interferometer is adjusted first to produce high contrast fringes directly on the surface of the sample and then through the microsphere on the surface by modifying the path length difference between the two arms to match the coherence planes.

(a) 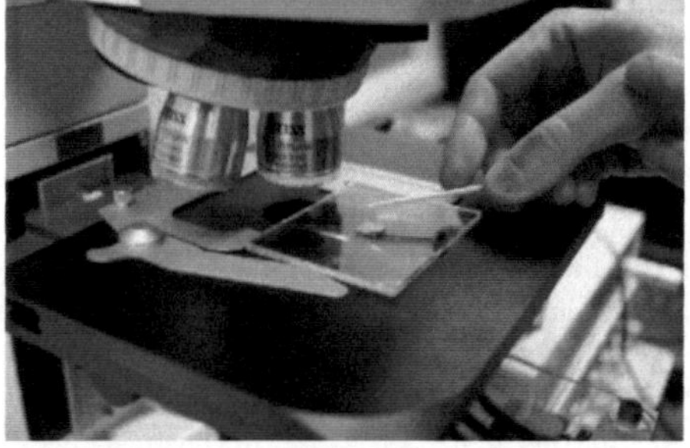

(b) 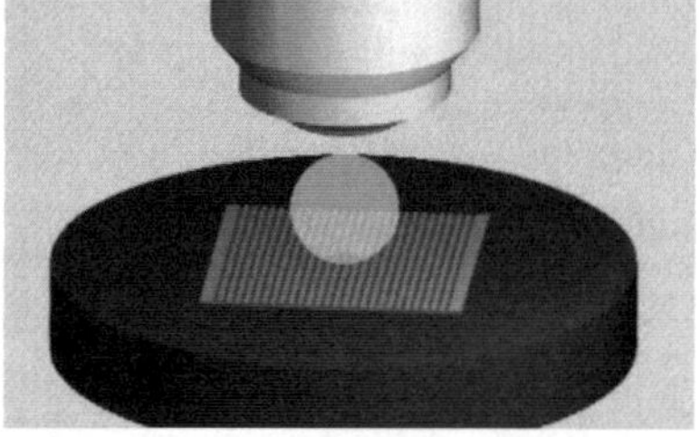

Figure 2. The nano-3D configuration (a) positioning of microsphere on sample surface in front of objective in the Zeiss classical optical microscope and (b) schematic layout of the microsphere placed on the sample, under the objective.

The algorithm used to make the phase measurement was a modified version of the 5 phase step algorithm ($\pi/2$), together with image averaging. The central region of interest was isolated and a phase discontinuity algorithm was used to unwrap the phase, followed by removal of the spherical form and light (3x3 pixels) median/low pass filtering to reduce the camera noise.

The surface shape was also measured using a Park XE70 AFM microscope in a vibration isolation chamber working in the non-contact mode. The tip used was a noncontact high frequency point probe with a tip radius of 2 nm and a width of 10 nm at a distance of 100 nm from the tip. The maximum field size is 50x50 μm for 256x256 pixels, with a lateral resolution of R_{lat} = 0.012 μm to 0.195 μm depending on the field size.

3. 3D resolution standard and fringe observation

For the demonstration of 2D super-resolution below 0.5 μm, several samples have been used in the literature: the periodic lines of a Blu-Ray DVD disk (spacing of 100 nm), the pores in an anodic aluminum oxide (AAO) sample (average diameter of 50 nm), individual adenoviruses (approximately 75 nm) [18, 19] and CD-ROM surface patterns [21].

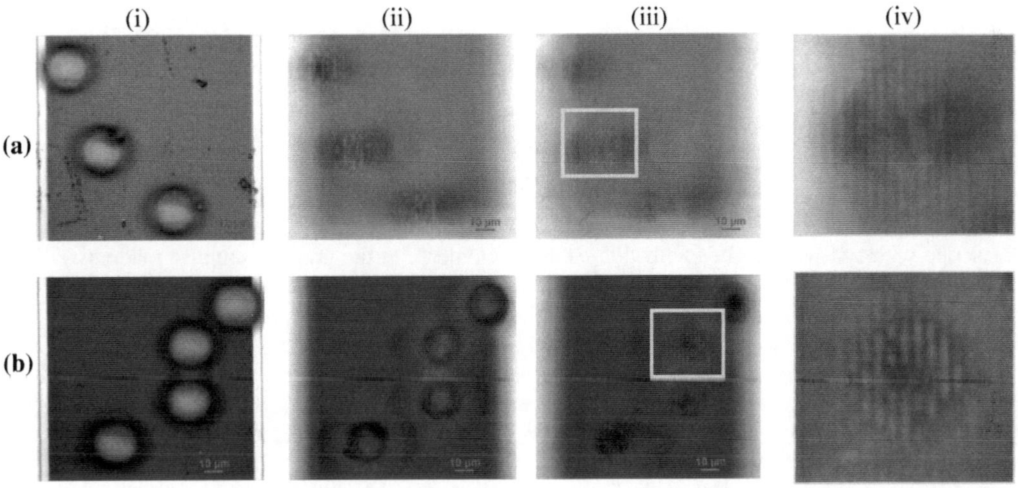

Figure 3. Reference gratings observed with Zeiss optical microscope with an objective of x50 (NA = 0.55) directly and through 24 μm diameter microspheres, changing focus from (i) to (iii) with a zoom in (iv) from the area of (iii) for (a) 1.2 μm pitch and (b) 0.6 μm pitch.

The choice of the sample to illustrate 3D super-resolution is not simple. 3D structures that have been used in the literature are the periodic line structures of a stripped Blu-Ray DVD disk (spacing of 100 nm) and a VLSI STR10-1000P silicon test pattern, composed of 5 μm wide, 100 nm high square structures with a pitch spacing of 5 μm [22].

In the present work we used a set of reference gratings (prototype RS-N gratings from SiMETRICS GmbH) consisting of several sub-gratings with different pitch values, dry etched in a 10 mm x 10 mm Si wafer up to a depth of 190 nm. The gratings measured had pitch values of 1.2 μm and 0.6 μm. The glass microspheres have a diameter of 24 μm (from Cospheric, size dispersion of 5%, n = 1.5). The microspheres placed on the 1.2 μm and 0.6 μm pitch gratings can be observed in figure 3, moving from the focus on the grating in figure 3(i) through to focus on the virtual image plane through the microsphere in figure 3(iii). The images in figure 3(iv) show a zoom from figure 3(iii) of the super-resolved grating structures.

The results in figure 4 show a 24 μm diameter sphere on the 0.6 μm pitch grating using the Linnik setup in air with white light. A 2D super-resolved image of the grating can be observed through the sphere in figure 4(a) and the fringes superimposed on the grating in figure 4(b). The latter image shows perturbations in the fringes due to the depth of the grating, showing that there is an influence of the phase of the light from the grating surface through the microsphere.

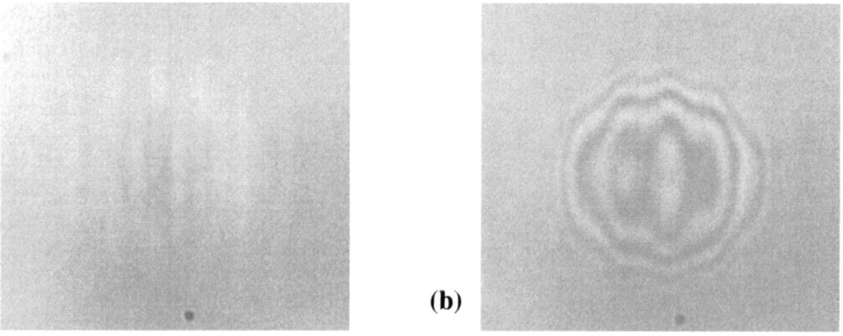

(a) **(b)**

Figure 4. Imaging through the 24 μm diameter microsphere in white light (a) directly, showing the 2D super-resolved image of a 0.6 μm pitch grating and (b) fringes superimposed on the image of the grating through the microsphere.

4. 3D measurements of Si gratings

The following results show measurements made on two of the different gratings with the Leitz-Linnik microscope with and without the microspheres. Measurements of the gratings are also made by AFM. A summary of the depth comparisons are given in Table 1 at the end of the section. Since the periods of the gratings are known, the magnification obtained through the microspheres is determined by the ratio between the period measured through the microspheres and the theoretical period value.

4.1. Measurements on 1.2 μm pitch grating

The measurements made on the 1.2 μm pitch grating under the Leitz-Linnik are possible directly without the sphere (figure 5(d),(e),(f)) but gives only a depth of 30 nm, compared with the nominal depth of 192 nm. Through the microsphere of 24 μm diameter, the grating is magnified by a factor of 4.5 and is much squarer, showing a depth of 150 nm (figure 5(a),(b),(c)), which is closer to that measured by AFM. Figure 6(b) shows the profile measured by AFM in the non-contact mode and reveals a pitch of 1.21 μm and a total depth of around 151 nm. It can be noticed that the depth profile measured is not square but that the sidewalls of the well appear inclined, together with a V shaped structure in the bottom of the well. The reasons why these details are not observed with the Leitz-Linnik could be due to a lack of lateral resolution in the direct measurements (figure 5(d),(e),(f)) or a difference between the positions observed in AFM and Leitz-Linnik through the microsphere (figure 5(a),(b),(c)).

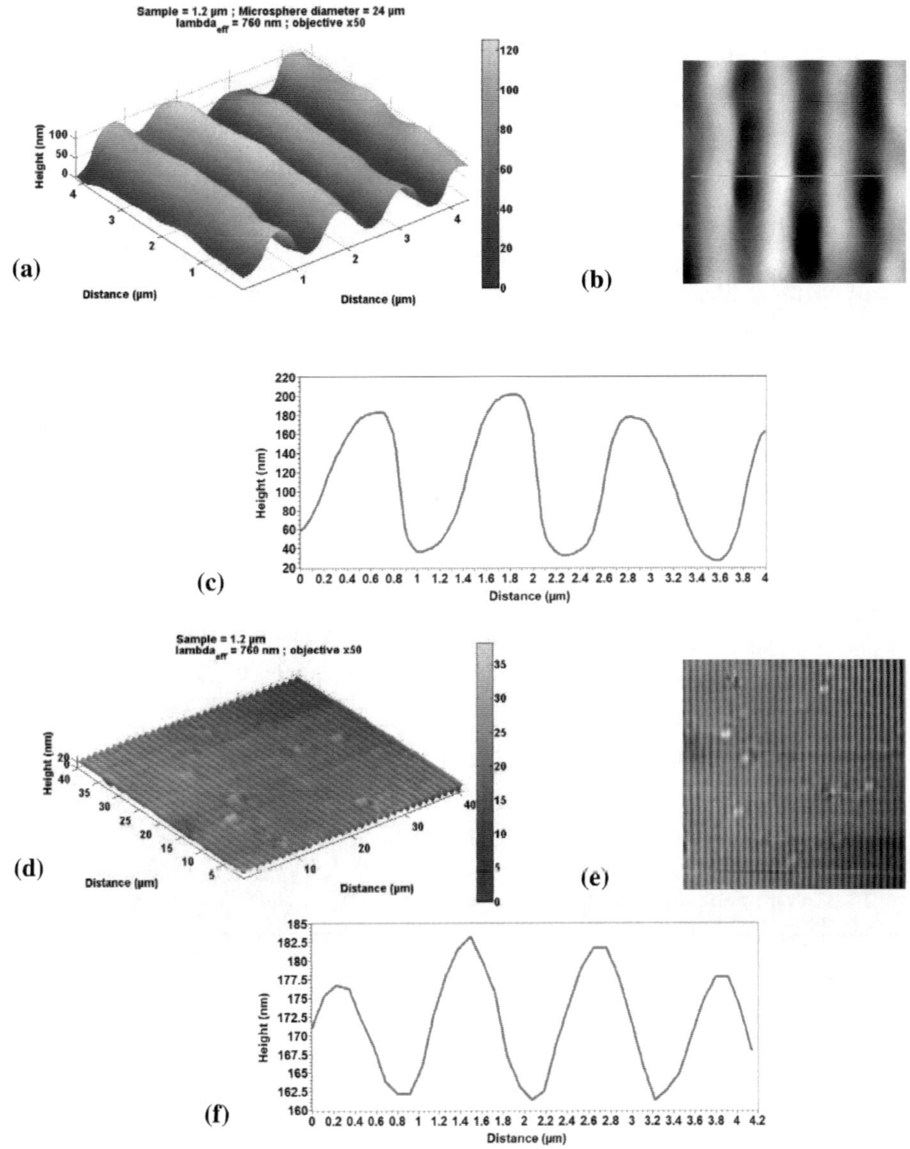

Figure 5. Comparison of measurements of the 1.2 µm pitch grating through the 24 µm diameter microsphere in white light (a) 3D view (b) grey level image of surface heights (c) line profile: average height = 150 nm and directly without sphere with the x50 Leitz-Linnik microscope (d) 3D (e) grey level surface heights and (f) line profile: average height = 30 nm.

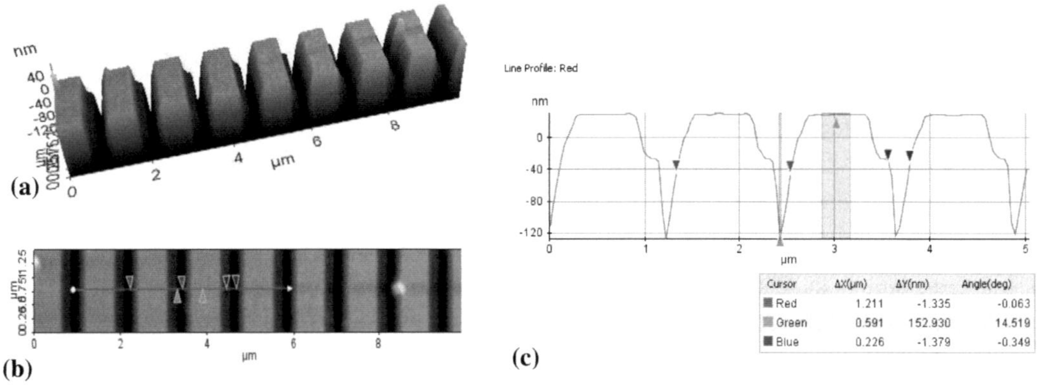

Figure 6. AFM measurements of the 1.2 µm pitch Si grating (a) 3D image of 10 µm x 1.5 µm area and (b) false colour image of surface heights and (c) line profile from (b); height between green arrows = 153 nm.

4.2. Measurements on 0.6 µm pitch grating

The measurements made on the 0.6 µm pitch grating under the Leitz-Linnik are not possible directly, being unresolvable, with R_c = 0.55 µm (Rayleigh criterion). Through the microsphere of 24 µm diameter, the grating is visible with a magnification of x3.65 and measurable (fringes resolved), giving a depth of 46 nm (figure 7). These results demonstrate sub-diffraction limited 3D imaging.

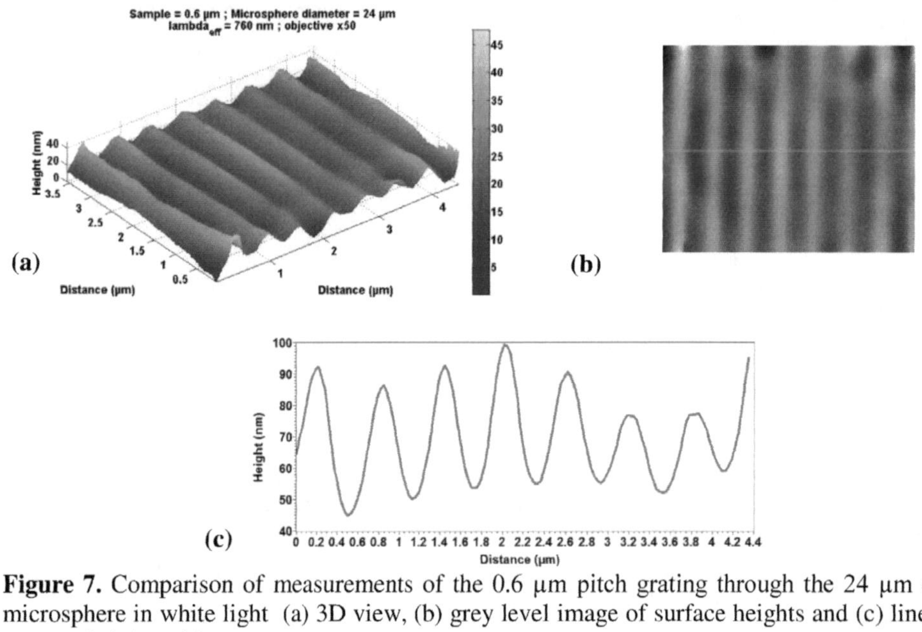

Figure 7. Comparison of measurements of the 0.6 µm pitch grating through the 24 µm diameter microsphere in white light (a) 3D view, (b) grey level image of surface heights and (c) line profile: average height = 46 nm.

The results of the AFM measurement are shown in figure 8(c), revealing a pitch of 0.633 µm and a total depth of around 58 nm.

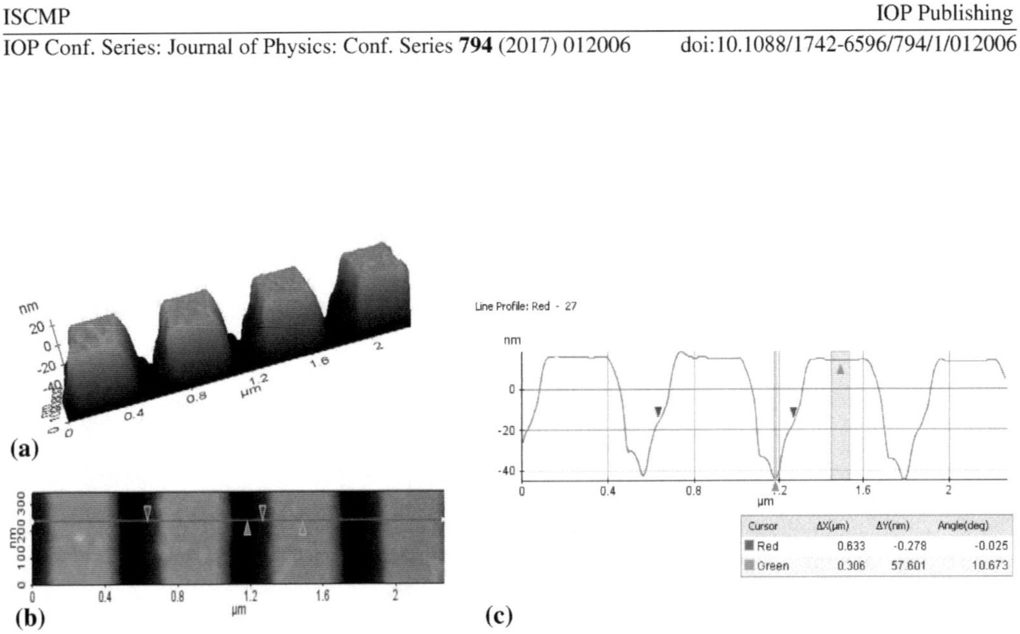

Figure 8. AFM measurements of the 0.6 µm pitch Si grating (a) 3D image of 2.3 µm x 0.34 µm area, (b) false colour image of surface heights and (c) line profile from (b); height between green arrows = 58 nm.

The results of the depth measurements with the different techniques are given in Table 1 for the two grating pitches.

Table 1. Comparison of depth measurements of different gratings of RS-N standard using different techniques. Legend: (xx) cannot be measured.

Grating pitch (µm)	Nominal depth (nm)	Depth AFM (nm)	Leitz-Linnik WL (λ = 760 nm)		
			Depth direct (nm)	Through 24 µm microsphere	
				Magnification	Depth (nm)
1.2	192	153	30	4.5	150
0.6	160	58	xx	3.65	46

5. Results of simulations

To better understand the phenomena involved, 2D FEM simulations were performed for a glass microsphere (n = 1.5) in air illuminated with a TE plane wavefront having a wavelength in free space of 500 nm coming from the top in figure 11. The boundary conditions for the right, left and bottom of the figure are Perfectly Matched Layer (PML) absorbing boundaries. A photonic nanojet is a propagated beam concentrated beyond the diffraction limit in the near field of a dielectric microsphere [29]. The formation of the photonic nanojet 3.5 µm below the microsphere can be clearly observed in figure 11(a), at which point (maximum intensity) the transverse Full Width at Half Maximum is around one wavelength. By analogy with a classical lens this could be considered as being the image focus point. When the sphere is on the grating surface, since the latter (the object) is between the lens and the image focus, the sphere may therefore create a virtual image with a magnification of larger than one. This is in agreement with the experimental observations but with an image position and magnification different to the ones predicted by classical optics, which can be easily understood because of the object scale compared to the wavelength.

The simulation for the case of the microsphere placed on top of the 1.2 µm pitch grating is shown in figure 11(b) with a close up of the zone of interest in figure 11(c). The parameters used for Si were $n = 3.42$, $\sigma = 10^{-12}$ Sm^{-1}, with a reflected ratio (power air/silicon) of 30%.

IOP Conf. Series: Journal of Physics: Conf. Series **794** (2017) 012006 doi:10.1088/1742-6596/794/1/012006

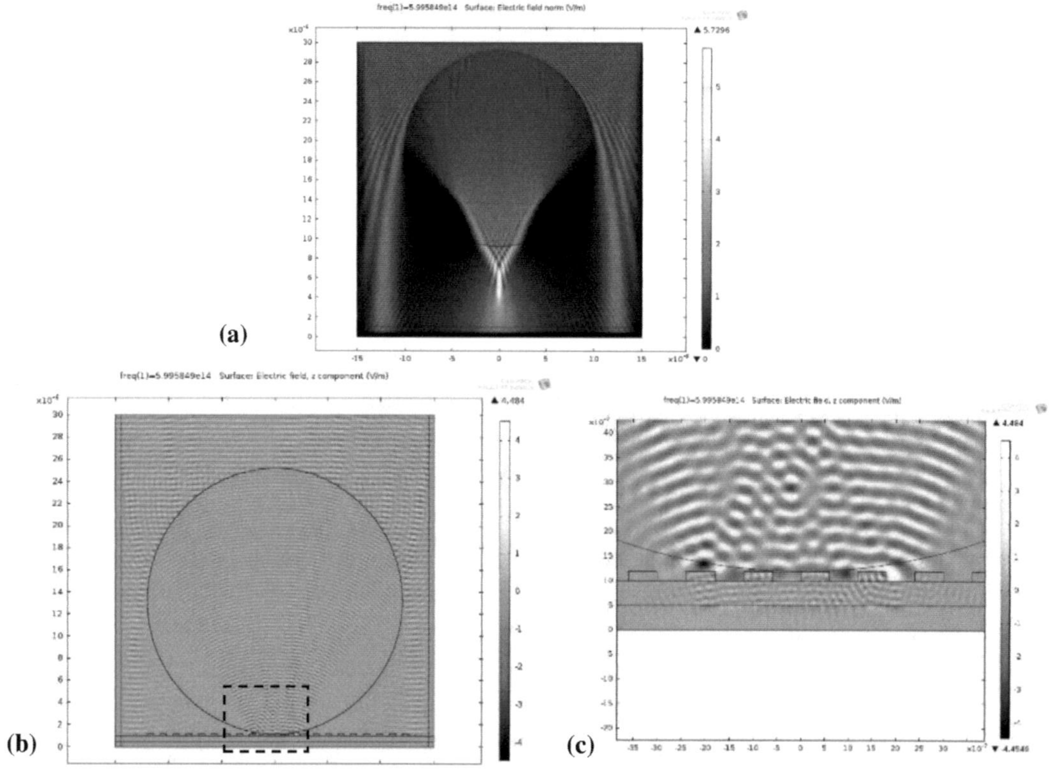

Figure 11. Determination of the imaging properties through a 24 μm glass microsphere illuminated with a TE plane wave at λ = 500 nm (from the top) using 2D FEM simulation in air (a) formation of the photonic nanojet below the microsphere |Ez|, (b) intensity of the wavefront emerging from the microsphere after interaction with the 1.2 μm pitch Si grating (real part of Ez) and (c) zoom from (b).

The results of these simulations show the interaction of the photonic nanojet with the silicon grating below the microsphere (figure 11(b) and (c)). Between 4 and 5 periods of the 1.2 μm grating are illuminated in accordance with the number of periods observed experimentally. The exact explanation of the super-resolution effect in 2D imaging is as yet unclear. Coherence must probably be taken into account, as must be the high spatial frequencies of the evanescent waves located in the near field [19, 25]. Independently, the simulations of figure 11(b) and (c) show that because of the contact between the sphere and the object, the light collected appears to correspond to a numerical aperture near to 1.

6. Conclusions

The aim of this work was to make an initial exploration of the 3D super-resolving properties of glass microspheres placed between the objective and sample of a Linnik interferometer. The first results using 24 μm diameter glass microspheres positioned on square profile gratings of different pitches in Si in a white-light Leitz-Linnik interferometer show that lateral super-resolution can be obtained and implemented with phase shifting to measure the 3D shape of surfaces. The smallest grating having a pitch of 0.6 μm could be measured through the microspheres whereas it could not be measured directly with the classical system, demonstrating sub-diffraction limited imaging.

Some initial simulations show the highly focused photonic nanojet in the near field of the microsphere. We have demonstrated how this can explain why an imaging process takes place with a magnification of larger than one. The perturbation of the electromagnetic field near to the contact between the microsphere and the grating also shows a numerical aperture of around 1 in air.

Both of the 3D measurement techniques displayed certain artefacts. For the super-resolved 3D images through the microspheres, there was notably regions of blur, which is possibly due to non-optimized illumination conditions or non-uniformity of the microspheres. For the AFM measurements, there are possible artefacts in the measurements of the grooves due to the interaction between the tip shape and grating edges. The results nonetheless confirm that nano-3D super-resolved imaging is possible in a far field optical imaging system using dielectric microspheres placed between the objective and sample. This new technique opens new possibilities for high resolution characterization in nanomaterials and the biological sciences.

7. Acknowledgements
Thanks are given to Julien Zelgowski for helping in positioning the microspheres.

8. References
[1] Schmit J, Reed J, Novak E and Gimzewski J K J 2008 *Opt. A: Pure Appl. Opt.* **10** 064001 1
[2] De Groot P 2015 *Handbook of Optical Metrology: Principles and Applications* (CRC Press, Boca Raton)
[3] Montgomery P and Montaner D 1999 *Microelectronic Engineering* **45** 291
[4] Huckabay H A, Armendariz K P, Newhart W H, Wildgen S M and Dunn R C 2013 Methods Mol. Biol. **950** 373
[5] Montgomery P, Leong-Hoi A Anstotz F Mitev D Pramatarova L and Haeberle O 2016 *J. Phys.: Conf. Series* **682** 012010
[6] Klar T A, Jakobs S, Dyba M, Egner A and Hell S W 2000 *Proc. Natl. Acad. Sci. USA* **97** 8206
[7] Betzig E, Patterson G H, Sougrat R, Wolf Lindwasser O, Olenych S, Bonifacino J S, Davidson M W, Lippincott-Schwartz J and Hess H F 2006 *Science* **313** 1642
[8] Xu K, Babcock H P and Zhuang X 2012 *Nature Methods* **9** 185
[9] Ehrenberg M 2014 (The Royal Swedish Academy of Sciences, Stockholm)
[10] Montgomery P C 1990 *Nanotechnology* **1** 54
[11] Fillard J P 1996 *Near Field Optics and Nanoscopy* (World Scientific, Singapore)
[12] Montgomery P C, Montaner D, Anstotz F and Serio B 2012 *J. Phys.: Conf. Series* **398** 012001
[13] Montgomery P C and Leong-Hoi A 2015 *J. Nanotechn., Sci. Appl.* **8** 31
[14] Gustafsson M G L, Agard D A and Sedat J W 1999 *J. Microsc.* **195** 10
[15] Bewersdorf J, Schmidt R and Hell S W 2006 *J. Microsc.* **222** 105
[16] Van Putten E, Akbulut D, Bertolotti J, Vos W, Lagendijk A and Mosk A 2011 *Phys. Rev. Lett.* **106** 193905
[17] Pendry J B 2000 *Phys. Rev. Lett.* **85** 18
[18] Wang Z, Guo W, Li L, Luk'yanchuk B, Khan A, Liu Z, Chen Z and Hong M 2011 *Nature Comm.* **2** 218
[19] Li L, Guo W, Yan Y, Lee S and Wang T 2013 *Light: Sci. & Appl.* **2** e104
[20] Gustafsson M G 2005 *Proc Natl Acad Sci U S A.* **02** 13081
[21] Debailleul M, Georges V, Simon B, Morin R and Haeberlé O 2009 *Opt. Lett.* **34** 79
[22] Liu H, Bailleul J, Simon B, Debailleul M, Colicchio B and Haeberlé O 2014 *Appl. Opt.* **53** 748
[23] Pecheva E, Montgomery P, Montaner D and Pramatarova L 2007 *Langmuir* **23** 3912
[24] Halter E, Montgomery P C, Montaner D, Barillon R, Del Nero M, Galindo C and Georg S 2010 *Appl. Surf. Sci.* **256** 6144
[25] Kassamakov I, Nolvi A and Hæggström E 2016 Conference on Lasers and Electro-Optics (CLEO), OSA Technical Digest AM4O.2
[26] Wang F, Liu L, Yu P, Liu Z, Yu H, Wang Y and Lib W J 2016 *Sci. Rep.* **6** 24703
[27] Wang Y, Guo S, Wang D, Lin Q, Rong L and Zhao J *Opt. Comm.* **366** 81
[28] Montgomery P C, Montaner D and Salzenstein F 2012 *Proc. SPIE* **8430** 843014
[29] Lecler S, Takakura Y and Meyrueis P 2005 *Opt. Lett.* **30** 2641

Boundary potential of lipid bilayers: methods, interpretations and biological applications

Yu A Ermakov

A N Frumkin Institute of physical chemistry and electrochemistry RAS, Moscow

E-mail: yury.a.ermakov@gmail.com

Abstract. The electric field distribution at boundaries of cell membranes consists of diffuse part of the electrical double layer and the potential drop over the polar area generally attributed to dipole effects. This report focuses on the molecular nature of dipole components of boundary potential and its relation to bilayer structure as it follows from different experimental approaches and molecular dynamic (MD) simulations. Alterations of the total boundary potential (BP) of planar bilayer lipid membranes (BLM) can be detected by the method of Intramembraneous Field Compensation, developed in our laboratory. When combined with traditional electrokinetic measurements in liposome suspension and Volta potential control at lipid monolayers it reveals alterations of the dipole potential induced by multivalent cations (Be^{2+}, Gd^{3+}) about 100-150 mV. It is related to the lipid phase transition detected by isotherm titration calorimetry (ITC) measurements. IFC method combined with perfusion of the cell show reversible electrostatic effects due to lysine adsorption and irreversible binding of polylysines accompanied by fast positive changes of BP as electrokinetic measurements, and slow negative ones attributed to BP dipole component. According to model proposed it related with changes of lipid hydration state varied by incorporation of ions or organic molecules into the lipid monolayer. Molecular dynamic simulations support this idea and relate dipole effects with H-bonded water molecules and lipid lateral coordination. Both effects are assumed as the principal reason of Gd^{3+} blocking effect on *E.coli* mechanosensitive channels.

1. Introduction

Electrostatic phenomena at cell walls are essential for membrane transport commonly studied by their lipid models – liposomes, planar bilayer lipid membranes (BLM) and lipid monolayers. Our studies are motivated by the idea that the dipole component of electric field distribution over a membrane/water interface (boundary potential, BP) reports on significant alterations of membrane structure and is accompanied by changes of membrane elasticity. The same effects in natural cells may modify the conformation state of membrane proteins and control their enzymatic activity. This idea was tested by experiments with bacterial mechanosensitive channels (MscL) of E-Coli incorporated into artificial lipid membranes of different composition (shown in figure 1).

Lipid membranes placed at the tip of glass pipet and exposed to varied hydrostatic pressure increased their lateral tension. Other technical details of the experiment published in the original paper [1]. Transmembrane current demonstrate pulses at high pressure as result of the channel opening by lateral tension applied to membranes. Experiment with membrane made from neutral phosphatidylcholine (PC) demonstrate, that MscL itself is not sensitive to Gd^{3+} (left panel). If some amount of negatively charged phosphatidylserine (PS) molecules are present in the membrane, MscL is completely blocked by Gd^{3+} cation but reactivated when this cation washed out from media (right panel). It proves a key role of phosphatidylserine in blocking effect of lanthanide on mechanosensitive channels. This fact motivates our special attention to anion phospholipids related to their interaction with inorganic cations and polycations of biological interest.

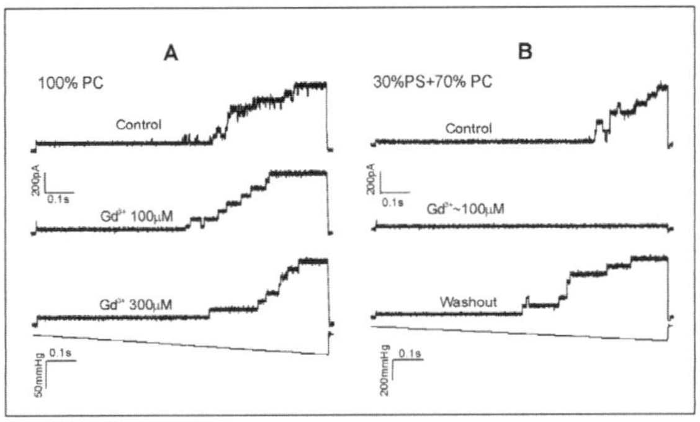

Figure 1. Current traces of MscL reconstituted to lipid membranes from soybean PC (A) and PS (B) placed at the tip of pipet and recorded at the linear pressure pump in the presence of different GdC13 concentration [1].

The current presentation focuses on the experimental approaches developed in our laboratory to control electrostatic field at membrane/water interface (boundary potential, BP), which consists of two components – electric field drop across the diffuse part of the electrical double layer, and electric potential in polar area of the lipid membrane [2]. Two principal components of the electric field at the membrane/water interfaces qualitatively illustrated by figure 1 where the electric field across the membrane is shown by a red line. The membrane composed of negatively charged lipids and facing to different ionic media with the same potential in the bulk (the short circuit condition). The negative surface potentials (φ_s^1 and φ_s^2) drops at both sides up to zero in the bulk; dipole components (φ_d^1 and φ_d^2) directed to high but different positive values inside of the membrane. The latter fact follows from many experimental works (i.e. [3,4]). Finally, the trans-membrane potential applied to hydrophobic area is defined by the difference between the boundary potentials.

$$\varphi_{in} = \varphi_b^1 - \varphi_b^2 \tag{1}$$

Because of the electrostriction effect on membranes, their electrostatic capacity is sensitive to this potential and becomes zero if compensated by external voltage applied between two sides of planar bilayer lipid membranes (BLM). The method of Intramembranous field compensation (IFC), the

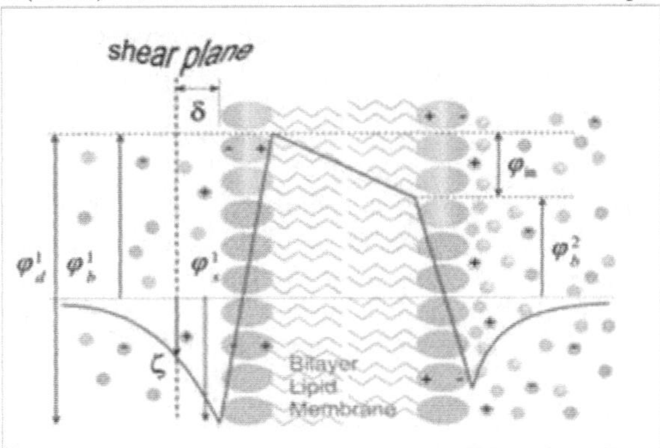

Figure 2. The electric field distribution (red line) at the lipid membrane interfaces.

IOP Conf. Series: Journal of Physics: Conf. Series **794** (2017) 012007 doi:10.1088/1742-6596/794/1/012007

experimental setup and the procedure of this compensation were developed in our laboratory and described in details in [2]. Two principal points are essential to note. The electric potential at the membrane surface can be found by electrokinetic measurements in liposome suspension. These measurements shown below were performed by dynamic scattering technique. It gives the sign of surface charge and potential but a slightly smaller magnitude, which corresponds to the potential value at some distance, δ, from the surface at the shear plane [5]. One needs alternative electrostatic methods to evaluate this parameter, for instance the IFC method.

2. Results and discussion

The surface potential, φ_s, related to screening the surface charge by ionic media as it well defined by formulas of Gouy-Chapman model. To describe the ion equilibria at the membrane surface it supplied with Langmuir type isotherms. We do not present here details and formulas of this theory, known as Gouy-Chapman-Stern (GCS) and presented in many original paper and reviews [6,7]. The second component of the boundary potential is generally attributed to dipoles oriented across the membrane interface. The dipole potential, φ_d, in spite of fact that electric field in this area is partially created by charged groups of substances or inorganic ions placed under Helmholtz plane assumed, which corresponds to membrane outer surface. Indeed, the molecular nature of dipole potential may differ for many real systems. Several examples of this presentation illustrate this situation by the help of BLM techniques and complementary methods.

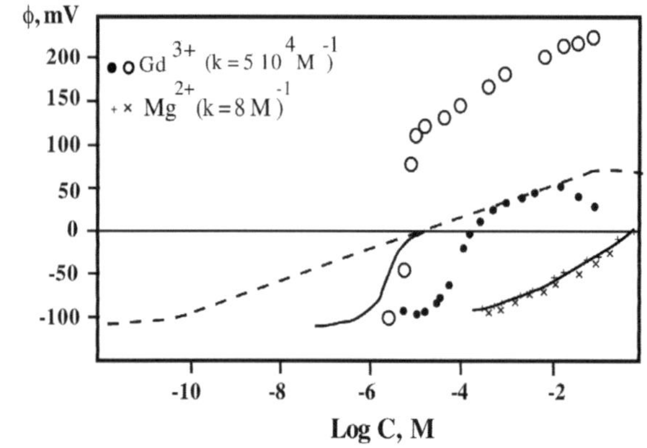

Figure 3. Boundary potential of planar BLM (opened points) and surface potential of liposomes (closed points) at different Gd^{3+} and Mg^{2+} concentration. Solid lines are for Gd^{3+} cations and GCS model with abscissa as total ion amount in the cell and dotted curve as a bulk ion concentration [8].

Electrostatic effects accompanied cation adsorption are illustrated in figure 3 by surface and boundary potentials measured in liposome suspension and by IFC method with planar BLM [8]. Experimental data related to membranes made from PS at varied concentration of Gd^{3+} (opened and closed points) and Mg^{2+} (crosses). Theoretical curves were calculated according to GSC model with binding coefficients shown in figure. Note, the adsorption of Mg^{2+} is well defined by the traditional version of the model with share plane position at δ=0.2 nm and experimental data for surface and boundary potentials coincide. The slope of the theoretical curve is about 30 mV per decade as it is predicted for the screening effect by divalent cations. It is not a case with Gd^{3+} cations. A high slope of experimental curves suggests a very high affinity of these cations to the lipid membrane surface, when their bulk concentration decreases (depletion effect). It means that mass balance condition has to be included in the GCS model. Dotted line in the figure corresponds to the traditional GCS model and has the predicted slope about 20 mV per decade for trivalent cations. Solid curve is for modified model with the same binding constant and abscissa for total ion amount in the experimental cell with undefined bulk concentration [8]. Because of high affinity of Gd^{3+} cation it became "potential determined" ion at bulk concentration of 10^{-10} M [5]. Independently on the depletion effect, a very significant difference between the surface and boundary potential was detected with these cations. They neutralize and overcharge the surface with a surface potential from -100 to +50 mV but naturally, they have no proper dipole moment to create the dipole effect about +150 mV. This effect is an intrinsic property of lipid membrane, which consists of negatively charged PS molecules. This conclusion follows from result of two experimental series depicted in figure 4 [8].

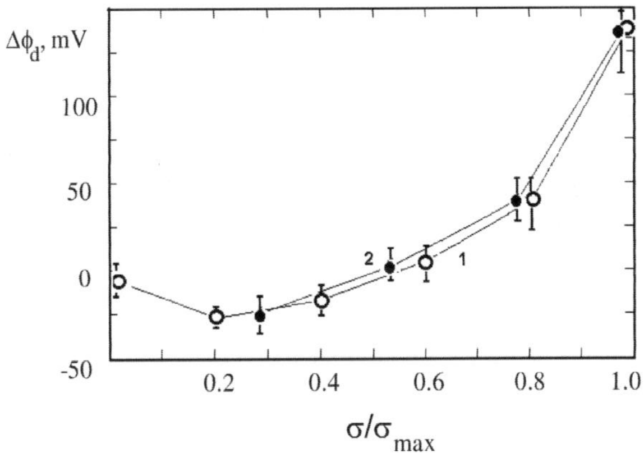

Figure 4. Dipole component of boundary potential induced by Gd^{3+} adsorption at membranes of varied PS/PC composition (opened points) or made from PS at different pH (black points) [8].

A series of measurements done with lipid membranes of varied surface charge density, σ, contains of PS in PS-PC mixture (opened points). Alternative variation of σ is realized by pH control in the media. Each point in figure 4 represents the surface potential of liposomes subtracted from the boundary potential of planar BLM measured at the same Gd^{3+} concentration in the bulk. This procedure removed all electrostatic effects due to changes in the surface charge and its screening by

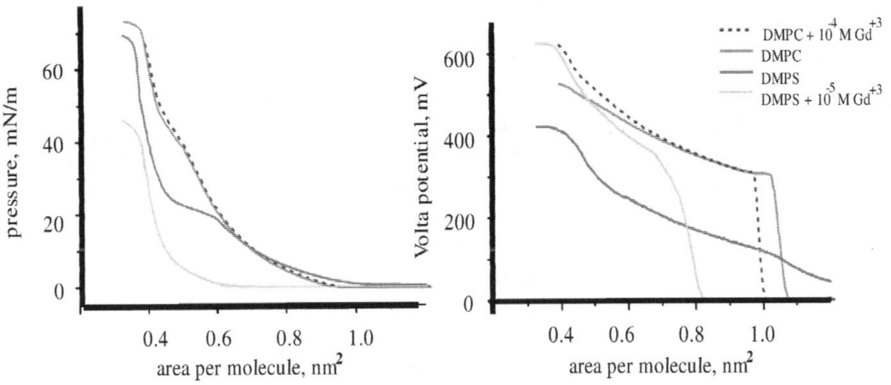

Figure 5. Lateral pressure (left panel) and Volta potential (right panel) measured simultaneously by compression of DMPC (blue lines) or DMPS monolayer at 10 mM of KCl and in the presence of sown $GdCl_3$ concentrations [1].

the electrolyte. The result is attributed to the change of the dipole component of the boundary potential $\Delta\phi_d$. An important conclusion follows from the data in figure 4, that negatively charged PS molecules, but not PC are a source of the dipole effect. This conclusion is supported by independent measurements with the technique of Langmuir monolayers (figure 5) [1].

Pressure-area diagrams and Volta potential of monolayers from dimyristoyl derivatives of PC and PS measured simultaneously are shown in the figure 5 by the same color. The shape of the diagrams with DMPC (blue lines) shows no sensitivity to adsorption of Gd^{3+}. Volta potential of these monolayers detects the cation adsorption as a positive shift about several tens of mV at the area below 0.6 nm per one PC molecule. The difference between "pure" PS monolayers and PS+Gd^{3+} is significant as in the pressure-area diagram and Volta potential measurements (compare red and green curves). Monolayers of PS over subphase of 10 mM KCl reveals a phase transition at a pressure about 20 mN/m with compressibility increased in solid (gel) state and Volta potential deflects to positive side at the same area. The same slope was observed in PS monolayer over subphase with Gd^{3+} concentration 10^{-5} M (green curve at left panel), which close to its bulk concentration at zero charge point in electrokinetic data for PS liposomes (solid theoretical curve in figure 3). The corresponding curve of Volta potential shifted to positive direction for about 200 mV as result of surface neutralization and the additional dipole effect. It leads to conclusion about gel state of lipid monolayer initiated by cation adsorption. Isotherm titration calorimetry reveals interesting details of this process [1].

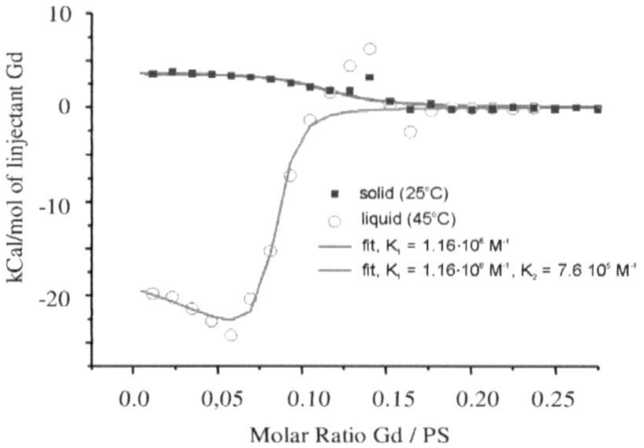

Figure 6. Integrated ITC data measured in DMPS suspension [1].

Integrated heat-production curves in figure 6 illustrate the result of $GdCl_3$ injection to suspension of DMPS liposomes performed by ITC technique at temperatures below and above the lipid phase transition (about 37° C). The molar ratio Gd/PS between 0.1 and 0.15 corresponds to zero charge point in the electrokinetic data. Note that Gd^{3+} binding to solid DMPS is an exothermic reaction, whereas it binding to liquid membranes causes massive heat release. However, at the end the heat effect changes its sign to positive because DMPS liposomes come to solid state. Simulation of this system by molecular dynamics reveals probable reasons of its behavior. It visualized the effect of multivalent cations (balls in figure 7, left panel) interaction with polar heads of lipids, which coordinates 2-3 lipid molecules into nano-clusters [9]. Some of them are colored in right panel and looks stable for long MD-trajectories (typical 100 ns). This fact explains their effect on the membrane elasticity revealed by experiments with MscL, incorporated in artificial membranes (figure1). The same effect is assumed important for any proteins imbedded to natural membranes.

It is necessary to mention that conclution about potential-elasticity relation does not depend on the presence of multivalent cations. We found correlation of these parameters for any systems studied by Langmuir monolayers [10]. One example is shown in figure 8.

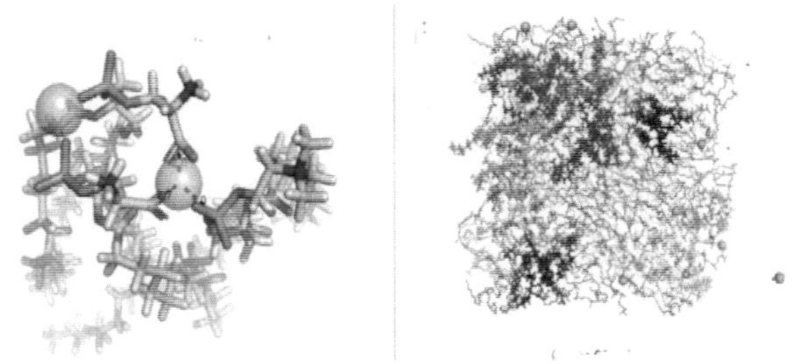

Figure 7. Coordination of PS molecules by multivalent cations (left) and their stable nanoclusters colored in right panel according to MD simulation.

Two derivatives namely the Area and Volta potential in respect of pressure defined for experimental data of figure 5 (red lines) and depicted by open and closed points in figure 8. The scales of both parameters rescaled to get a similar position of ordinates. It gives a coefficient E correlated these parameters as $\Delta\phi/\Delta p=E\ (\Delta A/\Delta p)$. The same correlation revealed for DMPS monolayers placed at different concentration of electrolyte KCl: 10^{-4}, 10^{-3}, 0.01, 0.1, and 1.0 M, and leads to values of E equal to 0.54, 0.55, 0.57, 0.49 and 0.88 V nm^{-2}, respectively. This fact clearly demonstrates the direct relationship between electrostatic and elastic phenomena in these systems. These experimental curves depicted in figure 9 in special scale as the work applied to compress monolayer (upper) and Volta potential (upper) versus to lateral pressure. It presents electrostatic-elastic relation in the clearest way [11].

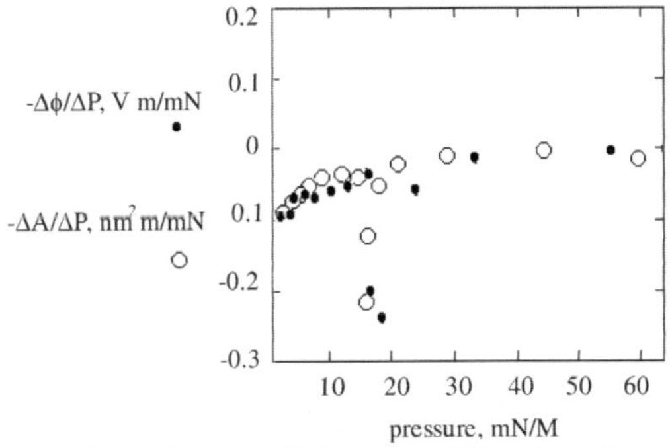

Figure 8. Increment of elasticity and Volta potential vs lateral pressure [10].

Data presented in figure 9 suggests a simple empirical description. The relative increment of monolayer compressibility, dA/A, assumed proportional to increment of lateral pressure with modulus k_p:

$$dA/A = dP/k_p\,. \tag{2}$$

This relation similar to Hook's law leads to integral form

$$A = A_0 + A_e \exp(-\frac{P}{k_p})\,. \tag{3}$$

It allows to determine the mechanical work applied to compress the monolayer of DMPS as:

$$W = A_0 P + A_e P \exp(-\frac{P}{k_p})\,. \tag{4}$$

Here A_0 and A_e – fixed and varied components of area per molecule. If electrostatic potential

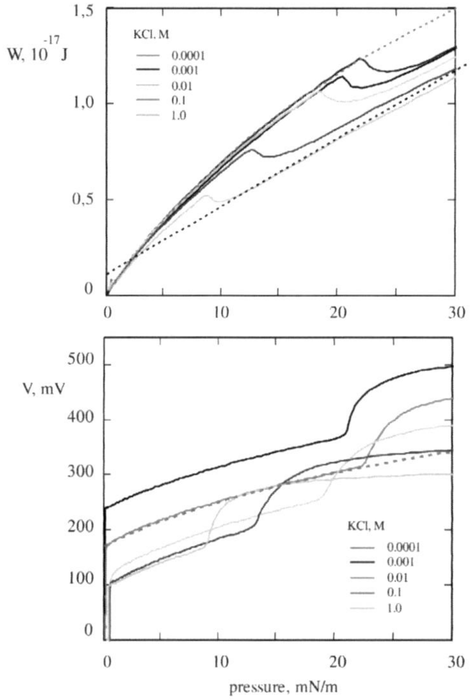

Figure 9. Data of compression DMPS monolayers at the subphase of different KCl concentration. Ordinate in upper panel is area per molecule multiplied by pressure (see equation (4)).

changes assumed proportional to increased compression energy, it gives the expression for changes of Volta potential with fitting coefficient N_d and f_0 for parallel shift in ordinate:

$$\Delta V = f_0 + N_d [A_0 P + A_e P \exp(-\frac{P}{k_p})]. \quad (5)$$

Dotted red curves in figure 9 corresponds to one of the experimental curve (violet) measured at 10^{-4} M KCl. It well described the shape as the pressure-area diagram and Volta potential below the point of phase transition with empirical parameters A_0=34 $Å^2$, A_e=58 $Å^2$, k_p= 20 m/mN and N_d = 0.12 V/N nm^2. After phase transition all experimental curves in upper panel move to the same asymptote with a linear dependence between W and lateral pressure:

$$W=109+35.5\ P \quad (6)$$

These parameters slightly vary for other curves. It means that the elasticity and electrostatics of DMPS monolayer nearly the same at different KCl concentration. In contrast, the presence of electrolyte is essential for critical value of pressure when monolayer transformed from liquid to solid state. This value became smaller when increased concentration of KCl and surface change of negatively charged PS molecules decreased due to K^+ adsorption. The same conclusion is true for series of experiments, when ionization state and surface charge of PS varied by pH of electrolyte (not shown). In both cases empirical relations listed above are well applied. All experimental curves presented in scale of upper panel positioned between two asymptotes/ which corresponds to liquid and solid states of lipids in monolayer.

3.General speculations and conclusions.

The biological importance of phosphatidylserines is related to their ability to control membrane elasticity and by this conformation mobility of proteins as result of PS interaction with inorganic actions in the media. Intrinsic properties of PS revealed in by Langmuir technique have the same molecular nature related with water molecules associated with the lipid polar heads. It depends on the cation type adsorbed at the membrane surface. Inorganic cations, especially multivalent ones, facilitate lipid phase transition to solid, gel state. Dipole component of boundary potential reports on this transformation and results from cation modification of lipid lateral interaction is accompanied by cluster formation in lipid membranes. The initial reason of this phenomenon may be the reorganization the net of hydrogen bonds between lipids. This hypothesis follows from the preliminary data of molecular dynamics and is currently under detailed studies.

Acknowledgements

The work is supported by Russian Foundation for Basic Research (No.16-04-00556a).

References:

[1] Ermakov Y A, Kamaraju K, Sengupta K and Sukharev S 2010 *Biophys. J.* **98** 1018
[2] Ermakov Yu A and Sokolov V S 2003 *Planar lipid bilayers (BLMs) and their applications* ed H T Tien and A Ottova (Amsterdam: Elsevier) chapter 3 p 109
[3] Wang L 2012 *Ann. Rev. Biochem.* **81** 615
[4] Zhan H and Lazaridis T 2012 *Biophys.Chem.* **161** 1
[5] Ermakov Yu A 2000 *Russian Colloid J.* **62** 389
[6] Cevc G, and Marsh D 1987 *Phospholipid Bilayers. Physical Principles and Models* (New York:Willey-Interscience)
[7] McLaughlin S 1983 *Physical chemistry of transmembrane ion motions* ed G Spach (Amsterdam: Elsevier) pp 69-76
[8] Ermakov Y A, Averbakh A Z, Yusipovich A I and Sukharev S 2001 *Biophys. J.* **80** 851
[9] Nesterenko A M and Ermakov Y A 2012 *Biochemistry (Moscow) Supplement Series A: Membrane and Cell Biology* **6** 320
[10] Ermakov Yu A 2011 *Biochemistry (Moscow) Supplement Series A: Membrane and Cell Biology* **5** 379
[11] Molotkovsky R J and Ermakov Yu A 2015 *Biochemistry (Moscow) Supplement Series A: Membrane and Cell Biology* **9** 48

IOP Conf. Series: Journal of Physics: Conf. Series **794** (2017) 012008 doi:10.1088/1742-6596/794/1/012008

Some calculational improvements in applying the Hubbard model to nanomaterials

Vladan Celebonovic

Institute of Physics,University of Belgrade,Pregrevica 118,11080 Belgrade, Serbia

E-mail: `vladan@ipb.ac.rs`

Abstract. John Hubbard proposed "his model" (HM) nearly sixty years ago, and one could be tempted to expect that everything (or almost so) about it is well known. Quite to the contrary, the HM is still arising interest and its applicability seems to be rising. The aim of this paper is to discuss several examples of new or already existing but improved results concerning the 2D HM and its applicability to nanomaterials. The following issues will be discussed: increasing the number of terms in various expressions used in calculating the conductivity; introducing the strain in these expressions (which is important for experiments under high pressure); defining the pressure in a $2D$ system and calculating the reflectivity but taking into account the imaginary part of the conductivity.

1. Introduction

Towards the middle of the last century, one of the outstanding problems of condensed matter physics was the metal to insulator transition (MI) . John Hubbard tried to contribute to finding an explanation of the MI transition by proposing what later became known as the Hubbard model - HM [1]. In spite of the time elapsed since Hubbard's proposal, the HM is still attracting considerable interest.

The publication of the HM greatly helped launch the study of correlated electron systems. There exists a semi-documented detail, illustrating this point: Mr. Hubbard was sitting and reading a book on many body theory, when his mother in law passed by and asked him what was the book about. To which Hubbard replied "A murder mystery"[2].

It is common knowledge that the behavior of a physical system is determined by its Hamiltonian H from which all the thermodynamical functions can be derived.The Hamiltonian of the HM is *apparently* simple. It contains only two terms,one describing the hopping of electrons within the lattice, and the other the interactions of a pair of electrons with opposite spins on a lattice node. The equations are avaliable in the literature, for example [3]. The fact that the HM is only *apparently* simple becomes clearly visible when attempting to solve it. The HM was solved in the $1D$ case [4], while work on solving the $2D$ case is ongoing within the Density matrix renormalisation group (DMRG) approach [5]. This kind of work is of course extremely interesting from the point of view of mathematical physics, but the obvious question is the applicability of $1D$ and $2D$ HM in material science.

In the years around 1980 and for some time after if was thought that the so called Bechgaard salts were materials which could be described by the $1D$ HM and that they were the simplest cases of strongly correlated electron systems. Applying the HM to these materials gave results agreeing with experiments, but they have not been as helpful as expected in understanding the high T_c materials. See, for example, [6].

Nanomaterials in $2D$ present a much wider field of possibilites for the application of the HM. To name just a few examples,these include organic electronics [7], stretchable electronic materials [8], photothermal cancer therapy [9].

There exist numerous publications on various aspects of the HM, including some by the present author [10] . What is the aim of another paper on the HM? The main aim of the present paper is improving the results obtained in previous papers. In practical terms, this means increasing the number of terms taken into account in various developments into series, and taking into account the fact that the imaginary conductivity is different from zero. It will be attempted to introduce a sitable definition of pressure in $2D$ systems. In that respect, it will be a distinct improvement over previous results. A detailed comparison of the experimental results with the predictions of these "improved" calculations will be performed in the future.

1.1. The method

The calculations to be discussed in the remainder of this paper were performed using the so called *memory function* method. It uses the idea, known in statistical mechanics, that knowledge of the Hamiltonian of a system gives the possibility of calculating its electrical conductivity. For a recent review see, for example, [11]. The essential equations of the method are

$$\chi_{AB}(\omega) = << A; B >> = -i \int_0^\infty \exp i\omega t < [A(t), B(0)] > dt \tag{1}$$

where $A = B = [j, H]$, j denotes the current operator, H the Hamiltonian and

$$\sigma(\omega) = i\frac{\omega_P^2}{4\pi z}[1 - \frac{\chi(z)}{\chi(0)}] \tag{2}$$

where σ denotes the electrical conductivity, z is the complex frequency, ω_P is the plasma frequency, $\chi(\omega)$ is the frequency dependent susceptibility and $\chi(0)$ the static susceptibility. As z is the complex frequency, it can be expressed as $z = z_1 + iz_2 = z_1 + i\alpha z_1$, where $\alpha > 0$. It can be shown that

$$\sigma_R = \frac{\omega_P^2 \chi_I}{4\pi z_1 \chi_0} \tag{3}$$

and

$$\sigma_I = \frac{\omega_P^2}{4\pi z_1}(1 - \frac{\chi_R}{\chi_0}) \tag{4}$$

2. Calculation of the electrical conductivity

2.1. Existing results

The Hamiltonian of the HM in $1D$, expressed in the formalism of second quantisation, has the following form

$$H = -t\sum_{l,\sigma}(c_{l,\sigma}^+ c_{l+1,\sigma} + c_{l+1,\sigma}^+ c_{l,\sigma}) + U\sum_j n_{j,\uparrow} n_{j,\downarrow} \tag{5}$$

where all the symbols have their standard meanings,and the current operator is

$$j = -it\sum_{l,\sigma}(c_{l,\sigma}^+ c_{l+1,\sigma} - c_{l+1,\sigma}^+ c_{l,\sigma}) \tag{6}$$

IOP Conf. Series: Journal of Physics: Conf. Series **794** (2017) 012008 doi:10.1088/1742-6596/794/1/012008

Using these two expressions it becomes possible to calculate the commutators $A = B = [j, H]$, to take into account their time dependence and obtain an expression for the current-current correlation function. All these steps are discussed in [12], and the final result is

$$
\begin{aligned}
\chi(z) = \sum_{p,g,k,q} & (32i(1/[(1+\exp[\beta(-\mu-2t\cos[g]))) (1+\exp[\beta(-\mu-2t\cos[k])))] - \\
& -1/[(1+\exp[\beta(-\mu-2t\cos[p]))) (1+\exp[\beta(-\mu-2t\cos[q])))] \times \\
& (Ut)^2(\alpha z_1 + i(z_1 + 2t(\cos[q] + \cos[p] - \cos[g] - \cos[k]))) \times \\
& (\cos(p+g)/2)(\cos((q+k)/2))(\cosh(g-p)-1)/ \\
& (N^4((\alpha z_1^2 + (z_1 + 2t(\cos(q) + \cos(p) - \cos(g) - \cos(k))))^2))
\end{aligned}
\tag{7}
$$

The symbol N denotes the number of lattice sites, μ is the chemical potential and the summations are limited to the first Brioullin zone. The chemical potential μ of the electron gas on a $1D$ lattice is given by [13]

$$
\mu = \frac{(\beta t)^6 (na-1)|t|}{1.1029 + 0.1694(\beta t)^2 + 0.0654(\beta t)^4} \tag{8}
$$

where β denotes the inverse temperature, a the lattice constant and n is the band filling. Summation of eq.(5) taking into account the smallest possible number of terms has been published in [12]. The result was an expression for the electrical conductivity of a $1D$ system of correlated electrons, which was applied to the Bechgaard salts, and the results obtained were in reasonable agreement with experiments.

The purely mathematical aim of the present calculation is to increase the number of terms taken into account in eq.(7),and in the subsequent calculation of the electrical conductivity. On the physical side, it will be attempted to take into account the dependence of the conductivity on the strain. This result will have applications in studies of nano materials under high external pressure.

2.2. New results for the conductivity

The starting point for this calculation is the summation in eq.(7). In order to take into account the influence of the external pressure (that is, the variability of a),one has at first to insert $a > 1$ in eq.(7), and after that introduce a change of variables in this expression. This has to be of the form $k \longrightarrow x/a$; $g \longrightarrow y/a$; $p \longrightarrow r/a$; $q \longrightarrow w/a$. Introducing all these changes in eq.(7), and summing over the variables $(k, y, r, w) \in (-\pi, \pi)$,one gets the following expession for the susceptibility

$$
\begin{aligned}
\chi \approx & (32i(\frac{1}{(1+\exp[\beta(2t-\mu])(1+\exp(\beta << 1 >>))} - \\
& \frac{1}{(1+\exp[\beta << 1 >>])(1+\exp(\beta << 1 >> << 1 >>))})(tU)^2 \\
& (\alpha z_1 + i(z_1 + 2t(\cos[1] - \cos[2]))) \cos[\frac{1}{2}(\frac{1-\pi}{a}) + (\frac{2-\pi}{a})a] \\
& (-1 + \cosh[(\frac{1-\pi}{a}) + (\frac{2-\pi}{a})a]))/(N^4(z_1^2\alpha^2 + (z_1 + 2t(\cos[1] - \cos[2]))^2)) + \\
& \frac{32i << 6 >> (-1 + \cosh(\frac{<<1>>}{a}) - << 1 >>)a}{N^4((z_1\alpha)^2 + (z_1 + << 1 >>)))^2}) + << 3736 >>
\end{aligned}
\tag{9}
$$

Numbers in parenthesis $<< .. >>$ denote the number of terms omitted. This long expression for χ is impractical in any real calculation.

An applicable expression can be obtained by a relatively simple procedure. Multiply out the products and powers in eq.(9), and express the result as a sum. This is a well defined operation, the end result of which is a huge sum of more than 30000 terms, which is far from applicable.

However, take the first 10 terms of this huge sum, expand it and take only the real part of he frequency into account (as done in [12] but for a small number of terms). One thus gets an expression for the real part of the susceptibility of a $1D$ HM. So, what is the improvement of this result compared to [12]? The number of terms taken into account in various developments into series has been doubled. The result for χ_R,taking into account only the first 7 terms is

$$\begin{aligned}\chi_R \cong & \frac{192\exp[3\beta\mu+2\beta t]t^3U^2\cos[1/2]}{(\exp[\beta\mu]+\exp[2t\beta])^2}\times\frac{1}{N^4(\omega+4t(\sin[1/2])^2)^2}\\ &\times\frac{1}{(\exp[\beta\mu]+\exp[2t\beta\cos[1]])^2}+\frac{128\exp[3\beta\mu+2\beta t](tU)^2\omega\cos[1/2]}{(\exp[\beta\mu]+\exp[2\beta t])^2}\\ &\frac{1}{N^4(\omega+4t(\sin[1/2])^2)^2}+\times\frac{1}{(\exp[\beta\mu]+\exp[2t\beta\cos[1]])^2}\end{aligned} \tag{10}$$

Inserting eq.(10) into expression (4) one can obtain the imaginary part of the electrical conductivity. Using the Kramers-Kronig relations [12],[14] it becomes possible to determine the imaginary part of the susceptibility as the following integral .

$$\chi_I(\omega_0) = -2\frac{\omega_0}{\pi}P\int_0^\infty \frac{\chi_R(\omega)d\omega}{(\omega^2-\omega_0^2)} \tag{11}$$

where P is the principal value of the integral. The final result for the electrical conductivity of a $1D$ HM is

$$\begin{aligned}\sigma_R \cong & \frac{\omega_P^2}{4\pi\omega\chi_0}\times[\frac{1536\exp[2t\beta+2\beta\mu]t^4U^2\omega_0^3\sin[1/2]\sin[2]}{(\exp[2t\beta]+\exp[\mu\beta])}]\\ &\times\frac{1}{N^4\pi}\times\frac{1}{(\exp[2t\beta]+\exp[\beta\mu])}\times\frac{1}{(\omega_0^2-16t^2(\sin[1/2])^4)^4}+\ldots\end{aligned} \tag{12}$$

The advantage of this expression, compared to the one previously presented in [12] is in the increased number of terms taken into account in various developments into series. This should allow a more precise comparison with experimental data on the conductivity.

A more complicated case is the $2D$ lattice, the simplest of which is a square lattice of side length a which can be treated as an extension of these results. If the lattice sides are denoted by x and y,the electric current flowing through such a system can be expressed as

$$\vec{j} = j_x\vec{e_x} + j_y\vec{e_y} \tag{13}$$

where $\vec{e_x}$ and $\vec{e_y}$ are unitary vectors of the two lattice axes.The total current is given by

$$j^2 = j_x^2 + j_y^2 \tag{14}$$

By definition $j = \sigma E$ where σ is the electrical conductivity and E the electric field,and assuming that $E_x = E_y = E$, it finally follows that

$$\sigma_{2D}^2 = \sigma_x^2 + \sigma_y^2 \tag{15}$$

The conductivities along the two axes are assumed to be mutually independent.

3. Compressing and stretching a material

The calculation of the electrical conductivity of the HM discussed in the preceeding section reffers to standard pressure. However, in nature, applications and various experiments, situations occur in which a material is subdued to variable pressure. The obvious effect which increased pressure has on a material is a decrease of its characteristic dimensions and of all the characteristics of the material which depend on it. Therefore the question is how can one take into account the influence of external pressure on the conductivity of the HM. The consequences of stretching on the conductivity of a material, within the HM, have been discussed to some extent in [3].

3.1. The 1D HM

Mathematically speaking, in the case of a $1D$ material, the function chracterizing the influence of high pressure on the conductivity would be the ratio

$$\frac{\partial\sigma}{\partial a} \tag{16}$$

where σ is the conductivity and a the lattice constant. The influence of external pressure on a material is characterized by the strain ϵ. The strain applied to an object of initial length l_0 so that it achieves length $l < l_0$ is defined by $\epsilon = (l_0 - l)/l_0$. Therefore, eq.(16) can be transformed in the form

$$\frac{\partial\sigma}{\partial a} = \frac{\partial\sigma}{\partial\epsilon}\left(\frac{\partial a}{\partial\epsilon}\right)^{-1} \tag{17}$$

The conductivity depends on the strain only indirectly, through the chemical potential, so that

$$\frac{\partial\sigma}{\partial a} = \frac{\partial\sigma}{\partial\epsilon}\left(\frac{\partial a}{\partial\epsilon}\right)^{-1} = \frac{\partial\sigma}{\partial\mu}\frac{\partial\mu}{\partial\epsilon}\left(\frac{\partial a}{\partial\epsilon}\right)^{-1} \tag{18}$$

Using expression (8) for the chemical potential and the definition of the strain, it turns out that the product $(\partial\mu/\partial\epsilon)(\partial\epsilon/\partial a)$ does not depend on a, which means that it does not depend on external pressure. However, the conductivity itself depends (among other factors) on the external pressure. It can be shown, performing some algebra on eq.(12), that

$$\frac{\partial\sigma}{\partial\mu} \cong \frac{1152\exp[2\beta(t+\mu)]t^8U^2\beta\omega_0\omega_P^2\ldots}{N^4\pi^2\chi_0\omega(\exp[2\beta t]+\exp[\beta\mu])^2\ldots} \tag{19}$$

Inserting eqs.(8) and (19) in full form into eq.(18), one could obtain an explicit expression for the strain dependence of the conductivity.

3.2. The simplest 2D case

A more complex, and more interesting case is the $2D$ HM. The simplest possible case of a $2D$ lattice is a rectangular lattice, with sides of length a and b. If the external pressure is acting within the plane of the lattice, its effect will be to reduce the surface of the elementary lattice cell. Denoting this surface by S, it follows that the influence of external pressure on the conductivity will be characterized by the derivative $\partial\sigma/\partial S$.

Changes of the dimensions of the lattice cell under pressure enter the calculation through the change of its sides. In the case of a rectangular lattice, the surface of the elementary lattice cell is given by

$$S = ab \tag{20}$$

As $a = a_0(1-\epsilon)$ and $b = b_0(1-\epsilon)$, where ϵ is the strain, it follows that

$$S = S_0(1-\epsilon)^2 \tag{21}$$

It follows that

$$\frac{\partial\sigma}{\partial S} = \frac{\partial\sigma}{\partial\epsilon}\frac{\partial\epsilon}{\partial S} = \frac{\partial\sigma}{\partial\mu}\frac{\partial\mu}{\partial\epsilon}\left(\frac{\partial S}{\partial\epsilon}\right)^{-1} \quad (22)$$

Clearly $\frac{\partial S}{\partial\epsilon} = -2S_0(1-\epsilon)$, so

$$\frac{\partial\sigma}{\partial S} = -\frac{1}{2S_0(1-\epsilon)}\frac{\partial\sigma}{\partial\mu}\frac{\partial\mu}{\partial\epsilon} \quad (23)$$

How do we approximate the chemical potential of the electron gas on a $2D$ rectangular lattice? This was solved in [15], where it was shown that

$$n = \frac{1}{2ab}\left[1 - \exp[-\mu/T] I_0\left(\frac{2t_x}{T}\right) I_0\left(\frac{2t_y}{T}\right)\right] \quad (24)$$

where n is the filling factor and I_0 the modified Bessel function of the first kind of order 0. Developing into series, taking t_x and t_y as small parameters, and limiting the development to lowest order terms, one gets the following result

$$\exp[-\mu/T] = \frac{1-2nab}{1+\left(\frac{t_x^2+t_y^2}{T^2}\right)+\left(\frac{t_x^2 t_y^2}{T^4}\right)} \quad (25)$$

Taking the natural logarithm of both sides, developing and retaining only the lowest order terms, one gets the following approximate result for the chemical potential of the electron gas on a $2D$ rectangular lattice:

$$\mu \cong k_B T[2abn + 2(abn)^2 + (\frac{8}{3})(abn)^3 + ((\frac{t_x t_y}{(k_B T)^2})^2 - \frac{t_x^2+t_y^2}{(k_B T)^2})^4 + ...] \quad (26)$$

Introducing the strain ϵ in eq.(26) by $a = a_0(1-\epsilon)$ and $b = b_0(1-\epsilon)$, it becomes possible to determine the derivative $\partial\mu/\partial\epsilon$. The derivative $\partial\sigma/\partial\epsilon$ can (in principle) be calculated from the previously defined and calculated conductivity of the $2D$ rectangular lattice. Combining those partial results, one gets the final approximation for $\partial\sigma/\partial S$, which describes the influence of high pressure on the conductivity of this particular $2D$ lattice.

Equation (24) gives the possibility of determining the region (or point) in phase space in which the chemical potential of a $2D$ electron gas becomes approximately equal to zero. In $1D$ problems, the point $\mu = 0$ corresponds to a half-filled band, and experimentally to a "'clean"' specimen. Inserting $\mu = 0$ into eq.(24) it follows that

$$1 - I_0(2t_x/T) I_0(2t_y/T) - 2S_0 n(1-\epsilon)^2 = 0 \quad (27)$$

which can be solved to give

$$n = \frac{1 - I_0(2t_x/T) I_0(2t_y/T)}{2S_0(1-\epsilon)^2} \quad (28)$$

This is the value of the filling factor for which $\mu = 0$, expressed as a function of various lattice parameters and the strain provoked by the application of pressure.

It would be interesting mathematically to determine if a point for which $\partial\sigma/\partial S = 0$ exists in principle, or for some particular combination of material parameters. From the experimental point of view, the existence of such a point would imply that a material is not sensitive to applied pressure.

4. Optics,invisibility and the HM

The electrical conductivity of a material is linked to its reflectivity. This theoretical result is known for decades and it has very interesting experimental consequences. Namely, the existence of this link opens up the possibility of calculating the conductivity from the measured values of the reflectivity.The only"'ingredient" of such a calculation is some theoretical model of the material under study; in this paper we shall discuss to some extent the HM. Full details of this calculation have recently been discussed in [10], but some possibilities for improvement shall be presented here.

Two parameters characterize the propagation of an electromagnetic wave through anon-magnetic material. These are the dielectric function $\epsilon(\omega)$ and the refractive index $N(\omega)$, with $N(\omega) = \sqrt{\epsilon(\omega)}$ [10]. It is known from theory of optics that

$$\epsilon = \epsilon_R(\omega) + i\epsilon_I(\omega)$$
$$N(\omega) = n(\omega) + iK(\omega) \tag{29}$$

$K(\omega)$ denotes the extinction coefficient. Reflectivity of a material is defined by the following quotient [10]

$$R(\omega) = \frac{(n-1)^2 + K^2}{(n+1)^2 + K^2} \tag{30}$$

Algebraic manipulation, discussed in [10], leads to

$$K^4 + K^2 \times \left[1 - \frac{4\pi\sigma_I}{\omega}\right] - (\frac{2\pi\sigma_R}{\omega})^2 = 0 \tag{31}$$

The final expression for the reflectivity is [10]

$$R(\omega) = \frac{(2\pi\sigma_R)^2 - K\omega \times [4\pi\sigma_R - K\omega(1+K^2)]}{(2\pi\sigma_R)^2 + K\omega \times [4\pi\sigma_R - K\omega(1+K^2)]} \tag{32}$$

This is the expression linking the reflectivity with the real and imaginary parts of the electrical conductivity of a material. Inserting the appropriate expressions for σ_R and σ_I would lead to a long equation expressing the reflectivity as a function of various material parameters, which are experimentally measurable. This equation would be the "bridge" between the transport and optical parameters of a material. In general form, without making any reference to the HM, this equation is

$$R \cong \frac{-4\pi(\sigma_I^2 + \sigma_R^2) + \omega[\sigma_I - \sigma_I\surd..]}{4\pi\sigma_I^2 - \sigma_I\omega + ...} \tag{33}$$

Equations (4),(10),(11),(12) and (15) lead to the functions σ_R and σ_I for the case of a $2D$ rectangular lattice. Inserting these results into the full expression for R, would give the dependence of the reflectivity on various measurable parameters of a material.

This opens up the possibility of calculating the value of the reflectivity for various values of material parameters. Turning the reasoning "'upside down", it also gives the opportunity of finding the values of material parameters for which reflectivity will have any given value. The final implication is that it thus becomes possible to find values of material parameters for which reflectivity will become zero,or have any arbitrarily small value.

The importance of this conclusion is clear. We see (by microscope, telescope or any other instrument) non-radiating objects due to the fact that they reflect light (from some source) which is incoming on their surfaces. Logically, if the reflectivity is small, the object in question will be hardly visible, and ultimately will become invisible. The conclusion is that invisibility, or visibility with difficulty of an object, can be achieved by a carfeul choice of the parameters of its material, and not only by various forms of cloaking.

The experimental parameter which can be controlled most easily is the temperature of the material. This fact opens another possibility of theoretically reaching the situation of $R \cong 0$. It is clear that the following expression holds:

$$R(T) \cong R(T_0) + (\frac{\partial R}{\partial T})(T - T_0) \tag{34}$$

with

$$\frac{\partial R}{\partial T} = \frac{\partial R}{\partial \sigma_R}\frac{\partial \sigma_R}{\partial T} \tag{35}$$

All the functions in his equation can be calculated using results presented in this paper.The reflectivity becomes zero at the temperature given by

$$T = T_0 - (\frac{\partial R}{\partial T})^{-1} R(T_0) \tag{36}$$

5. Conclusions

The preparation of this paper was started with a double aim: to present improvements in calculations on the HM performed by the present author (reviewed recently in [10]) , and to apply those improved results to some $2D$ nano materials.In the course of writing, this plan was somewhat changed, to "just" presenting several mathematical improvements to previously performed calculations.

The improvements discussed are as follows: In the calculation of the electrical conductivity, the number of terms which intervene in various developments into series was drastically increased compared to [12]. This was made possible by improvements in computing technology compared to what was avaliable back in 1997. The physical motivation for this calculation was enable the comparison between the experimental data and the theoretical predictions with ann increased precision. The calculation of the conductivity of a $2D$ rectangular lattice was made possible by a simple change of variables.

The influence of external pressure on the conductivity of this $2D$ lattice was characterized by the derivative $\partial\sigma/\partial S$. This function was calculated, and the mechanical strain was introduced.

Finally, the link between the HM and optics was discussed to some extent. In this discussion we have touched the problem of invisibility, and pointed out that it can be achieved without cloaking.

It is hoped that in the future all these improvements will be used in real calculations pertaining to nano materials.

Acknowledgments

Prepared within the research project 171005 of the Ministry of Education, Research and Technological Development of Serbia.

6. References

[1] Hubbard J 1963 *Proc R Soc London* **A 276** 238
[2] Editorial 2013 *Nature Physics* **9** 523
[3] Celebonovic V 2014 *J.Phys.: Conf.Series* **558** 012006
[4] Lieb E H and Wu F Y 1968 *Phys Rev Lett* **20** 1445
[5] http://www.itp.uni-hannover.de/ jeckelm/dmrg/
[6] Giamarchi T 2016 *C.R.Physique* **17** 322

[7] Anthony J E 2014 *Nature Materials* **13** 773
[8] Kim Y, Zhu J, Yeom B, Di Prima M, Su X, Kim J-G, Yoo S J, Uher C and Kotov N 2013 *Nature* **500** 59
[9] Chen Yu, Lianzhou W and Jianlin S 2016 *Nano Today*, in print
[10] Celebonovic V 2016 Hubbard Model in Materials Science: Electrical Conductivity and Reflectivity of Models of Some 2*D* Materials, in *Advanced 2D Materials*, A.Tiwari and A.Syvajarvi (eds.) Scrivener Publishing LLC pp.115-144
[11] Bhalla P, Das N and Singh N 2016 *Phys Lett* **A380** 2000
[12] Celebonovic V 1997 *Phys.Low-Dim.Struct.* **3/4** 65
[13] Celebonovic V 1996 *Phys.Low-Dim.Struct.* **11/12** 25
[14] Peiponen K E and Saarinen J J 2009 *Rep.Progr.Phys.* **72** 056401
[15] Celebonovic V 2002 *Phys.Low-Dim. Struct* **9/10** 105 and preprint cond-mat/0211640

Liquid crystal cell with graphene electrodes

V Marinova[1,2*], Z F Tong[3], S Petrov[3], S H Lin[3], M S Chen[2], Y H Lin[2], Y C Lai[2], P Yu[2] and K Y Hsu[2]

[1] Institute of Optical Materials and Technologies, Acad. G. Bonchev Str 109, Sofia, Bulgaria
[2] Department of Photonics, National Chiao Tung University, University Str., 1001, Hsinchu 30010, Taiwan
[3] Department of Electrophysics, National Chiao Tung University, University Str., 1001, Hsinchu, Taiwan

E-mail: veramarinova@nctu.edu.tw; vmarinova@iomt.bas.bg

Abstract. We demonstrate a graphene-based nematic liquid crystal cells fabricated by photo alignment technique. The modulation characteristics have been measured revealing that photo alignment method shows great potential for protecting the graphene layer on glass substrate during LC device fabrication.

1. Introduction

Liquid crystal displays (LCDs) are among the most important devices in modern life, intensively used in variety of photonics and optoelectronic applications [1]. Essential components for LCDs are transparent and electrically conductive layers. Among them, a solid solution from indium (III) oxide (In_2O_3) and tin (IV) oxide (SnO_2), known as ITO, is the dominant one. Recently, the demand for ITO thin-film transparent electrodes has dramatically increased. However, a stable supply of ITO may be difficult to be achieved for the recently expanding market for optoelectronic devices because of the cost and scarcity of Indium, the principal material of ITO. Therefore, the development of alternative materials is necessary to resolve this significant problem. Furthermore, ITO deposition is sensitive to the acidic and basic environments and requires complicated technical processing. In this aspect graphene, due to its excellent electrical conductivity, low sheet resistance and very high transparency in a broad spectral range [2,3] appears to be the perfect candidate to replace the ITO contacts.

After the pioneering work of Blake et al. [4] demonstrating the first LC display with one-side graphene electrode, several attempts have been proposed for complete replacement of ITO contacts. For example, high transmittance graphene-based LC cell with performances competitive to ITO-based LC cell have been reported [5], however, the fabrication of graphene-based LC device is problematic due to the mechanical rubbing which damage the graphene layer during the alignment procedure [6].

In this paper, we demonstrate a fabrication of nematic liquid crystal cell with graphene-based electrodes using photo alignment technique. The modulation characteristics of LC cell with graphene electrodes are measured and compared with the similar cell using conventional ITO electrodes. The results reveal that the photo-alignment technique opens a great potential for protecting the graphene during LC devices fabrication.

IOP Conf. Series: Journal of Physics: Conf. Series **794** (2017) 012009 doi:10.1088/1742-6596/794/1/012009

2. Experimental details and discussion

2.1. Graphene-based LC cell fabrication

Basically the LC device consists of LC layer sandwiched by a pair of glass substrates (figure 1 shows the schematic structure). In such configuration each glass substrate is covered with conductive layer and an alignment layer.

To realize the proposed structure, first the graphene was grown on a Cu foil, using a low-pressure chemical vapor deposition (LPCVD) method [7,8]. Generally, LPCVD is widely used approach to growth high quality graphene, in large area and with controlled thickness [9]. The growth process was performed in a quartz tube using 50 sccm high purity hydrogen and 20 sccm methane flow at a temperature of 1000 °C. After completing the growth process, the graphene was transferred on the glass substrates by spin coating a poly(methyl-methacrylate) PMMA on the top of graphene/Cu sheets. After PMMA deposition on the graphene site, the PMMA/graphene/Cu foil was floated on $Fe(NO_3)_3{\cdot}9H_2O$ solution in order to selectively etch the Cu foil. Subsequently, the resultant film with PMMA and graphene was transferred on several glass substrates, followed by removal of the PMMA using acetone at 60 °C.

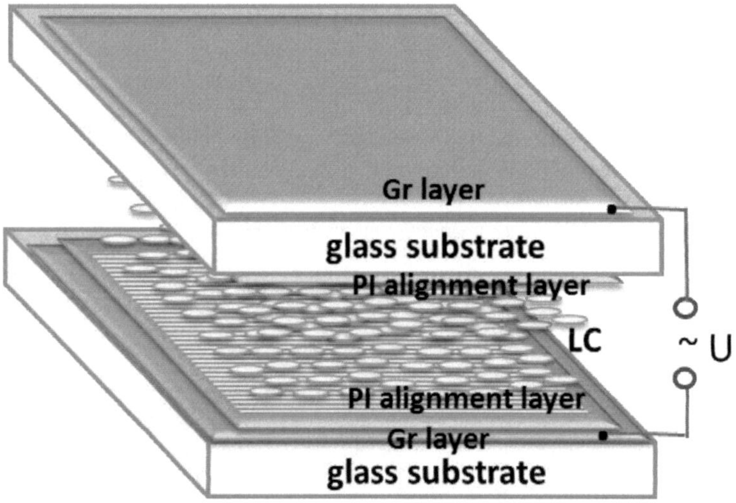

Figure 1. Schematic structure of LCD (Graphene or ITO layer stand as a conductive layer).

Before assembling the cells, the quality of graphene layer on a glass substrates were verified by Raman spectroscopy analysis. The Raman spectra were collected using a high resolution confocal Raman microscope (LabRAM HR Horiba-Jobin Yvon, HR 800) with 632.8 nm He/Ne-laser excitation under ambient conditions.

The Raman spectra of graphene transferred on a glass substrate is shown in figure 2. As it seen, the peak intensity of characteristics D-band is rather weak which indicates the fewer amounts of defects on graphene surface [10]. Furthermore, the 2D/G ratio is larger than 2 which figure out that over 90% of the area is single layer graphene.

IOP Conf. Series: Journal of Physics: Conf. Series **794** (2017) 012009 doi:10.1088/1742-6596/794/1/012009

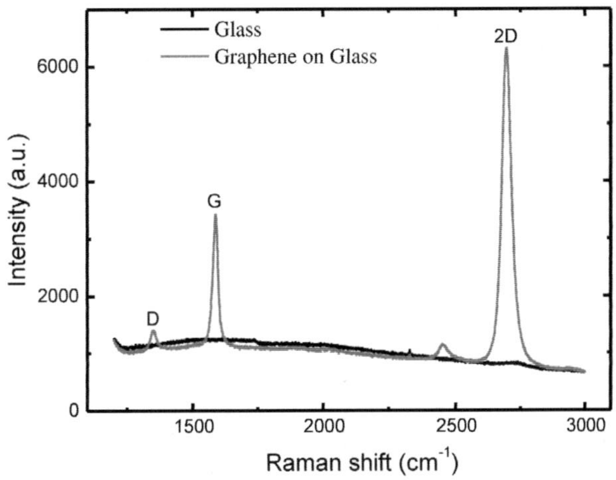

Figure 2. Raman spectra of graphene layer transferred on a glass substrate

Next, the alignment layer was prepared by spin coating a photo-curable polyimide (PI) layer (Innolux Corporation) on the above graphene-covered glass substrates and illuminated with linearly polarized UV light (wavelength of 365 nm and intensity of 1.6 mW/cm^2) for 10 minutes. We note that using the photo-alignment method the direction of alignment is parallel to the polarization of the UV light [11]. Finally, the empty cells are prepared using 12 μm Mylar spacer. The LC molecules (nematic phase E7, Merck) were injected into the empty cell.

In addition, a reference LC cell using conventional ITO contacts with the same thickness was prepared for comparison using standard mechanical rubbing procedure.

2.2. *Modulation characteristics*

The modulation characteristics were measured by inserting the LC cell between a pair of crossed polarizers as shown at figure 3(a). AC rectangular voltage with f=1 kHz was applied on the device and the power of the transmitted laser beam was monitored by a power meter.

The results are shown in figure 3(b). As it is seen, the transmitted light intensity (I) through the LC cell follows the sinusoidal function of the amplitude of applied voltage due to the phase retardation (Γ) provided by the LC molecules, given by the formula [1]

$$I = I_0 sin^2\left(\frac{\Gamma}{2}\right) \quad (1)$$

where *Io* is the initial output light intensity.

The modulation results for graphene-based LC cell are compared with the ITO-based cell, made as reference sample. As is illustrated in figure 3(b), both LC cells have similar modulation characteristics, with quite similar values of the driving voltages for both cells. Although the LC cell with ITO electrodes demonstrates higher contrast, the oscillation behaviour of the transmitted light intensity as a function of the driving voltage values of both reference cells is very similar.

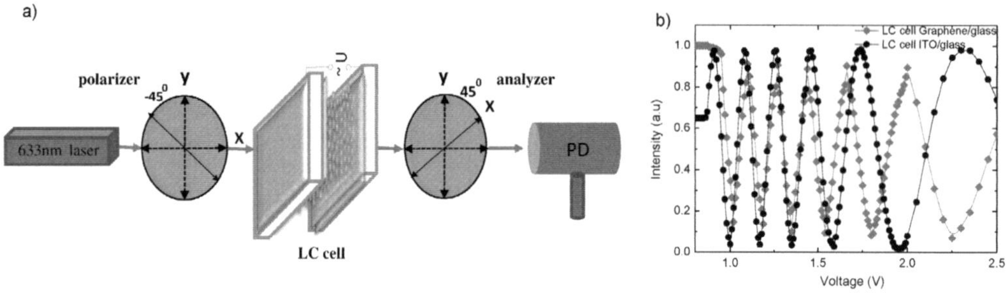

Figure 3. (a) Experimental set-up for voltage –transmittance measurements and (b) Transmitted light intensity versus applied voltage for LC cells using graphene contacts and photo-alignment method (rectangular symbols) and ITO contacts and mechanical rubbing method (circle symbols)

3. Conclusion

We demonstrated graphene contacts-based LC cells prepared by using photo-alignment method. The voltage-transmitted characteristics show similar behaviour compared to the ITO reference LC cell prepared using mechanical rubbing. The results indicate that the photo-alignment technique can prevent damage of the graphene layer during the alignment procedure.

Acknowledgements

The financial support by the Ministry of Science and Technology, Taiwan under contracts MOST 103-22221-E-009-079, 104-2221-E-009-164, 103-2911-I-009-516 and AUT Ministry of Education, Taiwan and Bulgarian Science Fund under the project FNI-T-02/26 are gratefully acknowledged. Thanks are due to the Innolux Corporation, Taiwan for providing materials for photoalignment layers deposition.

References

[1] Yeh P and Gu C 2010 *Optics of liquid crystal displays* (Wiley New Jersey)
[2] Zheng Q and Kim J K 2015 *Graphene for Transparent Conductors Synthesis, Properties and Applications* (Springer)
[3] Huang X, Zeng Z, Fan Z, Liu J and Zhang Hua 2012 *Adv. Mater.* **24** 5979–6004
[4] Blake P, Brimicombe P D, Nair R R, Booth T J, Jiang D, Schedin F, Ponomarenko L A, Morozov S V, Gleeson H F, Hill E W, Geim A K and Novoselov K S 2008 *Nanoletters* **8** 1704-1708
[5] Wahle M, Kasdorf O, Kitzerow H S, Liang Y, Feng X and Mullen K 2011 *Mol. Cryst. Liq. Cryst.* **543** 187-193
[6] Jung Y U, Park K W, Hur S T, Choi S W and Kang S J 2014 *Liquid Crystals* **41** 101-105
[7] Lai Y C, Rafailov P M, Vlaikova E, Marinova V, Lin S H, Yu P, Chi G C, Dimitrov D, Sveshtarov P, Mehandjiev V and Gospodinov M M 2016 *J. Phys. Conf. Ser.* **682** 012009
[8] Lai Y C, Yu S C, Rafailov P M, Vlaikova E, Valkov S, Petrov S, Koprinarova J, Terziyska P, Marinova V, Lin S H, Yu P, Chi G C, Dimitrov D and Gospodinov M M 2014 *J. Phys. Conf. Ser.* **558** 01205
[9] Muñoz R and Gómez-Aleixandre C 2013 *Chem. Vapor Depos.* **19** 297-332
[10] Ferrari C, Meyer J C, Scardaci V, Asiraghi C C, Lazzeri M, Mauri F, Piscanec S, Diang J, Novoselov K. S, Roth S and Geim A 2007 *Phys. Rev. Lett.* **97** 187401
[11] Yaroshchuk Y and Reznikov Y 2012 *J. Mater. Chem.* **22** 286-300

IOP Conf. Series: Journal of Physics: Conf. Series **794** (2017) 012010 doi:10.1088/1742-6596/794/1/012010

Independent adsorption of monovalent cations and cationic polymers at PE/PG lipid membranes

Daria A Khomich[1], Alexey M Nesterenko[1,2*], Andrei Yu Kostritskii[3], Diana A Kondinskaia[3], Yuri A Ermakov[4], Andrey A Gurtovenko[3,5]

[1] Biological Faculty, Lomonosov Moscow State University
[2] Belozersky Institute of Physico-Chemical Biology MSU
[3] Faculty of Physics, St. Petersburg State University
[4] Frumkin Institute of Physical Chemistry and Electrochemistry, RAS
[5] Institute of Macromolecular Compounds, RAS.

E-mail: comconadin@gmail.com

Abstract. Synthetic cationic polymers constitute a wide class of polymeric biocides. Commonly their antimicrobial effect is associated to their interaction with bacterial membranes. In the present study we analyze the interaction of various cationic polymers with model bacterial membranes comprised of a mixture of phosphatidylethanolamine (PE) and phosphatidylglycerol (PG). We describe a polymer-membrane interaction as a process of modification of the surface charge. It is well known that small monovalent inorganic cations (Na^+, K^+) cannot overcharge the surface of a bilayer containing anionic lipids. In contrast, polycations are able to overcharge anionic membranes and demonstrate a very large input to the electric field distribution at the membrane-water interface. We aimed here to study the electrostatic effects associated with the interaction of polycations of different types with a model lipid membrane whose composition closely resembles that of bacterial membranes (PE:PG = 1:4). Four different cationic polymers (polyvinylamine, polyallylamine, poly-L-lysine and polyethylenimine) were adsorbed at a model PE/PG bilayer in MD simulations. Adsorption of sodium cations was inspected separately for PE/PG bilayers of different composition and cation's binding parameters were determined. From computational experiments and consequent theoretical analysis we concluded that sodium adsorption at anionic binding sites does not depend on the presence of polycations. Therefore, we hypothesize that antimicrobial activity of the studied cationic polymers should depend on the ionic composition of the medium.

1. Introduction

There are many substances that are able to provide antimicrobial activity and a large number of these biocides are cationic polymers. Their bactericidal effect depends on the length of the polymer chain and is followed from their interaction to the bacterial membrane [1].

Prokaryotic membrane usually carry a negative surface charge as it contains anionic and zwitterionic lipids. Bactericidic effect of polycation can be associated with membrane neutralization, thus it is important to create a quantitative description of polymer-membrane interaction and corresponding change in membrane surface charge. Such interactions may be studied on model systems, especially on phosphatidilethanolamine/phosphatydilglycerol (PE/PG) mixtures in the case of prokaryotic membranes [2]. Molecular dynamic technique looks very attractive and powerful for such investigation.

Polycations, contrary to monovalent ions, demonstrate a great contribution to electric field distribution when adsorbed at the membrane-water interface [3]. Neutralization of negatively charged surface with a polycation can be described quantitatively considering irreversible nature of polymer adsorption. Monovalent cations that are always presented in background electrolyte are also able to affect the surface charge of a bilayer containing anionic lipids. When both polycation and monovalent cation interact with anionic surface it can puzzle the situation. We performed a series of computational experiments and proved that adsorption of polycation and sodium ions can be considered as independent. We analyzed the interaction of the membrane with monovalent ions individually before studying their joint action. First part of the work is devoted to interaction of sodium cations to PE/PG bilayers of different lipid compositions. Next we analyze interaction of four cationic polymers to PE/PG bilayer in the presence of sodium as a background electrolyte. As a result the quantitative description of polycation adsorption is introduced and verified over the set of computational experiments.

2. Adsorption of monovalent cations

2.1. Theoretical background

All of monovalent cations have a non-zero affinity to negatively charged phospholipids [4], whereas K^+ and Na^+ is almost not adsorbed on neutral (zwitterionic) phospholipids [5]. All the ions create an electric double layer (EDL) through their interaction to anionic lipids, wherein ions adsorbed at lipid/water interface penetrate the membrane at certain depth. Thickness of the dense layer is determined by the heads of phospholipids and their interaction to adjacent water [6]. Relation between ion adsorption, membrane surface potential and surface charge in such systems can be described with sufficient accuracy by means of Gouy-Chapman-Stern theory (GCS).

Following Ermakov [4] we describe cation equilibrium between membrane and adjacent solution with the Langmuir adsorption isotherm. Also we assume only anionic lipids to be the binding centers for cations. Surface filling θ can be expressed as a ratio of the surface density of these centers, free from the cations (σ), and a their total density (σ_{max}):

$$1-\theta = \frac{\sigma}{\sigma_{max}} = \frac{1}{1+KC(0)}\,, \tag{1}$$

where K is the adsorption constant and $C(0)$ is the concentration of the cation at the outer Helmholtz plane (the border beyond which ions have a Boltzmann distribution is called 'outer Helmholtz plane'). Note that the amount of ions adsorbed at the interface depends on the concentration of ions at the outer Helmholtz plane $C(0)$ and not on the ion concentration in bulk. Both center densities σ, σ_0 are expressed as charge densities.

Total density σ_{max} is linearly related to the concentration of the anionic lipid in the mixture (α):

$$\sigma_{max} = Q \cdot \alpha\,. \tag{2}$$

Next, we transform (1):

$$C(0) = \frac{\alpha}{\sigma}\cdot\frac{Q}{K} + \frac{1}{K}\,. \tag{3}$$

If the computational experiment data are represented in coordinates $C(0)$ and α/σ it is possible to determine the parameters of adsorption (Q and K) as parameters of linear fit.

2.2. Computational details

Simulation was performed using the MD method in all-atom force field SLipids [7]. Length of the trajectories was 1 microsecond. Last 0.3 μs of every trajectory was analyzed as a stationary state. Membranes for experiments with monovalent cations were modeled at various ionic strengths, for systems having an ionic strength of 50 mM size of membrane was 20 lipid per monolayer (a total of 40 molecules) for the rest of 10 lipid per monolayer. Size of the system averaged 20000 atoms for systems of 40 lipid molecules and 10000 atoms for systems of 20 lipid molecules.

We determined the localization of the outer Helmholtz plane in order to determine the potential and the charge on the surface of the membrane. According to distribution of components at the lipid-water interface, for our systems the position of this plane was defined at 9 Å from average C_α atoms of phospholipid glycerol.

The distribution of the electric potential in the Gouy-Chapman model described by potential of mean force (Ψ_{PMF}) [6]:

$$\Psi_{PMF} = \frac{zRT}{F} \cdot \frac{C(0)}{C(\infty)}, \tag{4}$$

where z — ion charge number, $C(0)$ — the concentration of ion at the outer Helmholtz plane defined above, $C(\infty)$ — concentration in the bulk solution.

2.3. Computational results

The adsorption of sodium cations to (PE/PG) model system was observed using molecular dynamics simulation. We compared the area per lipid and the thickness of simulated membranes to corresponding experimental values [8] in order to validate the molecular model. At room temperature the model membrane made of pure POPG has average area per lipid of 67.7 Å^2 whereas the natural membrane of the same composition has the experimental value of 66.1 Å^2. At 65°C corresponding values are 72.2 and 71.3 Å^2. Full thickness of membrane (D_{HH}) is defined by many authors as the distance between the peaks on the plot of electron density whereas the thickness of the layer of hydrophobic hydrocarbon chains of lipids ($2D_C$) can be calculated from the distance between peaks in the derivative of the last plot. The computed values (D_{HH} = 38 Å, $2D_C$ = 25 Å) are also in a good agreement with experimentally measured: 37.4 and 27.8 Å, respectively.

We use equation (3) to evaluate the adsorption parameters, such as adsorption constant K and the surface density of binding sites Q, in accordance with the selected model. We simulated series of lipid bilayers in different ionic solutions and determined $C(0)$ and the ratio α/σ. $C(0)$ was calculated from the concentration profile, averaged over the trajectory (example of such profile is presented at figure 1b). Surface charge was calculated with Graham equation from the surface potential Ψ_{surf}, which is well described by the potential mean force (PMF). Cation concentration near the surface $C(0)$ and its concentration in the bulk solution $C(\infty)$ were obtained from trajectories and substituted into (4) to find PMF (see 'Computational details'). α is known initially as the amount of anionic lipid in the mixture prepared for simulation. For each simulated lipid bilayer we put the point $\{C(0), \alpha/\sigma\}$ on the coordinate plane and then approximated all points by equation (3) (figure 1a). The result was: K = 2.16 M^{-1}, Q = -17.3 μC/cm^2 (for T = 300 K) and K = 7.15 M^{-1}, Q = -25.5 μC/cm^2 (for T = 340 K).

The adsorption of sodium was found to be more intensive under high temperature. Position of the main plane of the adsorption of sodium cations (figure 1b) was found on the distribution of the ions. The presence of the second peak on the chart the distribution of Na indicates the complex nature of sodium binding sites.

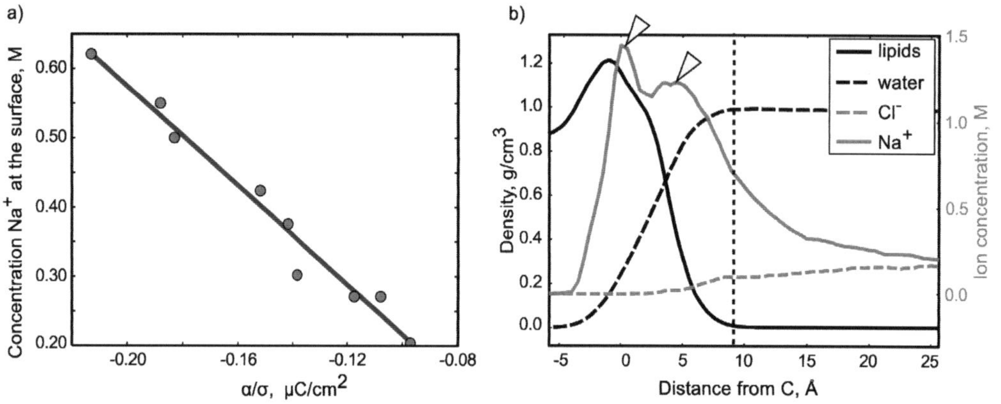

Figure 1. **(a)** Fitting the experimental values of charge density σ and near-surface cation concentration $C(0)$ with equation (3). Every marker corresponds to a separate simulation of PE/PG bilayer with PG content denoted by α. **(b)** The distribution of components at the lipid-water interface in the system POPE/POPG (2/3), I (in bulk) = 130 mM, T = 300 K. Black curves are plotted in density units (left axis), red curves are plotted in concentration units (right axis). Yellow arrows indicate the localization of the peaks which correspond to the planes of sodium ions adsorption. The vertical dashed line indicates the position of the membrane surface (outer Helmholtz plane).

3. Adsorption of polycations

3.1. Theoretical background

As we shown that GCS theory describes well the adsorption of sodium cations at pure membrane we can develop the similar theory in order to estimate sodium effect in the presence of polymer at the membrane surface. Let us derive the dependency between surface potential, adsorbed surface charge and concentration of monovalent electrolyte. The complete theory is presented in [9].

The near-surface concentration of monovalent cation, $C(0)$, is linked with the surface electric potential through the Boltzmann equation:

$$C(0) = C_{bulk} e^{-F\Psi_s/RT} . \tag{5}$$

Combining equations (1) and (5) provides us with an equation that links the binding site occupation θ and the surface potential Ψ_s:

$$\theta = \frac{K C_{bulk} e^{-F\Psi_s/RT}}{1 + K C_{bulk} e^{-F\Psi_s/RT}} \tag{6}$$

Ψ_s depends in turn on the overall surface charge density σ; the corresponding relation is known as the Grahame equation. In case of a symmetrical monovalent electrolyte one has:

$$\sigma = \sqrt{8 C_{bulk} \varepsilon \varepsilon_0 RT} \sinh\left(\frac{F}{2RT}\Psi_s\right) . \tag{7}$$

Next we should define surface charge dependency on occupation, $\sigma(\theta)$, and combine it with (6) and (7). The resulting expression will link the surface charge with the ionic concentration and adsorption parameters. First, let us consider a polymer-free membrane system. As we assumed previously (1), if the anionic lipids are the binding sites of sodium ions one has:

$$\sigma = -\sigma_{max} * (1 - \theta) \ . \tag{8}$$

Next, let us introduce the polymer adsorption into equation (8) (we assume that polycation adsorption is irreversible which is a good approximation for our systems). If an ion and a polymer do not compete for the same binding center, one has:

$$\sigma = -\sigma_{max} * (1 - \theta) + \sigma_p \ , \tag{9}$$

where σ_p is the density of the adsorbed positive charge (charge of a polycation divided on the membrane surface area).

A combination of (6), (7) and (9) gives:

$$\sqrt{8C_{bulk}\varepsilon\varepsilon_0 RT}\sinh\left(\frac{F}{2RT}\Psi_s\right) + \frac{\sigma_{max}}{1 + KC_{bulk}e^{-F\Psi_s/RT}} = \sigma_p \tag{10}$$

This equation can be validated in an experiment where polymer charge (and σ_p) is varied and surface potential Ψ_s is measured. In the next section we describe such a series of computational experiments.

3.2. *Computational results*

Several computational experiments were performed in order to test equation (10). Four polymers were adsorbed at the surface of pre-equilibrated dioleoyl-PE/dioleoyl-PG membrane (1:4): polyallylamine (PAA), polyethylenimine (PEI), polyvinylamine (PVA) and poly-L-lysine (PLL). In the experiments 1–3 molecules of a polymer were added to the bilayer in the presence of sodium chloride as a background electrolyte. Simulations were performed with systems of 128 lipids and about 10000 water molecules. The force field for polymers was developed recently [10]. The following protonation levels were used for the polymers: 15% (PAA), 50% (PVA), 50% (PEI) and 100% (PLL). One to three polymer molecule of 20 units were added to the bilayer for simulations. Detailed description of the simulation as well as resulting molecular picture of polymer-membrane interaction is provided in [9].

Polymer has 20 units in length and its protonation level corresponds to its charge at neutral pH. For each system surface potential Ψ_s was calculated as described above and in 'Computational details'. The resulting potential was plotted against polymeric charge (figure 2a) Theoretical curves are also presented at the same figure plotted for K = 2–6 M^{-1} and σ_{max} = 0.05 C/m^2 (this corresponds to 13 PG lipids per 41.5 nm^2). Experimental points could be fitted with equation (10) considering $K = 6\ M^{-1}$. Higher value of adsorption constant then it was obtained in the first section (6.0 vs 2.5 M^{-1}) can be explained as a result of using different acyl tails in these two simulation series (DOPG/DOPE vs POPG/POPE). Another reason could be the following: in our simple model we can not consider all complications of joint mono- and polycation binding.

4. Conclusion

As we found the adsorption of monovalent sodium cations at PE/PG lipid bilayer is perfectly described with Gouy-Chapman-Stern theory. It is the first time this theory is proved for phosphoglycerol containing bilayer. Simultaneous adsorption of monovalent cations and polycations could also be described with similar theory considering irreversible polymer adsorption and independent adsorption of mono- and polycations. The equation of joint

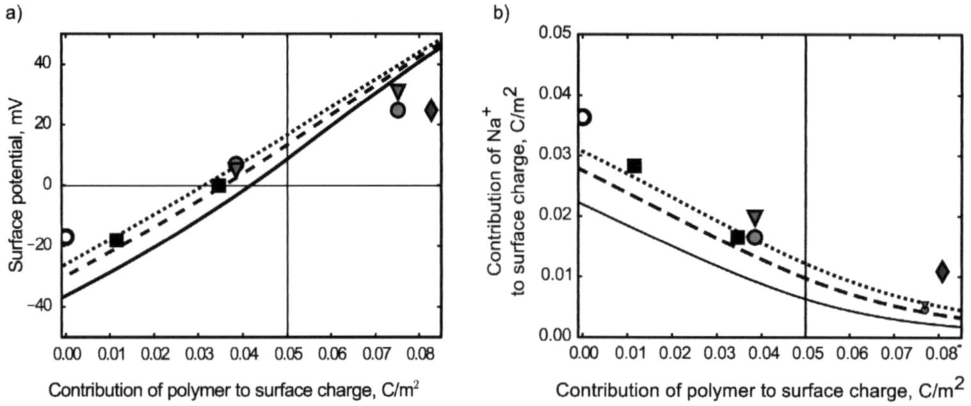

Figure 2. (a) Surface potential dependency on adsorbed polymer surface charge. Black theoretical curves are plotted with equation (10) using C = 100 mM, T = 310 K, $\sigma = 0.05$ C/cm^2 and $K = 2$ (dots), 4 (dashed), 6 (solid). (b) Contribution of sodium to the surface charge found by the same equation. Results of computational experiments are depicted by single markers: open circle (pure bilayer), black squares (PAA), green triangles (PVA), red circles (PEI) and blue diamond (PLL).

adsorption (10) was derived and seems suitable for describing the results of a series of numerical experiments.

According to the theory one may formulate the following predictions: (1) Point of zero charge (concentration of polymer required for neutralizing anionic liposome in a suspension) could be shifted to smaller polymer concentration by adding monovalent cations. (2) Type and amount of monovalent cations could affect biocidic effect of polycations. (3) Polycation binds irreversibly with any bilayer containing zwitter-ionic phosphatidylethanolamine in its composition.

Acknowledgements

The authors wish to acknowledge the use of the computer cluster of the Institute of Macromolecular Compounds RAS, the Lomonosov supercomputer complex at the Moscow State University. This work was partly supported by the Presidium of the RAS through the grant program "Molecular and Cellular Biology" and by Russian Foundation for Basic Research (No. 16-04-00556a).

References

[1] Panarin E F, Solovskii M V, Zaikina N A and Afinogenov G E 1985 *Die Makromol. Chemie* **9** 25–33
[2] Cullis P, Hope M, De Kruijff B, Verkleij A and Tilcock C 1985 *Phospholip. cell. regul.* **1** 1–59
[3] Finogenova O A, Filinsky D V and Ermakov Y A 2008 *Biochem. (Mosc.) Suppl. A* **25** 181–188
[4] Ermakov Y A 1990 *BBA - Biomembr.* **1023** 91–97
[5] Cevc G 1990 *BBA - Rev. Biomembr.* **1031** 311–382
[6] Nesterenko A M and Ermakov Y A 2013 *Biochem. (Mosc.) Suppl. A* **6** 320–328
[7] Jämbeck J P M and Lyubartsev A P 2013 *J. Chem. Theor. Comput.* **9** 774–784
[8] Pan J, Heberle F A, Tristram-Nagle S, Szymanski M, Koepfinger M, Katsaras J and Kučerka N 2012 *BBA - Biomembr.* **1818** 2135–2148
[9] Kostritskii A Y, Kondinskaia D A, Nesterenko A M and Gurtovenko A A 2016 *Langmuir* URL `http://dx.doi.org/10.1021/acs.langmuir.6b02593`
[10] Kondinskaia D A, Kostritskii A Y, Nesterenko A M, Antipina A Y and Gurtovenko A A 2016 *J. Phys. Chem. B* **120** 6546–6554

IOP Conf. Series: Journal of Physics: Conf. Series **794** (2017) 012011 doi:10.1088/1742-6596/794/1/012011

Surface characterisation of Ga^+ ion implanted ta-C thin films

M Berova

Institute of Solid State Physics, Bulgarian Academy of Sciences, 72 Tzarigradsko Chaussee, 1784 Sofia, Bulgaria

E-mail: mberova@issp.bas.bg

Abstract. Samples of thin film (d~40nm) tetrahedral amorphous carbon (ta-C), deposited by filtered cathodic vacuum arc (FCVA), have been implanted with Ga^+ at ion energy E = 20 keV and ion doses $D=3.10^{14}\div3.10^{15}$ cm^{-2}. The induced structural modification of the implanted material results in a considerable change of its optical properties, best manifested by a significant shift of the optical absorption edge to lower photon energies. This shift is accompanied by a considerable increase of the absorption coefficient (photo-darkening effect) in the measured photon energy range (0.5÷3.0 eV). These effects could be attributed both to additional defect introduction and increased graphitization, as well as due to Ga colloids formation, as confirmed by electron microscopy analysis. Further nano-scale structural and electronic properties characterisation of the Ga^+ implanted films has been carried out here using conductive atomic force microscopy (C-AFM) measurements. The observed properties modification results from the high concentration of introduced Ga^+ atoms, which is of the order of those for the host element. The obtained optical contrast (between implanted and unimplanted film material) could be made use of in the area of high-density optical data storage by using focused Ga^+ ion beams.

1. Introduction

Pure carbon can occur in many forms reflected by the nearest neighbour bonds, and by midrange and long-range ordering of the solid. Different forms of carbon are being exploited in newly engineered applications. The most well-known forms of elemental carbon found in nature are graphite and diamond. Besides the crystalline forms, amorphous form of carbon has been found to be having a lot of technological applications. The properties of these amorphous carbons sensitively depend on the relative concentration of *sp*3 and *sp*2 hybridized carbons. The resulting amorphous materials are variously referred to as tetrahedral amorphous carbon (*ta*-C), amorphous carbon (*a*-C), hydrogenated amorphous carbon (*a*-C:H), amorphous conducting carbon (*a*-CC), etc. The term tetrahedral is used to describe amorphous carbon films with a large percentage of *sp*3 bonding (up to 87%). The high *sp*3 content in the films results in unique properties that include extreme hardness (~70 GPa), chemical inertness, high electrical resistivity, and wide optical band gap [1–3].

Other important factors which make the films an attractive material for coatings include a smooth surface and low friction, thermal stability and transparency over a wide spectral range. These properties also offer advantages as compared to another wide optical bandgap material – silicon carbide (SiC) – for uses in nano-scale optical data recording for archival information storage using focused ion beams (FIB) techniques, where SiC thin films have found useful applications recently [4-10].

In the case of hydrogenated amorphous silicon carbide (a-SiC:H) films, computer operated FIB systems are used to both introduce irradiation defects and additionally chemically modify the amorphous structure of the films, thus reducing their optical bandgap in a very effective manner for the useful creation of optical contrast between implanted and non-implanted areas of the film material for applications in nano-scale optical data recording [11-13].

In the amorphous SiC film materials, a considerable part in the creation of useful optical contrast between irradiated and non-irradiated areas of the films is played by the transformation of substantial part of the present diamond-like (sp3) carbon bonds, before the irradiation, into graphite-like (sp2) carbon bonds, as a result of it [14-20]. It is expected, that a similar mechanism of the carbon bonds transformation would result when applying ion bombardment with different ions, e.g. gallium ions (Ga^+), in ta-C films, so that to achieve useful optical contrast between irradiated and non-irradiated areas of the films, which could possibly be of interest for further uses in archival information storage.

2. Experimental

Thin ta-C films (d~40 nm) were deposited on Corning glass substrates using a commercial FCVA system (Commonwealth Scientific Corporation). Carbon plasma is produced from the arc spot on the cathode, 99.999% pure graphite in high vacuum. Cathodic arcs are prolific generators of highly ionized carbon plasmas. With the FCVA technique, the plasma stream is steered through a magnetic filter to eliminate neutral particles generated at the cathode. At the filter exit, the fully ionized plasma, consisting of carbon ions and electrons, streams towards the substrate. The films were deposited at room temperature with an arc current of 120 A under floating conditions.

Ion implantation of Ga^+ was carried out at room temperature (RT) using a commercial broad-beam ion implanter. The ion-beam intensity was I~2 $\mu A/cm^2$, the ion energy was E=20 keV, and the ion fluencies used were D1=3.10^{14} and D2= 3.10^{15} cm^{-2}. The SRIM simulation program [21] was used to define the projected range Rp ~ 17 nm and the struggle ΔRp ~ 4 nm for the Ga^+ implanted ions into the ta-C film samples (d = 40 nm).

Conductive atomic force microscopy (C-AFM) measurements were carried out using scanning probe microscopy (MFP-3DTM, Asylum Research, Oxford Instruments) operating in contact mode. A Ti/Ir coated silicon probes (ASYELEC01) with a resonant frequency of 70 kHz and spring constant of 2 Nm^{-1} were employed. A bias of +5 V was applied to the sample surface during measurements.

3. Results and discussion

The surface morphology of pristine and gallium implanted ta-C samples was studied using AFM measurements in contact mode. In figure 1a, figure 1b and figure 1c are displayed the morphology changes of the ta-C samples with the gallium implantation. In general, the pristine films are homogeneous over large areas of the substrate (figure 1a). Figures 1b and 1c are showing roughness increase with the gallium fluence increase, particularly well expressed for the case with the highest gallium fluence (figure 1c). Presumably, this is due to the high concentration of the introduced Ga^+ atoms, which is of the order of those for the host element. The change of the structures could lead also to changes in the optical properties, which effect is a subject of some other publications of ours [22, 23].

The surface electrical conductivity change is presented in figure 1d, figure 1e and figure 1f, where the gallium fluence increase leads to apparent increase of the measured electrical conductivity. This is assumed to be due to the presence of electrically conductive graphitic nano-clusters, which are interconnected and grow with the gallium fluence. The conductivity increase is however limited as the graphitic clusters are limited in size with the growth of gallium metal colloids which also increase in size and concentration with the gallium fluence as reported in previous publications of ours [24, 25]. These graphitic clusters, though limited in size, are further increased in concentration and are redistributed with the gallium fluence increase (figure 3 and figure 4).

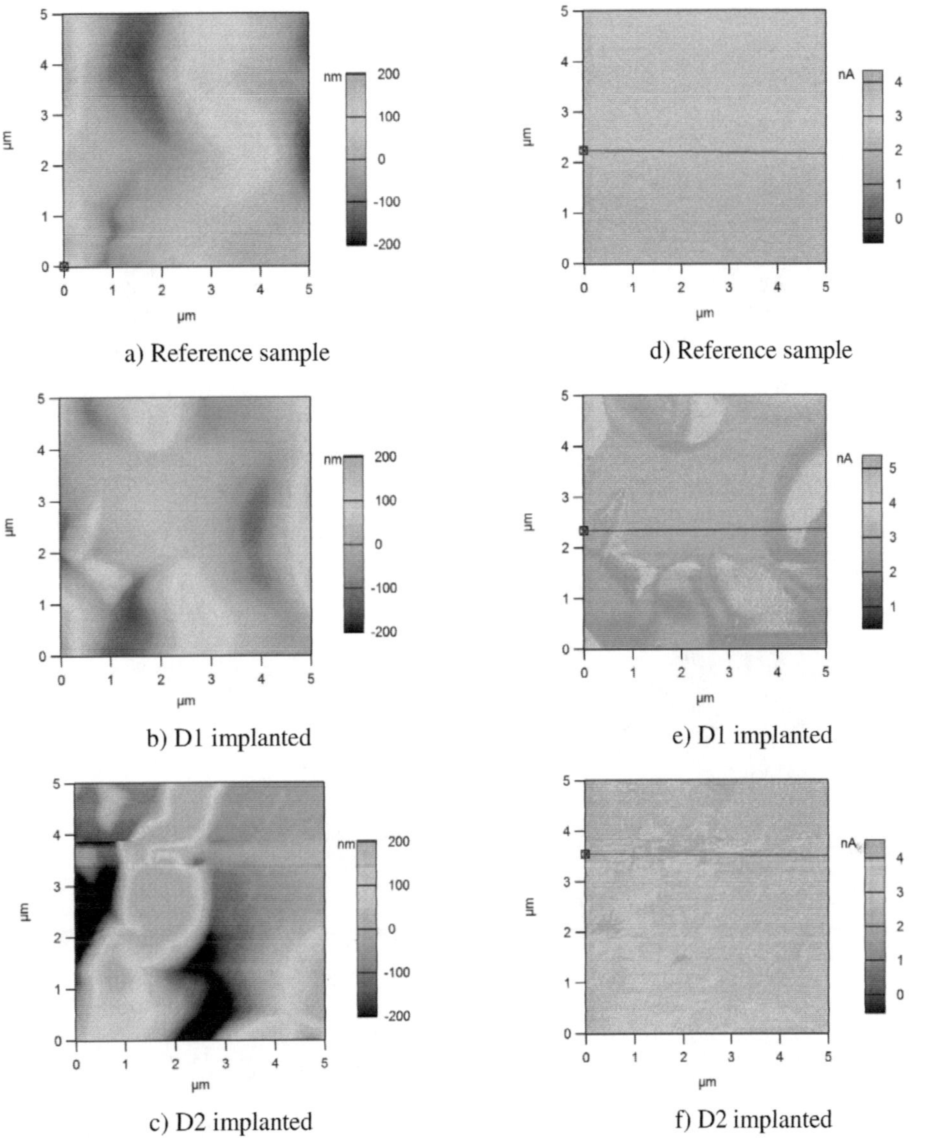

Figure 1. C-AFM images of topography: a) reference sample, b) D1 implanted and c) D2 implanted; and corresponding current: d) reference sample, e) D1 implanted and f) D2 implanted.

In the following figures, figure 2, figure 3 and figure 4, are displayed the three-dimensional current images, the current sections across the lines designated in the two-dimensional current images in figure 1 (figure 1d, figure 1e and figure 1f) and current histograms from 5x5 mm area, for the reference sample, the D1 sample and the D2 sample (figure 2, figure 3 and figure 4, respectively). Apparently, the electrical conductivity increases already for the lower gallium implantation fluence (D1), with only redistribution of the charge resulting from the higher gallium fluence (D2) and hardly noticeable any increase of current. As pointed out earlier, the conductivity increase appears limited as the graphitic clusters are eventually disrupted by the gallium colloids that increase in size and concentration with the fluence.

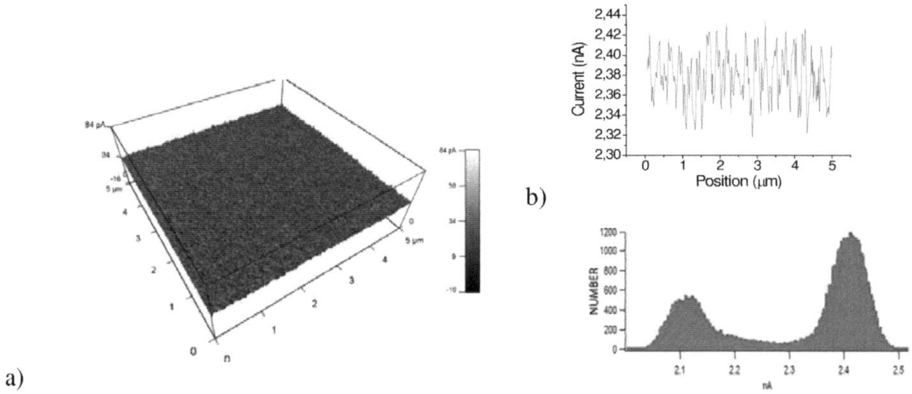

c) I_{aver}=2,3 nA, R_a=125 pA

Figure 2. Current image (a), current section across the line in figures 1d (b) and current hystogram from 5x5 μm area (c) for reference sample (I_{aver} – mean value of current; Ra – average deviation).

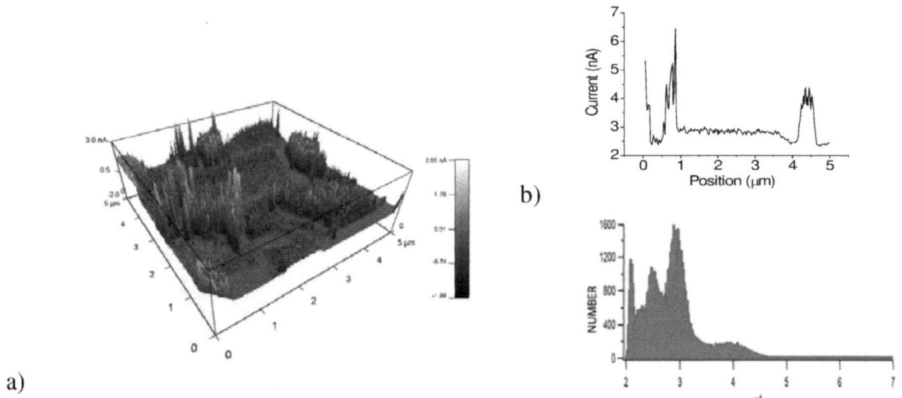

c) I_{aver}=2,867 nA, R_a=425 pA

Figure 3. Current image (a), current section across the line in figures 1e (b) and current hystogram from 5x5 μm area (c) for D1 sample (I_{aver} – mean value of current; Ra – average deviation).

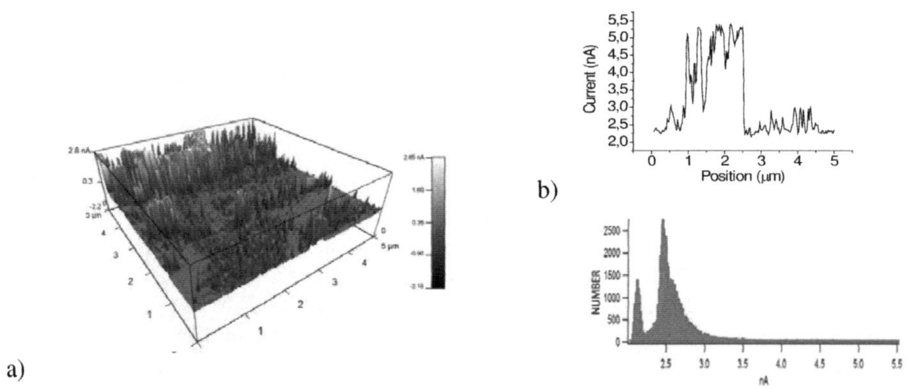

c) I_{aver}=2,78 nA, R_a=414 pA

Figure 4. Current image (a), current section across the line in figures 1f (b) and current hystogram from 5x5 μm area (c) for D2 sample (I_{aver} – mean value of current; Ra – average deviation).

4. Conclusion

The obtained results in the present work have shown that ion implantation of ta-C thin films with Ga^+ with relatively low ion fluence (D = 3.10^{14} cm^{-2}) has no significant effect on the film surface morphology. However, higher fluences of Ga^+ ion implantation (on the order of D = 3.10^{15} cm^{-2}) in such films results in considerable modification of the surface structure and morphology changes of the implanted ta-C films. Simultaneously, the surface electrical conductivity of the gallium implanted ta-C samples increases, presumably due to the growth of graphitic clusters in the implanted areas, which increase in concentration with the gallium fluence. The observed properties modification results also from the high concentration of introduced Ga^+ atoms, which is of the order of those for the host element. The change of the structures could lead also to changes in the optical properties, studied in earlier publications of ours, which effect could be made use of in high-density optical data storage.

Acknowledgements

The author would like to thank the Operational Program "Development of the Competitiveness of the Bulgarian Economy", project BG16 1P0003-1.2.04-0034-C0001, for the financial support, and the group of Prof. S. Kitova for the samples preparation and characterisation. The author would also like to thank Dr T.Tsvetkova and the staff of the ion beam center at the Helmholtz-Zentrum Dresden-Rossendorf e.V., a member of the Helmholtz Association, for planning and performing the ion implantation.

References:

[1] Veerasamy V S, Yuan J, Amaratunga G, Milne W I, Gilkes K W R, Weiler M and Brown L M 1993 *Phys. Rev.* B **48** 17954
[2] Robertson J 1991 Prog. *Solid State Chem.* **21** 199
[3] Xu S, Flynn D, Tay B K, Prawer S, Nugent K W, Silva S R P, Lifshitz Y and Milne W I 1997 *Philos. Mag.* B **76** 351
[4] Kalbitzer S 2004 *Nucl. Instrum. Methods* B **218** 343
[5] Tsvetkova T 1996 *Beam Processing of Advanced Materials*, Eds. J. Singh, S. Copley, J. Mazumder, ASM International, Metals Park, p 207
[6] Tzenov N, Tzolov M, Dimova-Malinovska D and Tsvetkova T 1993 "Ion bombardment effect on the optical properties of a-SiC:H films", *Electronic and Optoelectronic materials for the 21st century*, World Scientific, Singapore p 339
[7] Tsvetkova T, Tzenov N, Tzolov M, Dimova-Malinovska D, Adriaenssens G, Pattyn H and Lawerens W 1993 *J. Non-Cryst. Solids* **164-166** 897
[8] Tzenov N, Tzolov M, Dimova-Malinovska D, Tsvetkova T, Angelov C, Adriaenssens G and Pattyn H 1994 *Nucl. Instrum. and Methods* B **84** 195
[9] Bischoff L, Teichert J, Kitova S and Tsvetkova T 2003 *Vacuum* **69** 73
[10] Tsvetkova T, Angelov O, Dimova-Malinovska D, Sendova-Vassileva M, Bischoff L, Adriaenssens G J, Grudzinski W and Zuk J 2003 *Vacuum* **70** 467
[11] Tsvetkova T, Takahashi S, Zayats A, Dawson P, Turner R, Bischoff L, Angelov O and Dimova-Malinovska D 2005 *Vacuum* **79** 94
[12] Tsvetkova T, Takahashi S, Zayats A, Dawson P, Turner R, Bischoff L, Angelov O and Dimova-Malinovska D 2005 *Vacuum* **79** 100
[13] Takahashi S, Dawson P, Zayats A V, Bischoff L, Angelov O, Dimova-Malinovska D, Tsvetkova T and Townsend P D 2007 *J. Phys.* D*: Appl. Phys.* **40** 7492
[14] Tzenov N, Dimova-Malinovska D, Marinova Ts, Krastev V and Tsvetkova T 1996 *Nucl. Instrum. and Methods* B **112**, 342
[15] Tzenov N, Dimova-Malinovska D and Tsvetkova T 1996 *Mat. Res. Soc. Symp. Proc.* **396** 243
[16] Barancira T, Moons R, Koops G E J, Deweerd W, Pattyn H, Tzenov N, Tzolov M, Dimova-Malinovska D, Tsvetkova T, Venegas R, Zhang G L 1999 *J. Non-Cryst. Solids*, 244189

[17] Tsvetkova T, Tzenov N, Tzolov M, Dimova-Malinovska D, Adriaenssens G J and Pattyn H 2001 *Vacuum* **63** 749
[18] Tsvetkova T, Wright C D , Craciun M F , Bischoff L, Angelov O and Dimova-Malinovska D 2012 *J. Phys.: Conf. Series* **398** 012048
[19] Tsvetkova T, Wright C D, Hosseini P, Bischoff L and Zuk J 2013 *Acta Physica Polonica* A **123** 952
[20] Tsvetkova T, Wright C D, Kitova S, Bischoff L and Zuk J 2013 *Nucl. Instrum. & Meth.* B **307** 71
[21] Ziegler J F, Biersack J P, Littmark U *The Stopping and Range of Ions in Matter* 1985 (Pergamon, New York) Vol. 1
[22] Sandulov M, Berova M and Tsvetkova T 2014 *J. Phys.: Conf. Ser.* **558** 012044
[23] Sandulov M, Berova M, Tsvetkova T and Zuk J 2015 *Acta Physica Polonica* A **128** 953
[24] Berova M, Sandulov M, Tsvetkova T, Karashanova D, Boettger R and Bischoff L 2016 *J. Phys.: Conf. Ser.* **700** 012035
[25] Tsvetkova T, Berova M, Sandulov M, Kitova S, Avramov L, Boettger R and Bischoff L "Focused ion beam optical patterning of ta-C films", *Surface and Coatings Technology*, doi: 10.1016/j.surfcoat.2016.07.088

IOP Conf. Series: Journal of Physics: Conf. Series **794** (2017) 012012 doi:10.1088/1742-6596/794/1/012012

Improvement of growing of Ge QDs by the method of liquid phase epitaxy

I I Maronchuk[1*], D D Sanikovitch[1], A S Cherkashin[1], H Nitchev[2], D Dimova-Malinovska[2*]

[1] Laboratory of Applied Physics and Nanotechnologies for the Energy Application, Sevastopol National University of Nuclear Energy and Industry, str. Kurtchatova, 7, 299033-Sevastopol, Russia

[2] Central Laboratory of Solar Energy and New Energy Sources, Bulgarian Academy of Sciences, bulv. Tzarigradskao chaussee, 72, 1784 Sofia, Bulgaria

E-mail: dmalinovska@yahoo.co.uk, igimar@mail.ru

Abstract. This paper reports on improvement of the technological conditions for nano-heteroepitaxial structures (NHES) growth with Ge quantum dots (QDs) by liquid phase epitaxial (LPE) method applying impulse cooling on the substrate (ICS) The physical and mathematic modeling of the processes of growth and the analysis of the thermodynamic status has been carried out to optimize the construction of the thermal unit, the located in it graphite cassette and of the thermal conditions. For the analysis the Solid Works Flow Simulation program is applied, which has a satisfactory accuracy of calculations of heat-transfer simulation. The analysis has revealed shortcomings in the construction of the equipment. Having in mind these results the equipment is reconstructed and new different elements of the thermal block are installed. Good agreement of the experimental and calculated temperature distribution in the process of NHES with Ge QDs growing is obtained. The grown Ge QDs have improved structure with homogeneous distribution and size and depth of the Quantum Wells. The experiments carried out show good reproducibility of the growing process confirming the correctness of the mathematic modeling.

1. Introduction

Currently the development of technologies for growing of semiconductor nano-heteroepitaxial structures (NHES) with quantum wells (QW) and quantum dots (QDs) becomes important for application in solar cells [1].

In this context, the development of methods for growing of NHES without "wetting" layers between QDs by the Stranski-Krastanow mechanism [2], the method of liquid phase epitaxy (LPE) applying impulse cooling of the substrate (ICS) [3,4], used in this work, is actual and very promising. For its application in the devices technologies it is necessary the obtained nanostructures to be without defects, to have highly homogeneous in size and homogenously distributed QWs and QDs. The process should assure high reproducibility of the properties of the nanostructures being grown on large diameter substrates.

In our previous attempts it was not always possible to obtain reproducible results using the existing experimental equipment for growing epitaxial NHES with QDs. This observation required improvement of the construction of the working units of the equipment and optimization of the technological parameters of the process.

The main objective of this work has been focussed on scientific research for resolving the technical problems related to: development and manufacturing of essentially new working units of the equipment for growing NHES with Ge QDs by LPE method applying IC; optimization of thermal conditions of the process; reconstruction of the cassette and units of equipment by application the results of physical and mathematical modelling; development of the methodology of the thermal fields adjustments in LPE setup during the process of growth.

2. Experimental

From the previous study some disadvantages in the construction of the equipment and in the technological approaches of the processes that adversely affect the quality of the grown NHES with QDs by the method of LPE with ICS have been determinate [5-7]. The main shortcomings have been: poor reproducibility of the processes associated with the uncontrolled mixing of melt solution materials of different compositions and its uncontrolled removing from the substrate, which leads to deterioration of the quality of the grown material; unexpected changes in the thermal conditions of the process of growing related to shift of the position of the sliders of the working cassette relative to the stationary thermal field, which leads to uncontrolled change of temperature of the end surface of the heat absorber.

To overcome these (and other) shortcomings a new vertical type cylindrical drum cassette has been developed and manufactured. The diameter of the cassette is 90 mm and its length - 150 mm. Schematic diagram of the cassette (section along the length axis) is shown in figure 1.

In order to optimized the thermal conditions and the construction of the thermal units of the equipment and of located there graphite cassette, the physical and mathematical modelling of the processes of growing of NHES QDs by the method of LPE method with applied ICS have been carried, analysing the thermodynamic status. For the analysis the SolidWorks FlowSimulation program has been selected which has a satisfactory accuracy of calculations for simulation of heat-transfer process. For the modelling our experimental data obtained earlier [8] have been used and the boundary conditions for the mathematical model and to plane the performance of the work. For the modelling the difference in the temperatures on the area of the end of the heat absorber and the substrate, $\Delta T_{HA\text{-}S}$ 4°C, has been used and kept close to this value in the experiments. The temperature

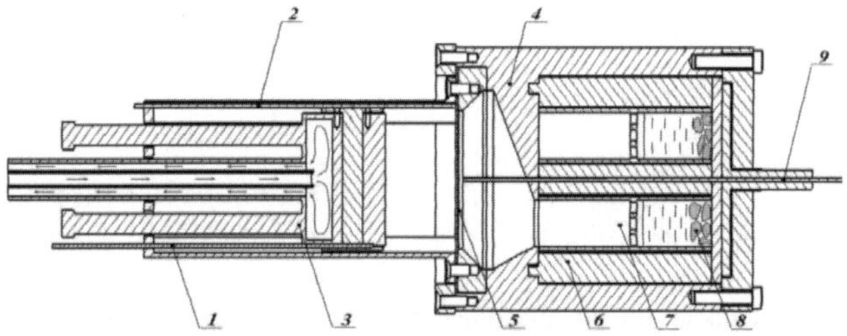

Figure 1. Schematic cross section diagram of the cassette with the water cooled heat absorber used for growing NHES QDs by the method of LPE method with applied ICS. 1- thermo-couple of the heat absorber, 2 - thermo-couple of the substrate, 3 - frame of the water-cooled heat absorber, 4 – corpus of the cassette, 5 – substrate, 6 – cylinder for the containers, 7 – containers. for the holders with melt solution materials, 8 – melt solution materials, 9 – slippery thermo-couple.

in the working space, T_{WS} = 425 °C, have been used for the modelling and during the process of growing. In the modelling of the conditions of the process of growing a c-Si substrate with diameter of 52 mm has been considered in the analysis. In the experiment the Ge QDs have been grown on the same type substrate - c-Si wafer with 52 mm diameter and orientation (111). The process has been carried out in the H_2 flow. A low melt temperature metal – Sn, has been added in the melt solutions used for formation of the buffer and spacer layers. The experiments of growing of NHES QDs and theoretical foundations of the method of LPE with ISC have been described in our previous publications [5-7].

Applying the physical and mathematical modelling has led to establishment of the negative factors affecting the process of growing of NHES QDs: the presence of inhomogeneous thermal field on the area of cassette of 4°C/cm axially and to 2°C/cm radially; the presence of temperature gradient up to 20°C/cm in the area of the working plane of the heat absorber; a high inertness of the process when the thermal conditions have changed; dependence of the quality, homogeneity in distribution of the Ge QDs; dependence of the reproducibility of the process of growth on the position of the heat absorber on the back side of the substrate. As a result of mathematical modelling and optimisation of the thermal units of the equipment and the design of the graphite cassette the shortcoming in the growing process have been eliminated and the construction of the equipment has been improved: materials with appropriated heat conductivity have been used for the heat absorber, new materials have been used for manufacturing of the cassette in order to obtain homogeneous heat field, new materials have been used for the new designed system of shields in the heating area around the Si substrate. From the analyse of the gas flow into the reactor it has been established homogeneous temperature distribution of the flow in the cross section of the reactor and the shortcomings in the construction of the gas supply in the input of the reactor have been pointed and overcame.

In order to establish the appropriated conditions for the process of growing according to the results of modelling the control of temperatures in the three-zones heater of the oven (with quasi-hermetic quartz reactor where the cassette is situated) has been carried out by different thermocouples disposed on different places: in the external side of the heat absorber in the "cold "zone, in the central area of the substrate and in the external side of the cassette in the "hot" zone. The temperatures have been measured to be 430° C, 460° C and 468° C, respectively. Using the slippery (moving) thermocouple (number 9 in figure 1) the temperature on the length of the cassette has been measured in direction from its external side to the substrate and the data have been used to obtain the temperature distribution on the surface of the cassette and in the different places of the oven unit of the equipment. The preliminary measurements have been carried out in the cases of vertical and horizontal positions of the cassette. As a result the negative influence of the gravitation effect on the temperature

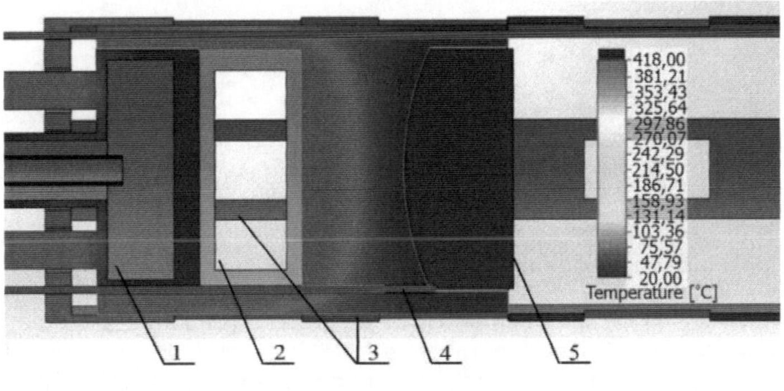

Figure 2. Scheme of the heat absorber used for growing NHES QDs by the method of LPE with applied ICS. The temperature colour legend is shown in the right side. 1 - water-cooled part of the heat absorber, 2 – technological opening, 3 - elements of the heat absorber, 4 - thermo-couple of the heat absorber, 5 – working area of the end surface of the heat absorber.

distribution has been determinate and the corresponding changes in the construction have been made to minimise it. The difference between the temperature of the surface of the cylindrical cassette and in its centre has been measured to be not higher than 0.5°C in the case of both positions of the cassette. This result shows homogeneous temperature distribution which demonstrates the advisability of the taken construction changes. These conditions have been applied for carrying out the process of growing open Ge QDs on c-Si substrate according the procedure described earlier [5,6]. The surface morphology of the obtained samples has been studied by AFM microscope type CMM-2000 (proton-Zelenograd, Russia) with soft cantilevel tip MSCT (Veeco, USA).

2. Results and discussion

The results obtained from physical-mathematical modelling of the thermal processes applied the values of the measured temperatures on different units of the equipment are compared to the corresponding calculated data from the simulation using SolidWorks FlowSymulation programme.

Figure 2 shows the optimised construction of the heat absorber and the distribution of the measured temperatures (the experimental data) in colours on different places on it. At the right side in the figure 2 the colour legend of the temperatures is presented, as well. The temperatures at the working area of the end of the heat absorber are very homogeneously distributed (red-brown colour of the area 5). The calculated temperature distribution along the radius of the working area of the edge of the heat absorber which has a contact with the substrate (having a circular shape) is displayed in figure 3. It is seen (curve 2 –calculated) that the temperature field along the diameter (respectively along the radius) of the working area of the heat absorber is distributed homogeneously and the difference in the temperature along the radius does not exceed 0.07° C. The average temperature on the working area of the heat absorber is 417.66° C which coincides with the corresponding measured temperature distribution presented in figure 2. It has to be noted that the rate of the cooling water has been kept at 60 l/h, and its average temperature has been 26° C. The measured temperature distribution along the radius of the head absorber (between its periphery and its centre) (figure 2, curve 1) has larger deviation - $\Delta T = 0.2°$ C, compared to the calculated one which is 0.1° C. This slight difference could be explained by the accuracy of the measurement by the thermocouples and the temperature gauges. Actually, the experimental measurements have been performed with an accuracy of 0.1° C.

Figure 4 demonstrates the calculated (curves 2) and the experimental (curves 1) temperature distribution on the diameter of the Si substrate (with diameter d= 52 mm). In figure 5 the calculated distribution of the temperatures over the working surface of the Si substrate (d = 52 mm) is shown in colour. It is observed that the average measured temperature of the working surface of the substrate is

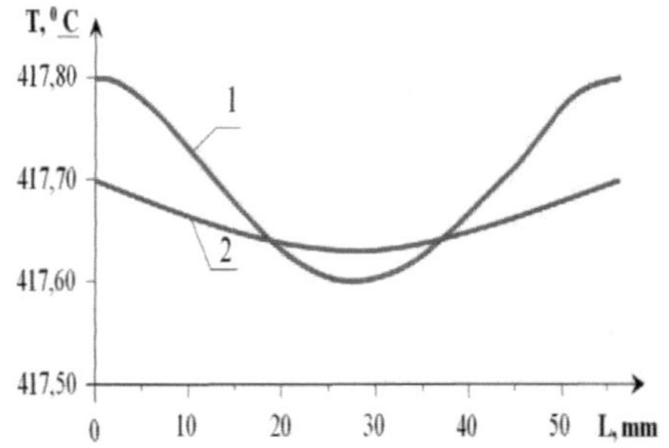

Figure 3. Temperature distribution on the diameter (respectively on the radius) of the working area of the end of the heat absorber using flow of cooling water with rate 60 l/h and temperature 26° C. Curve 1 - experimental data, curve 2 – calculated data.

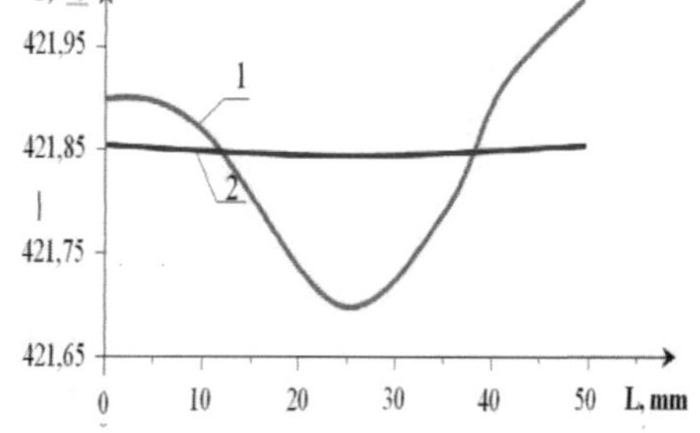

Figure 4. Temperature distribution on the diameter (respectively on the radius) of the Si substrate at average temperature in the working zone T_{WZ} = 425° C. Curve 1 - experimental data, curve 2 –. calculated data.

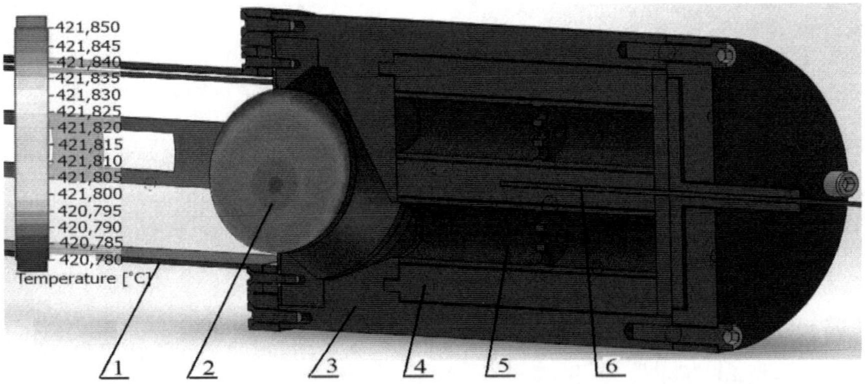

Figure 5. Scheme of the substrate with the calculated temperatures. The average temperature in the working zone $T_{WZ} = 425^{o}$ C and on the end surface of the heat sink $T_{HS} = 417.66^{o}$ C. The temperature colour legend is shown in the left side. 1 - frame of the heat sink, 2 - working area of the substrate, 3 - corpus of the cassette, 4 – drum for containers, 5 – container for the ware with the melt solutions, 6 – moving thermocouple.

421.85° C, the temperature distribution measured over the working surface of the substrate is homogeneous and the difference in the temperatures is very small, 0.22° C (between L = 0 mm and L = 26 mm) and 0.26° C (between L = 26 and L = 52 mm). The experimental data (figure 4, curves 1) and calculated values of temperature shown in figure 4 (curve 2) and in figure 5 coincide very well. The small difference between the experimental and calculated temperatures (0.1° C) could be associated with accuracy of the measurement by the corresponding devices. Also, the difference could be due to the fact that the calculations have been performed without taking into account the gravity component in the horizontal arrangement of the cassette, while the experimental measurements have been carried out in its vertical version arrangement. It should be noted that, as shown in figure 2 and figure 5, the average temperature difference between the working area of the end surface of the heat absorber (figure 2) and the substrate (figure 5) is measured to be $\Delta T_{HA\text{-}S} = 4.19^{o}$ C which is very close to the value used in the modelling - 4.0° C, as pointed earlier.

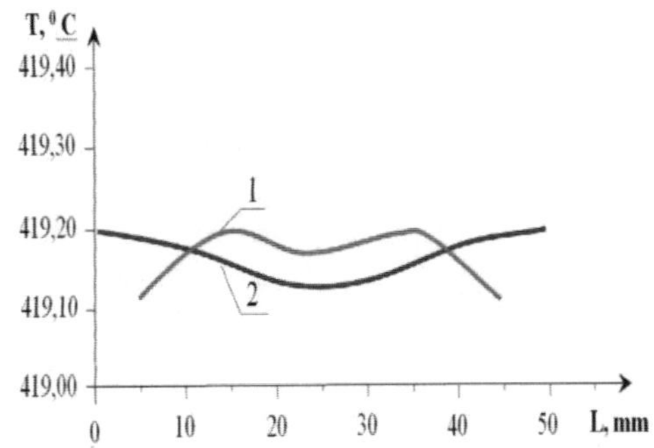

Figure 6. Temperature distribution on the diameter of the Si substrate at its overlapping with the end of the heat absorber and the melt solution, the average temperature in the working zone $T_{WZ} = 425^{o}$ C. Curve 1 - experimental data, curve 2 – calculated data.

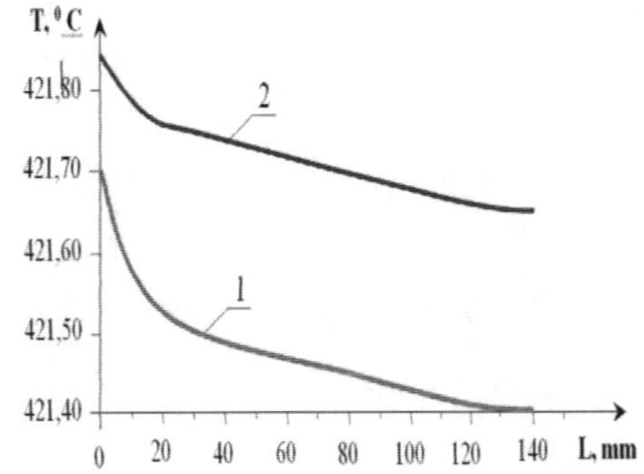

Figure 7. Temperature distribution on the length of the cassette (axial direction) at average temperature in the working zone $T_{WZ} = 425^{o}$ C. Curve 1 - experimental data, curve 2 – calculated data.

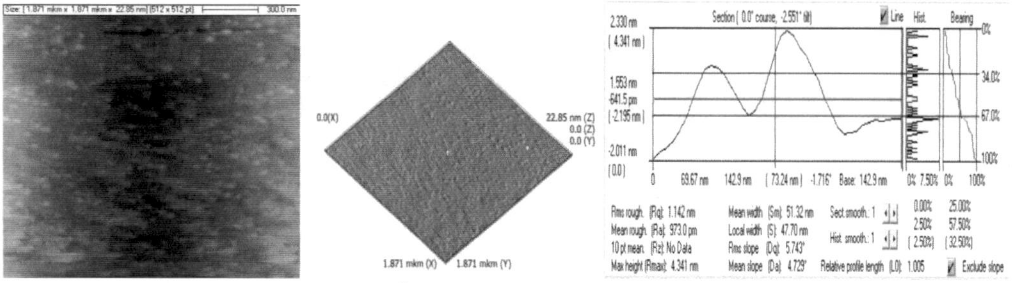

a b c

Figure 8. AFM surface view (a), three dimensional image (b) and profile (c) of the Ge QDs on the Si buffer layer on (111) Si substrate obtained after modernization of the equipment. The marker in(a) corresponds to 100 nm.

Figure 6 shows the distribution of temperature along the radius of the Si substrate at the average temperature in the working zone is $T_{wz} = 425°$ C when the substrate is aligned in the heat absorber and is put into contact with the melt solution. As can be seen from figure 6 the temperature profile is very smooth. The elimination of the shortcomings due to the asymmetric arrangement of the units in the construction of the cassette has resulted in very good temperatures inhomogeneity on the substrate. Some mismatches of experimental and theoretical curve in figure 6 can be caused not only by the factors described above, but also by the presence of some constructive special features during the process of measurement (the setting of the measuring thermocouple in the quartz capillary).

The calculated and the measured temperature profiles along the length of the cassette are demonstrated in figure 7. Both curves correlate well with each other. The shifts in the temperatures is due to the fact that the experimental data are measured with a sliding thermocouple located in the quartz capillary, which in its turn is located in the through hole made on the axis of the graphite cassette (number 4 in figure 2) which lead to the some difference (about 0.30^{o} C) in the calculated and measured data. The calculated values have been obtained in the case of the graphite cassette without a through hole and the thermocouple without quartz and metal housing which could explain the difference.

According to the theoretical calculation in the prevailing part of the cassette length (L is between 30 – 130 mm, figure 7) the distribution of the temperature has linear character with difference in the temperatures on the substrate and on the area at the end of the heat sink < 0.1^{o} C. Only at the areas of the cassette where the opening for the thermocouple (L is between 0 and 20 mm,

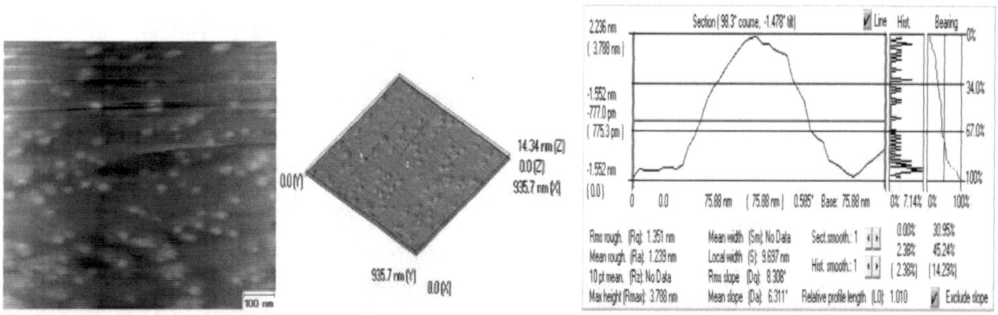

a b c

Figure 9. AFM surface view (a), three dimensional image (b) and profile (c) of the Ge QDs on the Si buffer layer on (111) Si substrate obtained before the modernization of the equipment. The marker in (a) corresponds to 100 nm.

figure 7) is situated there is small deviation from the linear curve probably caused by the difference in the thermo-conductivity of the materials: the substrate, melt solution, graphite cassette and the gas in the reactor camera. The temperature gradient in the linear part of the experimental data - 0.14° C is a little bit higher than in the calculated temperature distribution (0.1° C) which could be due to the gravitational effect, as discussed earlier, and it is suitable for the realization of the process. As a result, non-uniformity of the thermal field in the working area of the cassette in the axial direction is reduced to 0.3° C /cm and radial direction - to about 0.1° C /cm. In the working area the presence of the negative temperature gradient is reduced to 1° C/cm on a very large area of the end surface of the heat absorber.

In the figure 8 the AFM pictures of the open Ge QDs grown on buffer layer obtained on Si substrate (111) orientation at application of 1 cooling impulse after the modernization and optimization of equipment are shown. The Si substrate has been treated according the methods as described earlier in [5-7]. From the figure 8 b the sizes of the Ge QDs are determinate to be: the diameter is 30 nm and the height -5 nm. It is seen that the maximum difference in the size of the QDs is less that 20%. The surface roughness is < 1Å. The density of the grown Ge QDs is calculated to be 2.10^{10} cm^{-2}.

For the comparison the corresponding AFM pictures of the Ge QDs grown before the described above improvement and optimisation of the equipment are shown in figure 9. The Ge QDs are grown at the same conditions as those presented in figure 8. The diameter of the grown Ge GDs is in the range of 40-60 nm, and height is about 8-10 nm size. The difference in size of the Ge QDs and the surface roughness are higher and the density of their distribution is lower, less than 1.10^{10} cm^{-2}, compared to the structures grown in the optimized equipment.

4. Conclusion

The analyses of the results of physical-mathematical modelling of the thermal processes of the NHES growing of Ge QDs by the method of LPE with applying ICS have shown shortcomings in the constructed and used equipment. Applying the results of the thermodynamic analyse using the programme SolidWorks FlowSymulation the equipment for QDs growing has been reconstructed and improved. A new working cassette has been reconstructed and materials with appropriated thermal conductivities have been applied. The comparison of the result of the calculated and measured temperature distributions on the end of the heat absorber and on the substrate area has demonstrated good coincidence which confirms the correctness of the modernization of the equipment. The control experiments have been carried out and demonstrated better homogeneity in the surface distribution of the grown Ge QDs after optimisation of the construction of the equipment.

Acknowledgement
The work has been supported by the NSF of Bulgaria - grant DNTS/Ukraine - 01/0005 and by the grant №2015/702 from the Ministry of Education and Sciences of the Russian Federation.

References

[1] Pchelyakov O P, Bolkhovityanov Y B, Dvurechensky A V, Sokolov L V, Nikiforov A I, Yakimov A I and Voitlander B 2000 *Semiconductors* (in Russian) **34** 1281
[2] Ledencov N N, Ustinov V M, Shchukin V A, Kopiev P S, Alferov J I and Bimberg D 1998 *Semiconductors* (in Russian) **32** 385
[3] Kulyutkina T F, Maronchuk I E, Maronchuk A I and Naidencova M V *Epitaxial growth of structures with nano-dimensional features from liquid phase by pulse cooling of substrate United States Patent* US 7422632 B2 Class C30B 19/02 provisional application №60/688,769 Jun.8, 2005; date of patent Sep. 9, 2008
[4] Maronchuk I E, Kulyutkina T F and Maronchuk I I *Method of growth epitaxial nano hetero-junction structures with quantum dots array Patent Ukraine* UA № 94699 Class C 30B 19/00, C 30B 29/00, H 01L 21/20. 20.09.2010-10.06.2011 bulletin №5

[5] Maronchuk I I, Sanikovich D D, Nichev H and Dimova-Malinovska D *2016 J. Phys.: Conf. Series* **700** 012043
[6] Maronchuk I E, Maronchuk I I, Kulyutkina T F and Buikovsky S 2012 *Nanosystems, nanomaterials, nanotechnology* (in Russian) **10** 77
[7] Maronchuk I I, Maronchuk I E, Sankovitch D, Lovchinov K and Dimova-Malinovska D 2014 *J. Phys.: Conf. Series* **558** 012049
[8] Ginkin V P, Folomeev V I, Lyukhanova T M, Rakov V V, Maronchuk I I, Kartavykh A V and Egorov A V 2002 *J. Cryst. Growth* **236** 551

IOP Conf. Series: Journal of Physics: Conf. Series **794** (2017) 012013 doi:10.1088/1742-6596/794/1/012013

Optical properties of thick GaInAs(Sb)N layers grown by liquid-phase epitaxy

V Donchev[1,6], I Asenova[1,3], M Milanova[2], D Alonso-Álvarez[3], K Kirilov[1], N Shtinkov[4], I G Ivanov[5], S Georgiev[1], E Valcheva[1] and N Ekins-Daukes[3]

[1] Faculty of Physics, Sofia University, 5, J.Bourchier blvd., Sofia-1164, Bulgaria
[2] Central Laboratory of Applied Physics, 59 St. Petersburg blvd, 4000 Plovdiv, Bulgaria
[3] Department of Physics, Imperial College London, London, UK
[4] Department of Physics, University of Ottawa, Ottawa, ON, K1N 6N5, Canada
[5] Linköping University, Department of Physics, Chemistry & Biology, 581 83 Linköping, Sweden

E-mail: vtd@phys.uni-sofia.bg

Abstract. We present an experimental and theoretical study of GaInAs(Sb)N layers with thickness around 2 μm, grown by liquid-phase epitaxy (LPE) on n-type GaAs substrates. The samples are studied by surface photovoltage (SPV) spectroscopy and by photoluminescence spectroscopy. A theoretical model for the band structure of Sb-containing dilute nitrides is developed within the semi-empirical tight-binding approach in the $sp^3d^5s^*s_N$ parameterisation and is used to calculate the electronic structure for different alloy compositions. The SPV spectra measured at room temperature clearly show a red shift of the absorption edge with respect to the absorption of the GaAs substrate. The shifts are in agreement with theoretical calculations results obtained for In, Sb and N concentrations corresponding to the experimentally determined ones. Photoluminescence measurements performed at 300K and 2 K show a smaller red shift of the emission energy with respect to GaAs as compared to the SPV results. The differences are explained by a tail of slow defect states below the conduction band edge, which are probed by SPV, but are less active in the PL experiment.

1. Introduction

Dilute nitrides form a novel class of semiconductor alloys, which has emerged from the conventional III-V compounds by doping with small amount of nitrogen. The addition of nitrogen in small concentrations has proven to have a significant influence on their electronic properties, in particular bandgap narrowing and shift of the absorption edge to longer wavelengths. This allows extended band structure engineering and makes the dilute nitrides materials of great interest for a variety of applications in advanced optoelectronic devices [1]. The addition of In in the alloy facilitates the

[6] To whom any correspondence should be addressed

incorporation of N, while Sb act as a surfactant that reduces the strain-induced defect formation, favours 2D growth and further decreases the bandgap [2].

While the conventional fabrication methods for dilute nitrides are molecular beam epitaxy (MBE) and metalorganic chemical vapour deposition (MOCVD), liquid phase epitaxy (LPE) has recently shown promise in obtaining high-quality dilute nitride layers with lower defect concentrations for optoelectronic applications [3,4]. This is related to the fact that LPE growth is carried out under near equilibrium conditions which provide high quality material in terms of lifetime, mobility and freedom from defects of the epitaxial layers. In spite of this the LPE technique has been rarely used for growing dilute nitrides alloys and the data on structural, optical and electronic properties of LPE-grown dilute nitride alloys are still scarce. Therefore, these materials need to be extensively studied.

Among the experimental techniques applied to dilute nitrides, surface photovoltage (SPV) spectroscopy has been relatively rarely used [5–7] although it provides information on the optical absorption spectrum and carrier transport in the sample [8]. Besides, the few reported SPV studies on dilute nitrides concern MBE or MOCVD grown samples. The present contribution partially fills this gap by presenting an original study of the electronic and optical properties of LPE-grown GaInAsN and GaInAsSbN layers based on SPV spectroscopy, combined with photoluminescence (PL) and semi-empirical tight-binding calculations.

2. Experimental

A series of $Ga_{1-y}In_yAs_{1-x}N_x$ and $Ga_{1-y}In_yAs_{1-x-z}Sb_zN_x$ epitaxial layers were grown by the horizontal graphite slide-boat technique for LPE on (100) n-type GaAs:Si (~10^{18}cm^{-3}) substrates. A flux of Pd-membrane purified hydrogen at atmospheric pressure was used for the experiments. No special baking of the system was carried out before epitaxy. The starting materials for the solutions consisted of 6N pure solvent metals Ga, In, Sb and of polycrystalline GaAs and GaN with purity of 5N. Quaternary $Ga_{1-y}In_yAs_{1-x}N_x$ layers with low In content were fabricated from mixed In+Ga melt, while for $Ga_{1-y}In_yAs_{1-x-z}Sb_zN_x$ layer growth a few percent of Sb were added to the melt. The N content in the melt was 0.5 at.% for all grown layers. Thick epitaxial layers were grown from initial epitaxy temperature varying in the range 590–550°C at a cooling rate of 0.8°C/min for 2–3 minutes. The layer thicknesses (1.5-2.5 μm) and compositions were determined using a scanning electron microscope LYRA I XMU (Tescan) equipped with an energy dispersive X-ray (EDX) microanalyzer (Quantax, Bruker). The In content in the layers varies in the range $0.02 \leq y \leq 0.03$ depending on the melt composition and epitaxy temperatures. Nitrogen content in the layers determined from HRXRD measurements using Vegard's rule is $0.002 \leq x \leq 0.003$.

SPV spectra were recorded at room temperature using the metal-insulator-semiconductor operation mode of the SPV technique [8]. The sample was illuminated by means of a 250 W halogen lamp along with a SPEX grating monochromator (f = 0.25 m, 600 gr/mm) and an optical chopper (94 Hz). The probe electrode was a semitransparent SnO_2 film evaporated on the bottom surface of a quartz glass separated from the sample by a sheet of mica (15 μm). The probe signal with respect to the ground was fed to a high-impedance unity gain buffer and then measured by an SR830 lock-in amplifier. The photon flux was kept constant ($\approx 1.5\times10^{14}$ cm^{-2}s^{-1}) within ± 0.5 % for all wavelengths, by positioning a neutral density filter with variable optical density. More details about the SPV experimental set-up and measurement procedure can be found in [9].

PL measurements at 300 K were performed using laser excitation at 532 nm (2.33 eV) with 4 mW (~50W/cm^2) power and at 405 nm (3.06 eV) with 50 mW (~4.5kW/cm^2). The signal was detected by a Si detector in the former case and by a CCD camera in the latter case. PL at 2K was also measured with 532 nm excitation. In this case a much lower excitation power (~100 mW/cm^2) was enough to obtain a strong PL signal, which was detected by a CCD camera.

The microscopic lattice structure related to the incorporation of N in InGaAsN alloys was studied by micro-Raman spectroscopy using Jobin Yvone LabRAM HR800 spectrometer and laser excitation at 632 nm.

3. Theoretical calculations

The electronic structure of $Ga_{1-y}In_yAs_{1-x}N_x$ and $Ga_{1-y}In_yAs_{1-x-z}Sb_zN_x$ was calculated using the semi-empirical tight-binding (TB) method for dilute nitrides in the $sp^3d^5s^*s_N$ parameterization, including the spin-orbit coupling [10]. The effects of the addition of nitrogen on the band structure were considered independent from those of In and Sb [11], allowing us to use the virtual crystal approximation for the TB parameters of the GaInAsSb quaternary. Quadratic corrections were introduced to account for the large GaAsSb band gap bowing parameter of 1,43 eV [12]. The N-related TB parameters previously derived for the $Ga_{1-y}In_yAs_{1-x}N_x$ and $GaP_{1-x}N_x$ dilute nitride alloys were used [11], with valence band offsets with respect to GaAs of 0.21 eV, 0.77 eV, and 0.80 eV for the unstrained InAs, GaSb, and InSb, respectively [12]. The bandgap reduction with respect to pure GaAs was calculated for N, In, and Sb concentrations in the expected experimental range ($x \leq 0.01$; $y \leq 0.10$; $z \leq 0.02$).

4. Results and discussions

In figure 1a the SPV amplitude spectrum of a $Ga_{1-y}In_yAs_{1-x}N_x$ sample is compared with its PL spectrum and the SPV spectrum of the GaAs substrate. The small undulations in the SPV spectrum are due to the interference effects caused by the mica sheet. A clear red shift of the absorption edge with respect to GaAs is evidenced by the comparison of the SPV spectra. To assess the bandgap of the material we plot the square of the SPV vs photon energy, which for the ideal case of a bulk semiconductor should give a straight line resulting from the shape of the 3D combined density of states. In the present case we consider the linear part of the obtained curve and extrapolate the straight line to find its intersection with the abscissa, which is assumed to be the bandgap, E_g. For the current sample E_g = 1.328 eV and therefore the bandgap shift ΔE_g with respect to GaAs (E_g=1.423 eV) is 95 meV. The same result (within the experimental error) is obtained by using Tauc plot assuming that SPV is proportional to the absorption coefficient. On the other hand the PL peak position is at 1.362 eV, i.e. 34 meV higher that the bandgap determined from the SPV results.

In figure 1b the SPV amplitude spectrum of a $Ga_{1-y}In_yAs_{1-x-z}Sb_zN_x$ sample is compared with its PL spectrum and the SPV spectrum of the GaAs substrate. The bandgap determined from the SPV curve is 1.312 eV. Therefore the bandgap shift ΔE_g with respect to GaAs is 111 meV in this case. The PL peak appears at 1.357 eV, i.e. 45 meV higher than the bandgap determined from the SPV results.

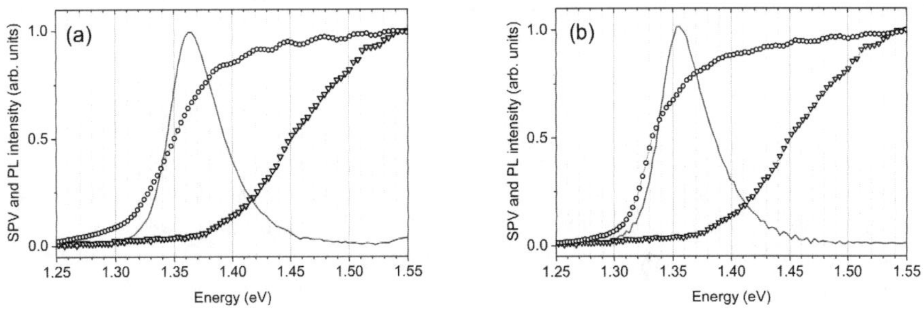

Figure 1. SPV (circles) and PL (line) spectra of a $Ga_{1-y}In_yAs_{1-x}N_x$ (a) and a $Ga_{1-y}In_yAs_{1-x-z}Sb_zN_x$ (b) sample measured at room temperature. The SPV spectra of the GaAs substrate are represented by triangles.

The above described comparisons between SPV and PL were performed for a number of $Ga_{1-y}In_yAs_{1-x}N_x$ and $Ga_{1-y}In_yAs_{1-x-z}Sb_zN_x$ samples. The results show a clear decrease of the bandgap of the dilute alloy with respect that of GaAs. The addition of Sb further reduces the bandgap, which is in accordance with literature data [2]. On the other hand there is a clear discrepancy between the bandgap energy determined from SPV and the position of the PL peak – the values found by SPV are lower than that from PL by some 30-45 meV depending on the sample. This is a kind of "anti-Stokes shift" and is opposite to the Stokes shift between SPV and PL reported in [6] for $Ga_{1-y}In_yAs_{1-x}N_x$ samples with $y = 0.72$ and x ranging from 0.009 to 0.017.

A few $Ga_{1-y}In_yAs_{1-x}N_x$ samples were measured by PL at 2K. A typical 2K PL spectrum is represented by solid line in figure 2, together with the room temperature SPV (circles) and PL (dashed line) spectra. For the purposes of comparison the origin of the energy axis is set at the GaAs bandgap at the corresponding temperature, namely 1.519 eV for 2K and 1.423 eV for 300 K. This way we compare the shifts with respect to the GaAs bandgap rather than the bandgaps themselves. The 2K PL spectrum reveals a peak at –51 meV and a tail from its low-energy side with many small peaks on it. The PL peak obtained at 300 K is much broader but is located at the same energy position as the 2K PL peak. Therefore the PL peaks at 2K and 300 K show the same bandgap red shift of 51 meV with respect to the GaAs bandgap at the corresponding temperature. The SPV spectrum reveals the usual step. Applying the procedure described above gives a bandgap shift with respect to GaAs of 83 meV. Thus, again a discrepancy between SPV and PL results is observed (32 meV in the present case) indicating higher bandgap transition energies for PL.

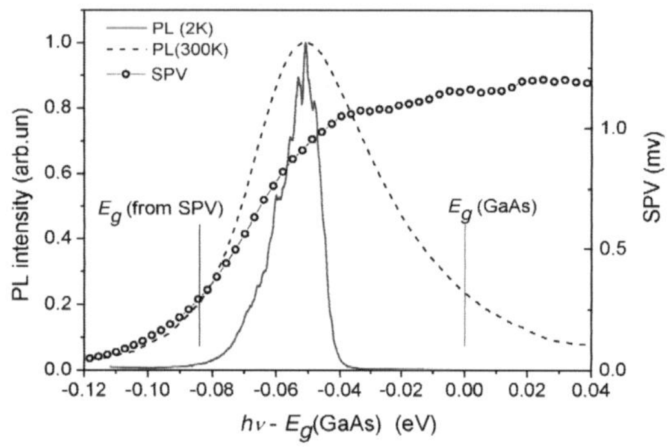

Figure 2. PL of a $Ga_{1-y}In_yAs_{1-x}N_x$ sample measured at 300 K (dashed line) and 2 K (solid line) compared to its 300 K SPV spectrum (circles). The energy is measured from the GaAs bandgap at the corresponding temperature.

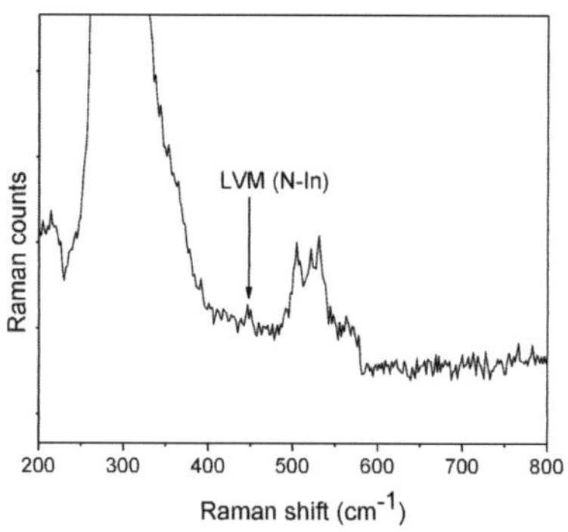

Figure 3. Raman spectrum of as grown InGaAsN layer showing local vibration modes of N-In.

A tentative explanation of these discrepancies is presented below. In an ideal case all N atoms introduced in the crystal take places of As atoms. In figure 3 by means of micro-Raman spectroscopy N-induced local mode is observed in the range 440 to 460 cm^{-1} providing evidence for the formation of In-N bonds. In practice the N atoms take various positions in the crystal lattice as well as form clusters. This results in a series of defect states, some of which are in the bandgap. We assume that these states are slower with respect to recombination (e.g because the electron has to overcome an energy barrier before recombination) than the extended states in the conduction band. The light intensity in the SPV experiment is much lower than that in the PL one. So, with increasing photon energy, SPV probes first these states and as a result gives a relatively low optical transitions energy. On the other hand under laser excitation, which is the case of the PL experiment, these states are filled with electrons. As they are less effective for radiative recombination, the PL mainly probes states with higher energies, which are faster and more effective for emission. Thus, the PL peak appears at higher energies than the bandgap obtained by SPV. The existence of a tail of defect states in the bandgap is evidenced by the low-energy tail of the 2K PL spectrum. If there is an activation barrier for radiative recombination via the slow centres, the low temperature makes further difficult the overcoming of this barrier. Similar PL results were obtained in [13] for $GaAs_{1-x}N_x$ (x ~ 0.25%) and explained by localized and delocalized defect states below the GaAs conduction band. A continuous distribution of shallow defect states located just below the conduction band edge of $Ga_{1-y}In_yAs_{1-x}N_x$ is reported in [14].

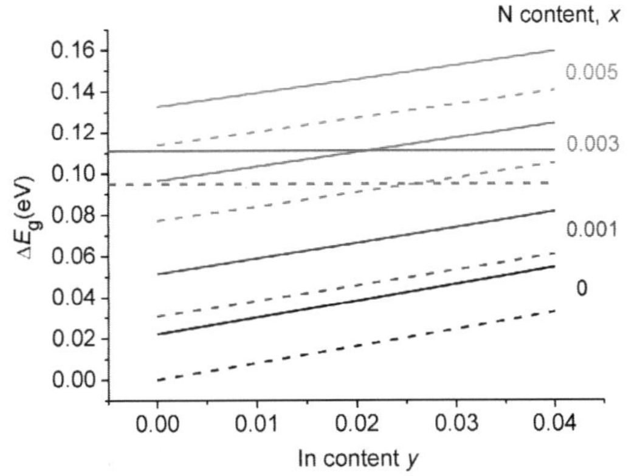

Figure 4. Calculated bandgap shifts of $Ga_{1-y}In_yAs_{1-x-z}Sb_zN_x$ as a function of the In content, for Sb concentrations of 1% (solid lines) and 0% (dotted lines). Curves of different colours correspond to different N fractions as shown. The horizontal lines indicate the experimental shifts obtained by SPV for the samples from figure 1.

Figure 4 shows the calculated dependences of the bandgap shift with respect to GaAs in the expected range of N, In, and Sb concentrations. The addition of Sb reduces the bandgap with 22 meV/at. %, which is in agreement with recent experimental results showing a value of ≈20 meV/at. % for both pure ternary [15] and dilute nitride [16]. This demonstrates the validity of the $sp^3d^5s^*s_N$ model for calculating the bandgap energies of dilute nitrides containing Sb, and validates the use of universal N-related TB parameters for all dilute III-V nitrides. Remarkably, the behaviour of the pentanary $Ga_{1-y}In_yAs_{1-x-z}Sb_zN_x$ is described without introducing new parameters to the standard $sp^3d^5s^*s_N$ parameterization [10,11]. Figure 4 also shows that the calculated bandgap shifts corresponding to the experimentally found In content of 0.02–0.03 and N content of around 0.003 are 91–97 meV for the case without Sb and 110–117 meV for the case of 1% Sb. These values are in good agreement with the experimental values of 95 meV and 111 meV, respectively determined by SPV.

5. Conclusion

An original study of the optical properties of thick LPE grown GaInAs(Sb)N layers is presented. The absorption features, measured by SPV spectroscopy, are compared with results obtained by PL and tight-binding calculations in terms of bandgap shift with respect to GaAs. The bandgap shifts found by SPV are of the order of 80–95 meV for the samples without Sb and around 100–110 meV for the samples containing Sb. They are in good agreement with the calculation results, while the bandgap shifts determined by PL are 30–45 meV lower. This is explained by the existence of a tail of slow defect states just below the conduction band edge, which are probed by SPV, but are less active in the PL experiment. This work demonstrates the ability of the SPV technique as an alternative of the optical absorption spectroscopy for studying dilute nitride materials. The obtained results contribute to better understanding of the physical properties of dilute nitrides grown by LPE and their potential for optoelectronic applications.

Acknowledgements

This work was supported by the research fund of the Sofia University (contract 114/2016) and COST action 1406 MutiscaleSolar. The authors thank A. Tsonev for the EDX and SEM measurements and J. W. Gerlach (Leibniz Institute for Surface Modification) for the HRXRD measurements.

References

[1] Erol E A, ed. 2008 *Dilute III-V Nitride Semiconductors and Material Systems. Physics and Technology* (Berlin: Springer)

[2] Fehse R and Adams A R 2004 Recombination Processes in GaInNAs Long-Wavelength Lasers, in: *Physics and Applications of Dilute Nitrides*, ed I.A. Buyanova and W. Chen (New York, Taylor & Francis) pp. 362–394

[3] Dhar S, Halder N, Kumar J and Arora B M 2004 Observation of a 0.7 eV electron trap in dilute GaAsN layers grown by liquid-phase epitaxy *Appl. Phys. Lett.* **85** 964–66

[4] Milanova M, Vitanov P, Terziyska P, Popov G and Koleva G 2012 Structural and electrical characteristics of InGaAsN layers grown by LPE *J.Crystal Growth* **346** 79-82

[5] Kudrawiec R et al. 2014 Surface photovoltage and modulation spectroscopy of E_- and E_+ transitions in GaNAs layers *Thin Solid Films* **567** 101–104

[6] Bansal B, Kadir A, Bhattacharya A, Arora B M and Bhat R B 2006 Alloy disorder effects on the room temperature optical properties of GaInNAs quantum wells *Appl. Phys. Lett.* **89** 032110

[7] Donchev V, Milanova M, Lemieux J, Shtinkov N and Ivanov I G 2016 Surface photovoltage and photoluminescence study of thick Ga(In)AsN layers grown by liquid-phase epitaxy *J. Phys. Conf. Ser.* **700** 012028

[8] Kronik L and Shapira Y 1999 Surface photovoltage phenomena: Theory, experiment, and applications *Surf. Sci. Rep.* **37** 1–206

[9] Donchev V, Ivanov T, Germanova K and Kirilov K 2010 Surface photovoltage spectroscopy – an advanced method for characterization of semiconductor nanostructures *Trends in Applied Spectroscopy* **8** 27–66

[10] Shtinkov N, Desjardins P and Masut R 2003 Empirical tight-binding model for the electronic structure of dilute GaNAs alloys *Phys. Rev.* B. **67** 081202

[11] Turcotte S, Shtinkov N, Desjardins P, Masut R and Leonelli R 2004 Empirical tight-binding calculations of the electronic structure of dilute III–V–N semiconductor alloys *J. Vac. Sci. Technol.* A **22** 776

[12] Vurgaftman I, Meyer J R and Ram-Mohan L R 2001 Band parameters for III–V compound semiconductors and their alloys *J. Appl. Phys.* **89** 5815

[13] Bhuyan S, Das S K, Dhar S, Pal B and Bansal B 2014 Optical density of states in ultradilute GaAsN alloy: Coexistence of free excitons and impurity band of localized and delocalized states *J. Appl. Phys.* **116** 023103

[14] Kaplar R J, Ringel S A, Kurtz S R, Klem J F and Allerman A A 2002 Deep-level defects in InGaAsN grown by molecular-beam epitaxy *Appl. Phys. Lett.* **80** 4777–4779

[15] Bian L F, Jiang D S, Tan P H, Lu S L, Sun B Q, Li L H , et al. 2004 Photoluminescence characteristics of GaAsSbN/GaAs epilayers lattice-matched to GaAs substrates *Solid State Commun.* **132** 707–711

[16] Wang T S, Tsai J T, Lin K I, Hwang J S, Lin H H and Chou L C 2008 Characterization of band gap in GaAsSb/GaAs heterojunction and band alignment in GaAsSb/GaAs multiple quantum wells *Mater. Sci. Eng. B* **147** 131–135

Magneto-optic manganese doped ZnO thin films

P Petkova[1], V Altunyan[1], P Vasilev[1], Tz Tzenkov[1] and R Mimouni[2]
[1]Shumen University "Konstantin Preslavsky", 115 Universitetska Street, 9712 Shumen, Bulgaria
[2]Unité de physique des dispositifs à semi-conducteurs, Faculté des sciences de Tunis, Université Tunis El Manar, 2092 Tunis, Tunisia

E-mail: Petya232@abv.bg

Abstract. This investigation deals with the magneto-optical properties of pure and manganese doped ZnO thin films in the spectral region 600 – 1200 nm. The spectra of Verdet constant (V) are determined in the same spectral region for these thin films. The magneto-optic anomaly factor γ has been expressed. The spectral dependence of the spin-spin exchange interaction constant K has been studied. The change in energy splitting due to magnetic field for the investigated doped samples is also presented.

1. Introduction

Transparent oxides thin films have been widely investigated for their interesting structural, optical and electrical performances. Due to the decreasing scale of electronic devices using these oxides raises a growing interest in producing self-assembled micro and nano-structured materials. ZnO has attracted much attention as a promising material in recent years due to its unique properties such as a direct wide band gap on the order of 3.3 eV [1], for short wavelength optoelectronic devices, such as emitting diodes and laser diodes, especially as TCO in solar cells, gas sensors [2–7] and acoustic devices [8–10]. Transition metal doped dilute magnetic semiconductors (DMS), exhibiting room temperature ferromagnetism (RTFM) have attracted wide-spread interest for their promising applications in spintronics. In these materials, the charge and spin state of electrons are accommodated into a single material leading to interesting magnetic, magneto-optical and magneto-electric properties [11,12]. The magnetic properties of II–VI-based DMS have attracted much attention following Dietl's prediction [13] of RTFM in Mn-doped ZnO as well as its high Curie temperature due to strong p–d hybridization [14].

Therefore, in the present work, ZnO:Mn thin films have beene investigated in the magnetic field.

2. Experimental details

ZnO:Mn thin films were deposited on glass substrates at 460°C, using 10^{-2} Maqueous solution of zinc acetate di-hydrate ($Zn(CH_3COOH)_2.2H_2O$). The solution contained a mixture of water and propanol with fraction volumes of 1/4 and ¾ respectively. It was acidified with acetic acid (pH = 5). Magnesium chloride tetrahydrate ($MnCl_2$–$4H_2O$) was used as a source of Mn. The Mn-to-zinc molar ratios ([Mn]/[Zn]) were: 0%, 1%, 2% and 3%. Nitrogen was used as the gas carrier (pressure at 0.35 bar) through a 0.5 mm-diameter nozzle. The nozzle to substrate plane distance was fixed at the optimal value of 27 cm [15]. During the deposition process, the precursor mixture flow rate was taken constantly at 4 mL/min throughout the thin films deposition.

3. Results and Discussion

The magneto-optic Faraday effect presents the connection between optics, magnetism and atomic physics. Faraday rotation manifests itself as a rotation of the polarization plane of the light passing through the sample in the presence of a magnetic field, and is characterized by the Verdet constant (V) of the investigated sample. The rotation angle φ can be expressed by the formula [16]:

$$\varphi(\lambda) = A/(\lambda^2 - \lambda_0^2), \tag{1}$$

where A is a constant determined from the matrix elements of the interband transitions, λ is the wavelength, and λ_0 is the wavelength related to the interband transitions and corresponding to the natural frequency $\omega_0 = 2\pi/\lambda_0$ of an effective harmonic oscillator. The relationship between the rotation angle and the Verdet constant is φ = VBd, where B is the magnetic induction of the field and d is the sample thickness (figure 1).

The magneto-optic anomaly factor γ can be taken as a measure of the degree of covalency that exists in the bonds connecting the ions and atoms [17].

$$\gamma = \frac{\varphi}{\frac{e}{2mc^2}\lambda D}, \tag{2}$$

where φ is the rotation angle in the magnetic field, e is the charge of electron, λ is the wave length, m is the mass of the electron and c is the velocity of light.

In the paramagnetic materials, the anomaly factors γ can vary with the wavelength of the incident light, even if there is only one a absorption frequency contributing to dispersion (figure 2). The dispersivity of the investigated crystal can be presented by the following equation:

$$D = -\frac{\lambda}{c}\frac{d^2n}{d\lambda^2}. \tag{3}$$

$$\gamma = -\frac{VBl}{\frac{e\lambda^2}{2mc^2}\frac{d^2n}{d\lambda^2}}, \tag{4}$$

where V is the Verdet constant, B is the magnetic induction, l is the thickness of sample and n is the refractive index.

$$m^* = \frac{e\omega}{\omega_0^2 - \omega^2}\sqrt{\frac{Ne}{2n\varepsilon_0 cV}}, \tag{5}$$

where m* is the effective mass of electrons, N is the number of d shell electrons in Mn^{2+} ions, ω_o is the natural frequency of an effective harmonic oscillator and ε_o is the dielectric permittivity in the vacuum.

$$E_d = \frac{13.52Z^2}{\varepsilon^2}\left(\frac{m^*}{m}\right), \tag{6}$$

where Z is the serial number of element in the Mendeleyev's table and ε is the dielectric permittivity.

The magnetic induction is B = 0.97 T at the room temperature T = 300K and the paramagnetic sample can be treated has having a linear susceptibility given, to an accurate approximation [18], by

$$\chi = \frac{Ng^2\beta^2}{kT}\frac{s(s+1)}{3} = 2.64 * 10^{-4}\,cm^3/mol, \tag{7}$$

where N = 5, g = 2, $\beta = 34.21*10^{-5}$ and S = 5/2 for Mn^{2+}.

The spectral dependence of the spin-spin exchange interaction constant K in the case of pure and manganese doped ZnO thin films is presented in figure 3. It can be calculated by the following formula [19]:

$$K(\lambda) = \frac{V\left[\left(\left(\frac{\lambda}{\lambda_g}\right)^2 - 1\right)\right]^{3/2}}{\chi\lambda}, \tag{8}$$

where V is the Verdet constant, χ is the magnetic susceptibility of the sample and λ_g represents the wave length which corresponds with the band gap of the material.

The following formula

$$\Delta Eij = [(\hbar c)/(Ed)](\partial E/\partial n) \tag{9}$$

describes the parameter ΔEij [20]. This is the change in energy splitting due to magnetic field for a d^5 electron transitions in the Mn^{2+} ion (figure 4).

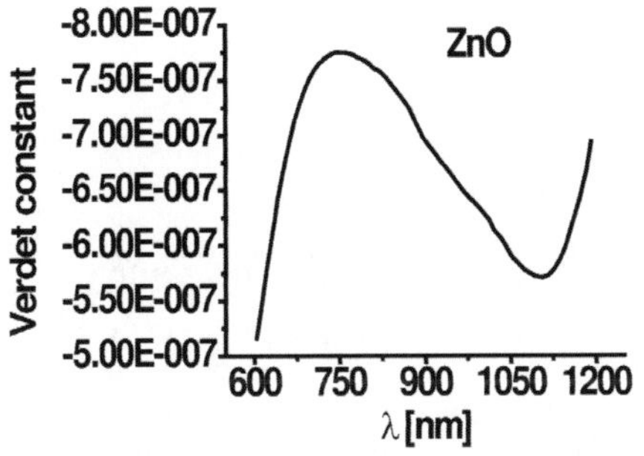

Figure 1a. The dependence V(λ) in the spectral region 600 – 1200 nm for pure ZnO.

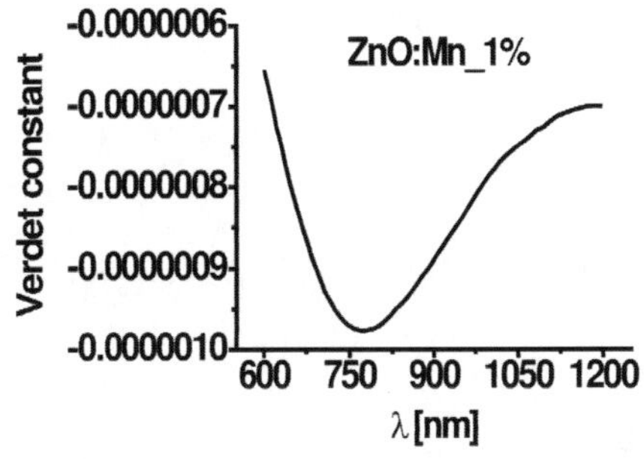

Figure 1b. The dependence V(λ) in the spectral region 600 – 1200 nm for ZnO:Mn_1%.

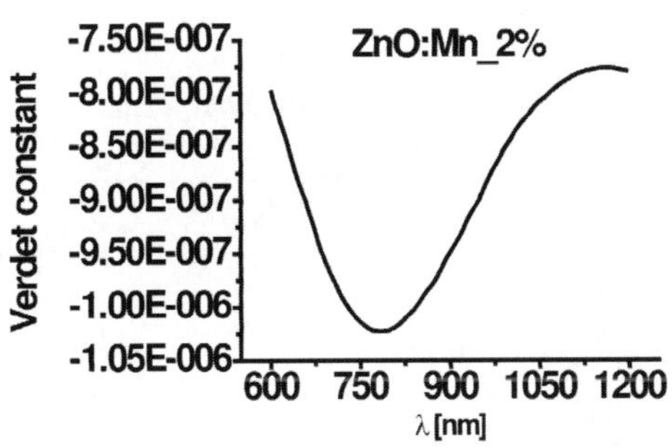

Figure 1c. The dependence V(λ) in the spectral region 600 – 1200 nm for ZnO:Mn_2%.

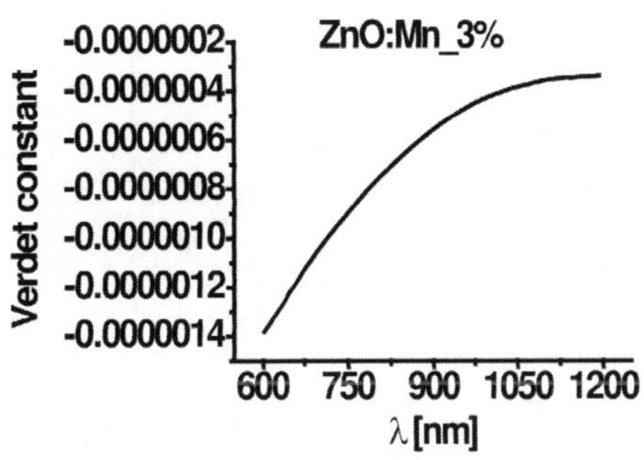

Figure 1d. The dependence V(λ) in the spectral region 600 – 1200 nm for ZnO:Mn_3%.

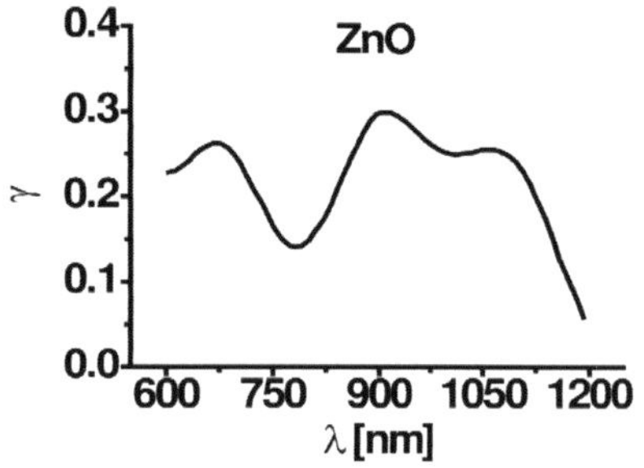

Figure 2a. The magneto-optic anomaly factor γ in the spectral region 600 – 1200 nm for pure ZnO.

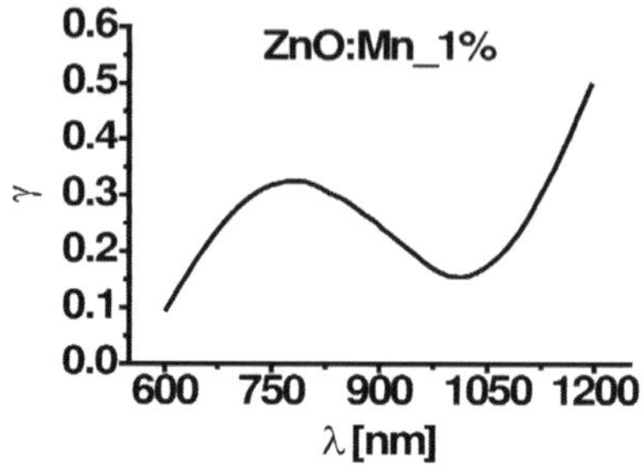

Figure 2b. The magneto-optic anomaly factor γ in the spectral region 600 – 1200 nm for ZnO:Mn_1%.

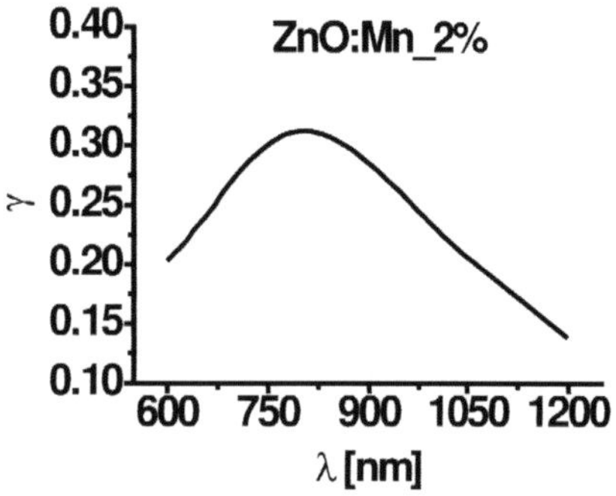

Figure 2c. The magneto-optic anomaly factor γ in the spectral region 600 – 1200 nm for ZnO:Mn_2%.

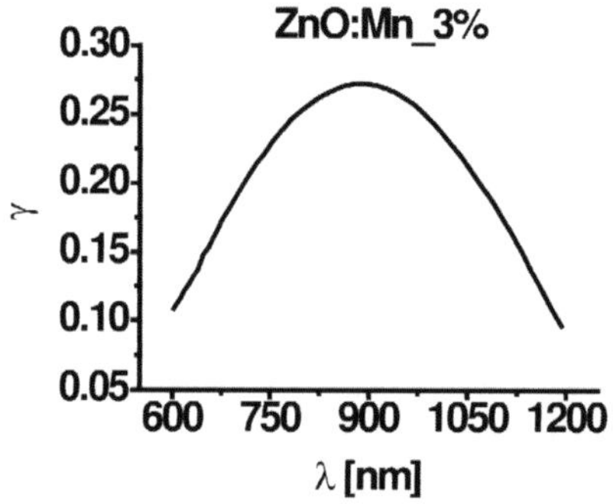

Figure 2d. The magneto-optic anomaly factor γ in the spectral region 600 – 1200 nm for ZnO:Mn_3%.

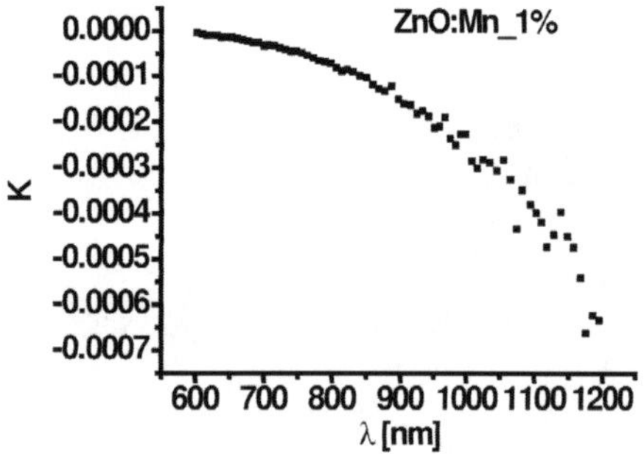

Figure 3a. The constant K(λ) of spin-spin exchange interaction for ZnO:Mn_1%.

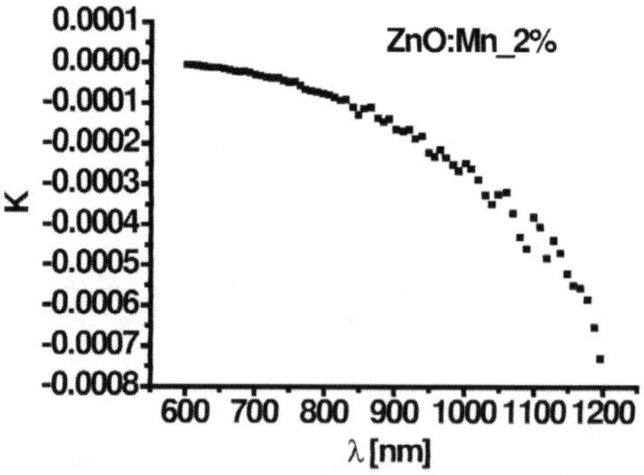

Figure 3b. The constant K(λ) of spin-spin exchange interaction for ZnO:Mn_2%.

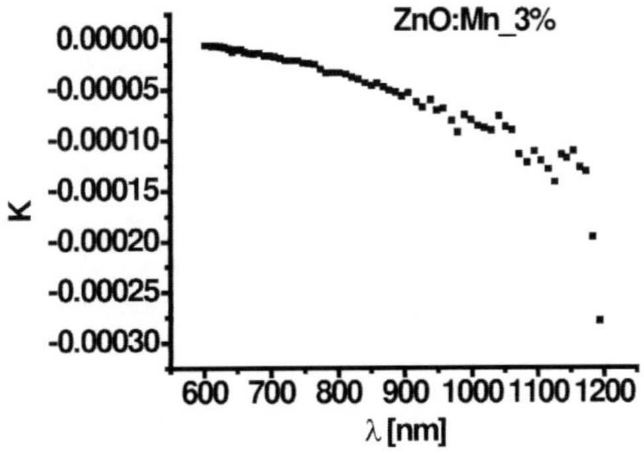

Figure 3c. The constant K(λ) of spin-spin exchange interaction for ZnO:Mn_3%.

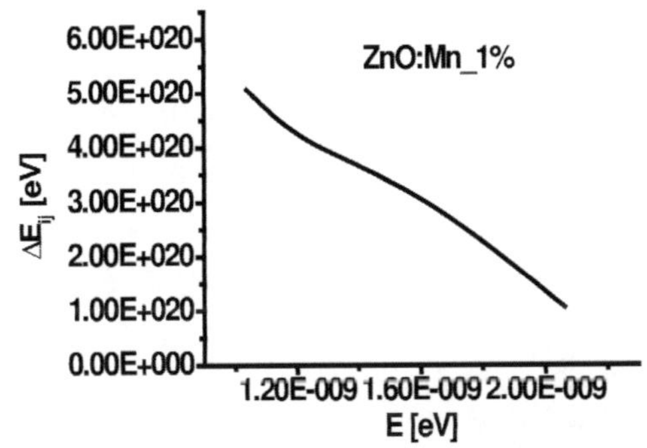

Figure 4a. The change in energy splitting ΔEij for ZnO:Mn_1%.

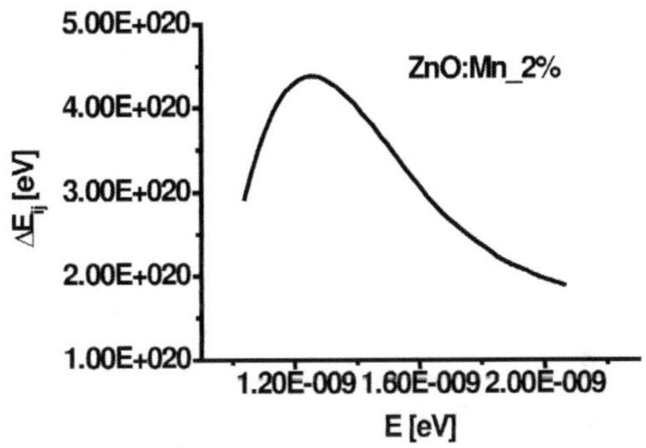

Figure 4b. The change in energy splitting ΔEij for ZnO:Mn_2%.

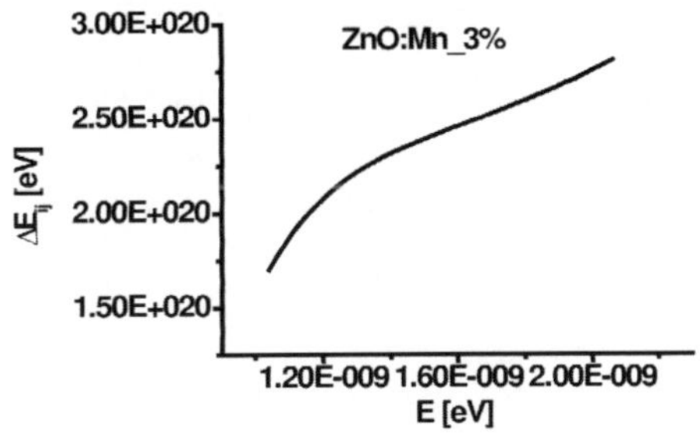

Figure 4c. The change in energy splitting ΔEij for ZnO:Mn_3%.

The photoluminescence spectra of pure and manganese doped ZnO thin films are presented in [21]. The broad visible bands with the increase of the manganese doping are observed at about 750 nm. This is attributed to the red emissions related to oxygen defects [22]. The values of Verdet constant at about 750 nm decrease when the concentration of Mn content is 1% and 3% and increase when the Mn content is 2% (Figure 1). This means that the effective mass of manganese electrons increases in the first case and decreases when the Mn content is 2% (see eq. 5). When we know the values of m_{ef} at about 750 nm, we can determine the ionization energy E_d of Mn^{2+} ions (see eq. 6). The degree of covalency of bonds between manganese ions at about 750 nm is biggest in the case when the Mn content is 1% (Figure 2).When the Mn content is 3%, the degree of covalency is the same as in the case of pure ZnO thin film. The spin-spin exchange interaction is the smallest, when the Mn content is 3 % (Figure 3). This means that the room temperature ferromagnetism of the manganese doped ZnO thin films [21] will be better expressed, when the Mn content is 1or 2 %. The change in the energy splitting is bigger when the Mn content is 1 or 2 % (Figure 4).

4. Conclusions

The values of Verdet constant are negative. This means that the pure and manganese doped ZnO thin films are paramagnetic. The magnetic anisotropy is proved by calculation of spin-spin exchange interaction constant. All investigations in this work are useful for the application of manganese doped ZnO thin films as the emitting and laser diodes.

Acknowledgements

The authors would like to thank for the financial support of the project RD-08-109/08.02.2016 of Konstantin Preslavsky University of Shumen.

References

[1] Bougrine A El Hichou A Addou M Ebothe´ J Kachouane A Troyon M 2003 *Mater. Chem. Phys.* **80** 438–445
[2] Arakelyan M V Galstyan E V Shahnazaryan G E 2008 *Armenian J. Phys.* **1** 138–141
[3] Morales J Solis J Estrada W Paraguay F L Miki-Yoshida M D 1999 *Superficies YVacı´ o* **9** 245–247
[4] Tubtimtae A Gardchareon A Mangkorntong P Mangkorntong N Santhaveesuk T Choopun S 2008 *CMU J. Nat. Sci. Spec. Issue Nanotechnol.* **1** 153–157
[5] Chou M S Teoh G L Lai H W Su H Y Hon H M 2006 *Sensors* **6** 1420–1427
[6] Chang F J Kuo H H Leu C I Hon H M 2002 *Sens. Actuators* **B84** 258–264
[7] Jones A Jones A T Mann B Firth G J 1984 *Sens. Actuators* **5** 75–88
[8] Shih C W Wang J M Nan Lin I 2008 *Diam. Relat. Mater.* **17** 390–395
[9] Fu Q Y Luo K J Du Y X Flewitt J A Li Y Markx H G et al. 2010 *Sens. Actuators* **B 143** 606–619
[10] Krishnamoorthy S Iliadis A A 2008 *Solid-State Electron.* **11** 1710–1716
[11] Wolf A S Awschalom D D Buhrman A R Daughton M J Molnar V S Roukes L M Ctchelkanova Y A Treger M D 2001 *Science* **294** 1488
[12] Ohno H 1998 *Science* **281** 951
[13] Dietl T Ohno H Matsukura F Cibert J Ferrand D 2000 *Science* **287** 1019
[14] Pearton J S Norton P D Ivill P M Hebard F A Zavada M J Chen M W Buyanova A I 2007 *J. Electron. Mater.* **36** 462–471
[15] Belgacem S Bennaceur R Saurel J Bougnot J 1990 *Rev. Phys. Appl.* **25** 1245–1258
[16] Kizel A V Krassilov I J Burkov I V 1974 *Uspehi fizicheskih nauk* **114** 295
[17] Sivaramakrishnan V 1954 *Magneto-optic study of some inorganic acids and their salts*
[18] Drago R 1981 *Physical Methods in Chemistry* (Urbana: University of Illinois)

[19] Syed M Siahmakoun A 2005 Optical Materials **27** 1629-1636
[20] Martin A Bravo D Dieguez E and Lopez J F 1996 *Phys. Rev. B* **54** 12915
[21] Mimouni R Kamoun O Yumak A Mhamdi A Boubaker K Petkova P Amlouk M 2015 *Journal of Alloys and Compounds* **645** 100–111
[22] Kohan F A Ceder G Morgan D 2000 *Phys. Rev. B* **61** 15019

As_2Se_3 thin films deposited by frequency assisted thermal evaporation – morphology and structure

T Hristova-Vasileva[1], I Bineva[1], A Dinescu[2], M Danila[2] and D Arsova[1]

[1] Bulgarian Academy of Sciences, Institute of Solid State Physics, 72 Tsarigradsko Chaussee blvd., 1784 Sofia, Bulgaria
[2] National Institute for R&D in Microelectronics – IMT Bucharest, 126 Erou Iancu Nicolae str., 077190 Bucharest, Romania

teddie@issp.bas.bg

Abstract. Thin As_2Se_3 films with thicknesses of 60 and 250 nm were deposited from the initial compound by frequency assisted thermal evaporation in vacuum – a new approach for shaping thin films surface. Frequencies of 0, 50 and 4000 Hz were applied on the substrates during the deposition process. The morphology was evaluated using AFM and SEM techniques. Some main optical properties and specifics of the films were determined by the means of spectroscopic ellipsometry analysis. The structure of the samples was investigated by XRD. A possible mechanism for explanation of the observed peculiarities is proposed on the basis of the classical materials science and molecular thermodynamics.

1. Introduction

In order to satisfy the ever rising requirements of the contemporary electronics, in particular optoelectronic and sensing technologies, the scientists focus their efforts towards development of new methods and approaches for quality increase and price decrease of materials, which are already implemented and used in practice, as well as towards obtaining of materials with new properties.

The frequency assisted thermal evaporation in vacuum (FATEV) is new, cheap and simple approach for topography shaping of thermally deposited films. The technology is based on application of audible frequency vibrations on the substrates during thermal deposition in vacuum. For this purpose a source of mechanical vibrations with sonic wave frequency is mounted in standard equipment for thermal evaporation.

The physical meaning of the FATEV approach is the following: The vibrations with audible frequency are directly transferred from the source to the substrate and corresponding waves are distributed on its surface by forming a standing waves network. Each instant a small portion of the evaporated substance condenses on the excited substrate and nuclei formation takes place at the nodes and anti-nodes of the standing wave network. With the increase of the nuclei size islands are formed which coalesce and a continuous film is formed which copies the substrate topology.

Until now this technology was used for deposition of selenium and tellurium films [1], and change of their topography depending on the applied frequency was observed – undulated surfaces were obtained at near subsonic frequencies with rms roughness (Rq) maximum at 50 Hz. The excitement with frequency of 4 kHz led to formation of extremely smooth surfaces. The quantification of these changes depends on the degree of crystallinity of the deposited material – the Rq deviations for amorphous selenium films are from 2.3 nm for frequency of 50 Hz to 0.2 nm for 4 kHz oscillations, while the Rq of crystalline tellurium decreases from 40 nm (at 50 Hz) to 13 nm (at 4 kHz). The mechanical distortions do not lead to any change in the crystallographic structure of the materials.

In this sense the aim of the present report is to investigate the influence of the mechanical vibrations with audible frequencies of 50 and 4000 Hz during deposition on the morphology, structure and optical properties of films of As_2Se_3 (a typical glassy chalcogenide semiconductor).

2. Experimental procedures

The source material, As_2Se_3, was synthesized by conventional melt quenching technique. Appropriate quantities of As and Se (purity of 5 N) were melted in evacuated to a residual pressure of 1.10^{-3} Pa quartz tubes. Two isothermal steps (300 and 550 °C) were performed during the synthesis, conformed to the physicochemical peculiarities of the source elements and the final phase. The melt was tempered for 2 hours at the last step and then rapidly quenched in ice-cold water (cooling rate of ~10-15 °C/s).

The frequency assisted thermal evaporation in vacuum (TEV) was performed in modified TEV equipment. A source of mechanical vibrations with acoustic wave frequency (2 inch 180 mW closed loudspeaker with frequency response in the interval 20-20000 Hz, situated normally to the evaporation flow) was set up in the vacuum chamber. The desired sine signal was generated using a computer program and translated through an amplifier towards the loudspeaker. Frequencies of 0, 50, and 4000 Hz were applied on a 2 inch Si (100) wafer during deposition. The amplitude of the vibrations was set to 80 % of the maximum loudspeaker output power.

Series of As_2Se_3 films with thicknesses of 60 and 250 nm were deposited at the three frequencies of the mechanical vibrations with deposition rate of ~0.5 nm/s. The residual pressure in the vacuum chamber was 7.10^{-3} Pa. The distance between the crucible and the substrate was 20 cm. The deposition rate and the thickness of the films during preparation were controlled using a quartz-crystal microbalance device Miki MSV 1841/A.

AFM measurements were performed on scanning probe microscope Multimode V (Veeco, Santa Barbara, CA). The images were taken in tapping mode, as each sample has been investigated on multiple points of the surface. Measurements in scale of 1, 3 and 10 μm have been performed with scanning rate in the interval 0.5-2.0 Hz and images resolution of 512 lines per scan direction. Al-coated silicon cantilevers TAP150-Al-G (Budget Sensors Innovative Solutions Bulgaria Ltd., Bulgaria) with nominal resonant frequency of ~150, and spring constant of 5 N/m have been used. The radius of the cantilever's tip is smaller than 10 nm. The root mean square roughness (Rq) of the samples was determined in scale of 10 μm. The images have been just flattened before the analysis.

SEM observations have been made using e-Line EBL equipment in SEM mode (Raith GmbH, Germany), working with accelerating voltage of 10 kV. The As_2Se_3 films were observed on cross-section with magnification from 20 to 150 thousand times.

The structure of the films was evaluated through grazing incidence X-ray diffraction (GIXRD) using Rigaku SmartLab X-ray thin film diffraction system (Rigaku Corporation, Japan) at $CuK_{\alpha1,2}$ irradiation (λ=1.5418 Å) and angle of incidence of 0.35 °. The XRD patterns were taken in Bragg angle (2θ) range from 10 to 100 ° with step of 0.01 °.

Ellipsometric measurements in the UV-VIS-NIR region were carried out with an automatic M2000D spectroscopic ellipsometer (J.A. Woollam, Germany, bought on FP7 INERA REGPOT-2012-2013-1 NMP project) within the spectral range of 193-1000 nm and angles of incidence from 45 to 65 ° by 5 °. The films thicknesses and optical constants were determined by the means of Tauc-Lorentz oscillator model using Woollam's CompleteEASE program.

3. Results and discussion

Cross-section SEM microphotographs of as-deposited As_2Se_3 films are presented in figure 1.

The SEM observations show homogeneous and smooth surface, without cracks or other inclusions. Only some shell-like formations (cracks), typical for glassy materials, were observed.

The films thicknesses (table 1) were evaluated using the SEM microphotographs. The obtained values were confirmed during the spectroscopic ellipsometry (SE) investigations, the results of which are also present in table 1. The determined values are in good agreement.

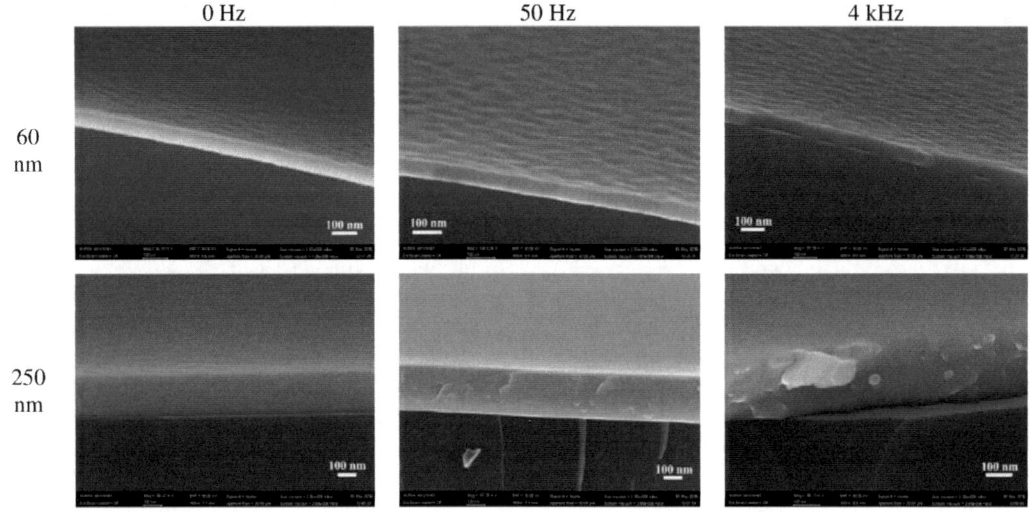

Figure 1. Cross-section SEM microphotographs of as-deposited As_2Se_3 films.

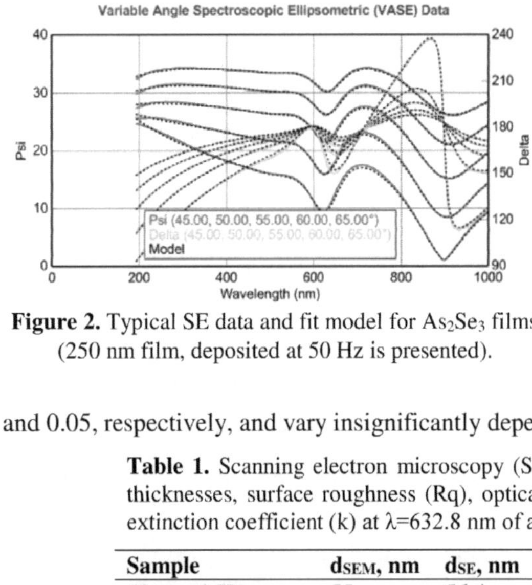

Figure 2. Typical SE data and fit model for As_2Se_3 films (250 nm film, deposited at 50 Hz is presented).

A representative SE data, as well as the corresponding fit, used for determination of the thickness and the optical constants of the films are presented in figure 2. The SE study showed a fair agreement of the optical constants with the literature data (optical band gap E_g=1.8 eV [2], refractive index n≈3 [3], and extinction coefficient k≈0.04 [3]). The results are presented in table 1.

The applied frequencies do not influence the optical properties of the films. As it can be seen in table 1 the optical band gap $E_g(As_2Se_3)$≈1.78 eV, and n and k at He-Ne laser wavelength (λ=632.8 nm) are around 3.2 and 0.05, respectively, and vary insignificantly depending on the frequency of the applied vibrations.

Table 1. Scanning electron microscopy (SEM) and spectroscopic ellipsometry (SE) thicknesses, surface roughness (Rq), optical band gap (E_g), refractive index (n) and extinction coefficient (k) at λ=632.8 nm of as-deposited As_2Se_3 films.

Sample	d_{SEM}, nm	d_{SE}, nm	Rq, nm	E_g, eV	n	k
60 nm, 0 Hz	55	56.4	1.21	1.78	3.18	0.05587
60 nm, 50 Hz	57	57.9	1.70	1.80	3.16	0.04621
60 nm, 4000 Hz	53	53.4	2.09	1.79	3.17	0.04823
250 nm, 0 Hz	236	235.0	0.509	1.75	3.22	0.07655
250 nm, 50 Hz	247	246.1	0.449	1.78	3.19	0.05887
250 nm, 4000 Hz	240	239.1	0.448	1.78	3.19	0.05723

The results from the X-ray diffraction measurements at grazing incidence (GIXRD) are presented in figure 3. Because of the small thickness, some peaks of the Si wafer appear on the XRD pattern.

Three X-ray amorphous halos are observed on the XRD patterns (figure 3), the positions of which correspond to the pattern of monoclinic As_2Se_3 [4], but are shifted towards lower 2θ angles, most probably due to internal stress in the films.

The possibility for formation of nanocrystalline objects, caused by the frequency excitement, in the as-deposited films was examined. For this purpose the main peaks of the XRD patterns were fitted using Lorentz model in order to obtain the exact position of the peak and its full width at half maximum (FWHM) – table 2. Using the obtained values, the average crystallite sizes (D) and the microstrain (ε) of the samples were calculated using Eq. (1), and Eq. (2), respectively:

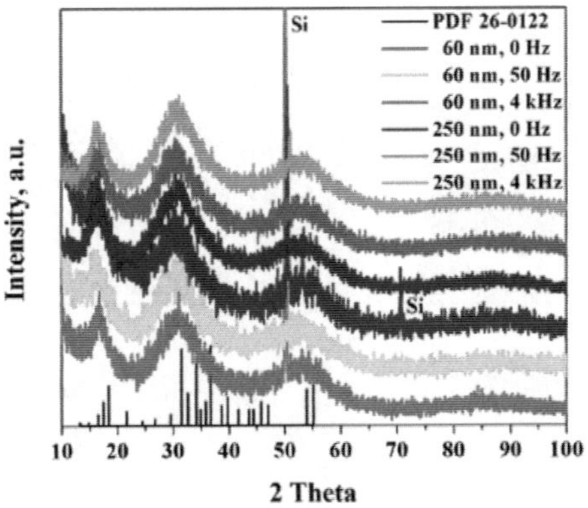

Figure 3. GIXRD of as-deposited As_2Se_3 films. PDF 26-0122 - JCPDS card of bulk As_2Se_3 [4].

$$D = \frac{K\lambda}{\beta\cos\theta} \quad (1)$$

$$\beta = 4\varepsilon\frac{\sin\theta}{\cos\theta} \quad (2)$$

where K - Scherrer constant (shape factor, K=0.94 [5]); β – line broadening (FWHM in radians); θ – Bragg angle; λ – X-ray wavelength (λ=0.15418 nm). The results are presented in table 2.

Table 2. Position (2θ) and full width at half maximum (FWHM) of the main XRD peak, average crystallite size (D) and microstrain (ε) of as-deposited As_2Se_3 films.

Sample	2θ, deg	FWHM, deg	D, nm	ε
60 nm, 0 Hz	30.69	8.42	1	0.13385
60 nm, 50 Hz	30.43	8.60	1	0.13798
60 nm, 4000 Hz	29.81	10.87	0.8	0.17827
250 nm, 0 Hz	30.76	8.64	1	0.13712
250 nm, 50 Hz	30.55	9.78	0.9	0.15624
250 nm, 4000 Hz	30.93	8.55	1	0.13485

The estimated crystallites size value is with a fair approach towards accuracy due to the following factors: 1) the investigated peak is too complex and the obtained FWHM could differ a little, 2) the value of the Scherrer constant could be different since we do not know the exact shape of the crystallites. However, the obtained values are around 1 nm in diameter or below, which roughly corresponds to the cell parameters of the As_2Se_3 JCPDS file (a=6.69 Å, b=13.86 Å, c=10.00 Å, β=113.8 °) [4], and are a proof for the glassy nature of the deposited films. Other proof for the disordered structure of the films is the relatively high strain, observed in them.

AFM images of the as-deposited samples are presented in figure 3.

The topology of the samples, revealed from the AFM measurements, is in good correlation with the SEM observations – the films are homogeneous and no crystalline formations were seen.

The root mean square roughness Rq of the samples was determined and the results are presented in table 1.

Two facts are of particular interest: 1) the rms surface roughness of the films decreases with the increase of their thickness, and 2) the effect, caused by the mechanical excitement during deposition, is significantly smaller or even absent in comparison to the results obtained during the investigation of Se and Te films, deposited by the FATEV approach at the same conditions [1].

These deviations could be explained taking into account the following facts:

- The stable state of As_2Se_3 is glassy and it is hardly obtained in crystalline form by thermal evaporation in vacuum [6].

- The evaporated As_2Se_3 molecules and structural fragments tend to attach to any place forming a disordered glassy network due to their high viscosity in gas and liquid state and thus to reach unimpeded the glass-transition temperature and freeze as a supercooled liquid [7].

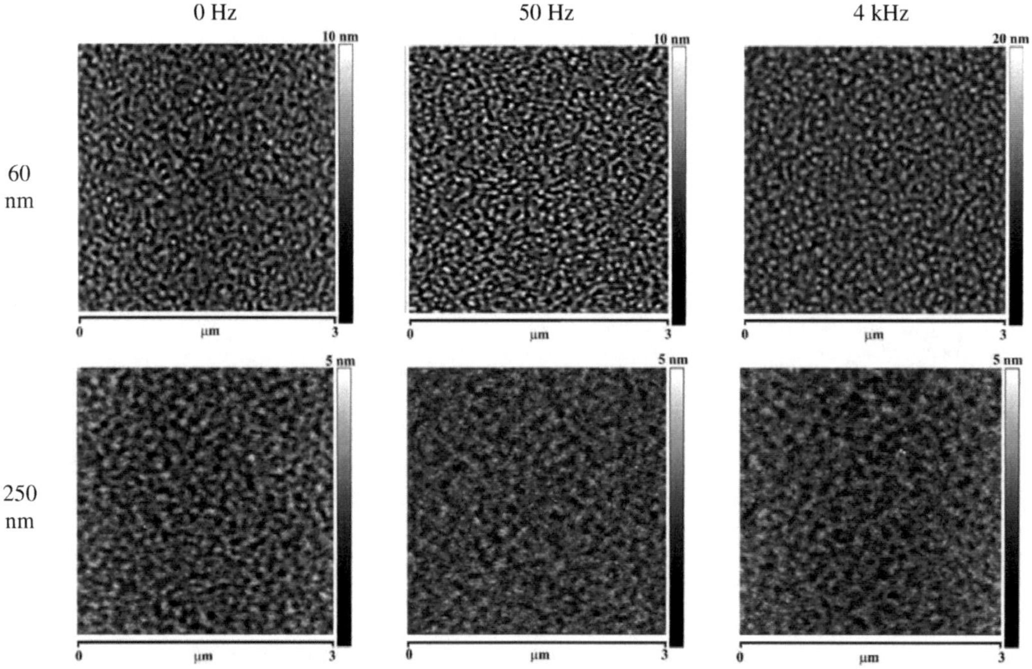

Figure 3. AFM images at scale of 3 μm of as-deposited As_2Se_3 films. The maximum height scale is 5 nm for all 250 nm thick films, while for the thin ones it is 10 nm for 0 Hz and 50 Hz, and 20 nm for 4 kHz.

- Recently it has been theoretically predicted during molecular dynamics simulation of the effect of a standing surface acoustic wave on adatom diffusion on face centred crystal surface, that the main mechanism responsible for a structuring effect is the inertia [8].

- The same authors state that the diffusion of atoms from the substrate could also take place during the excitement [8].

In our case, the impurities present in the glassy matrix of As_2Se_3 worsen their optical properties, but do not cause their crystallization [9]. Therefore, as the influence of the inertial momentum decreases with the increase of the thickness, and since there is no typical crystallographic orientation of the deposit, only the glassforming ability of the material takes place. This should be the reason for the registered flat surfaces of the As_2Se_3 films with thickness of 250 nm and the significantly rougher ones of the 60 nm thick films.

The observed unusual increase of the Rq at excitement with 4 kHz wave in the 60 nm films can be explained by the high viscosity of As_2Se_3 [10] and its lack of crystallization tendency – the dynamic curvature of the substrate' surface gives an additional momentum to the deposit, which is not related to the initial frequency, but to the mobility of its adatoms, i.e. the deposited film starts to resonate at its own frequency. Thus when the condensate "freezes" an undulated surface is formed. It has to be noted that the films, deposited at 0 Hz could also be subjected to a certain extend to uncontrollable vibrations, caused by equipment imperfections. However, it can be said that the increase of the frequency of the substrate mechanical excitement during deposition leads to formation of rougher films (than these obtained by conventional thermal deposition in vacuum), whose curvature is caused not directly by the standing waves network, but by the viscous movement of the material adatoms, excited by the vibrations. In the case of 250 nm films, when there is enough quantity of the deposited material, and taking into account the above statement that the excitement effect decreases with the thickness, it is not surprising that the thicker films are significantly smoother and not influenced by any excitement frequency.

Taking into account the results obtained from the investigation of the FATEV deposited selenium and tellurium films [1] it can be concluded that the vibration excitement of the substrate during thermal deposition is highly influenced by the viscosity and the tendency towards crystallization of the condensates – the high viscosity [10] and priority trend towards formation of disordered structures (in the case of As_2Se_3) dull the excitement effect; the almost equal viscosity [11] but existence of certain crystallization affinity even at room temperature if crystallization centres are present [12] (in the case of Se) lead to formation of undulated surfaces at near sub-sonic frequencies and extremely smooth ones at 4 kHz, while the typically crystalline material (in the case of Te) follows the last path but the changes are significantly larger.

4. Conclusions

Glassy As_2Se_3 films were deposited by frequency assisted thermal evaporation in vacuum at excitement frequencies of 0, 50 and 4000 Hz. The thicknesses of the films were determined to be 60 and 250 nm by cross-section scanning electron microscopy and spectroscopic ellipsometry. The optical properties of the samples coincide with the literature data about the same compound and do not change depending on the applied vibrations. A profound study of the films structure was preformed by the means of XRD analysis. The results show that the films are amorphous with particle sizes below the lattice parameters of the corresponding As_2Se_3 crystal. This, together with the SEM observations (appearance of shell-like cracks in cross-section) makes us believe that the deposited samples possess glassy nature. The excitement during deposition does not lead to noticeable changes of the crystalline structure. The AFM measurements reveal an unusual deviation of the surface roughness both with the thickness and the excitement frequency increase. The films surface roughness decreases with the increase of the thickness and stays equal for all 250 nm thick films, while the excitement of the 60 nm films leads to increasing curvature with the increase of the vibrations frequency. These specifics were explained by the means of molecular thermodynamics taking into account the physical properties of As_2Se_3. A comparison with other materials, previously deposited by the same method, was preformed.

Acknowledgements

This work was financed by the Program for career development of young scientists of the Bulgarian Academy of Sciences (contract DFNP-160/13.05.16).

References

[1] Hristova-Vasileva T, Bineva I, Dinescu A, Arsova D and Nesheva D 2016 “Cymatics” of Selenium and Tellurium films deposited in vacuum on vibrating substrates, *Surf. Coat. Technol.*, in press
[2] Saitoh A and Tanaka K 2011 *J. Optoel. Adv. Mater.* **13**, 65
[3] Todorov R, Paneva A and Petkov K 2010 *Thin Solid Films* **518**, 3280
[4] International Centre of Diffraction Data, Joint Committee on Powder Diffraction Standards, Powder Diffraction File 26-0122
[5] Langford J and Wilson A 1978 *J. Appl. Cryst.* **11**, 102
[6] Borisova Z 1983 *Halkogenidnie poluprovodnikovie stekla* (Izd. LGU-Leningrad) – in Russian
[7] Zallen R 2004 *The physics of amorphous solids* (Wiley-VCH)
[8] Taillan C, Combe N and Morillo J 2015 *Eur. Phys. J. B* **88**, 317
[9] Danto S, Thompson D, Wachtel P, Musgraves J D and Richardson K 2013 *Int. J. Appl. Glass Sci.* **4**, 31
[10] Málek J and Shánělova J 2005 *J. Non-Cryst. Solids* **351**, 3458
[11] Koštál P and Málek J 2010 *J. Non-Cryst. Solids* **356**, 2803
[12] Csarnovics I, Nemec P, Nazabal V, Veres M, Csik A, Allix M and Kokenyesi S 2014 *Mat. Chem. Phys.* **143**, 889

Effect of oxygen concentration and metal electrode on the resistive switching in MIM capacitors with transition metal oxides

D Spassov[1], A Paskaleva[1], K Fröhlich[2] and Tz Ivanov[1]

[1] Institute of Solid State Physics, Bulgarian Academy of Sciences, TzarigradskoChaussee 72, Sofia 1734, Bulgaria
[2] Institute of Electrical Engineering, SAS, Dubravska cesta 9, 841 04 Bratislava, Slovakia

E-mail: d.spassov@issp.bas.bg

Abstract. The influence of the oxygen content in the dielectric layer and the effect of the bottom electrode on the resistive switching in Au/Pt/TaO_x/TiN and Au/Pt/TaO_x/Ta structures have been studied. The sputtered TaO_x layers have been prepared by using oxygen concentrations of 10 or 7% O_2 in the Ar+O_2 working ambient as well as by a gradual variation of the O_2 content in the deposition process from 5 to 10%. Two deposition regimes for TiN electrodes have been investigated: reactive sputtering of Ti target in Ar+N_2 ambient, and sputtering of TiN target in pure Ar. Bipolar resistive switching behavior is observed in all examined structures. It is demonstrated that the resistive switching effect is affected by the oxygen content in the working ambient as well as by the type and the deposition conditions of the bottom electrodes. Most stable effect, with ON/OFF ratio above 100 is obtained in TaO_x deposited with variable O_2 content in the ambient. The obtained switching voltage between the high resistive and low resistive state (SET) is about -1.5 V and the reverse changeover (RESET) is ~2 V. A well pronounced resistive switching is achieved with reactively sputtered TiN while for the other bottom electrodes the effect is negligible.

1. Introduction

The wide spread in the recent years of portable digital devices as well as the general trend in the mainstream computer industry for increasing the speed of the data transfers of internal mass storage devices and their durability defined the significant interest of the semiconductor industry to non-volatile memory (NVM) technology. The current NVM devices are based on the flash technology incorporating floating gate field effect transistor cell. To satisfy the ever increasing demands for higher data volumes stored, lower power consumption and faster data exchange, the dimensions of floating gate cell are constantly shrunk. Although the flash technology can fulfill the requirements to NVM currently, in the long term its inherent limitations will hamper future scaling [1]. Several alternative NVMs involving different memory effects were proposed including: charge trapping memory, phase change memory, ferroelectric field effect transistor; resistive switching memories; nanoelectromechanical memory; spin transfer torque devices. The use of resistive switching (RS) effect for information storage offers some attractive advantages in respect to the floating gate technology: i) simpler memory cell based on metal insulator metal (MIM) structure, the so-called memristor; ii) smaller cell size, hence high scalable potential; iii) power consumption, high speed and better endurance/retention; iiii) lower production costs.

RS effect, found in the 1960s, is a phenomenon of abrupt non-volatile change of the resistance of the dielectric from high resistive state (HRS) to low resistive state (LRS) and vice versa. At specific electrical stress conditions a reversible dielectric breakdown can be obtained, altering the insulating properties of the material. The memory effect is achieved by transferring the structure in HRS (logical "0") or LRS (logical "1") through applying to the two terminals voltage with certain value and polarity. The switch between HRS to LRS is referred to as "SET" operation and the reverse process is called "RESET" operation. RS is obtained either by a creation of conductive filament (the nature of the filament depends on the used material) in the dielectric or by the barrier height modification at the interface with the electrode(s) [2]. RS can also be classified by the voltage polarity required to execute SET and RESET operations. Two types of RS can be distinguished in this case: unipolar where SET and RESET do not depend on the polarity of V but only on its value; and bipolar where SET and RESET operations occur at different voltage polarity.

The resistive switching is observed in many materials, but its existence in the binary transition metal oxides is of particular interest for the semiconductor industry [3,4]. These oxides have dielectric constant higher than that of SiO_2, and as high-k dielectrics they were extensively investigated as a replacement of SiO_2 in the advanced ultra-high density integrated circuits. It was found that they exhibit some unexpected properties which could be implemented in other microelectronics devices. One of these effects is RS. Among the various high-k oxides, Ta_2O_5 attracted a lot of attention as a dielectric in the storage capacitors of dynamic random access memories [5]. The resistive switching was also observed in Ta_2O_5 offering a unique possibility for combining the dynamic random access memory with non-volatile resistive random access memory in a single device.

In this work we present experimental results on RS in TaO_x-based MIM structures with different electrodes in dependence on the oxygen content in the deposition ambient.

2. Experimental procedure

The test MIM structures were fabricated on p-type (100) Si wafers. After a standard chemical cleaning TiN or Ta bottom electrodes were deposited on Si. Two types of sputtered TiN bottom electrodes were prepared by: i) reactive sputtering of Ti target in $Ar+N_2$ ambient at substrate temperature of 200 °C; and ii) magnetron sputtering of TiN target in Ar without substrate heating. Ta bottom electrodes were also fabricated by magnetron sputtering of Ta target in Ar. The thickness of the bottom electrodes were 70 nm. Next, TaO_x layers with a thickness of about 10 nm (as determined by ellipsometry) were deposited through reactive magnetron sputtering of Ta target in a mixture of Ar and O_2. During the deposition substrates were not intentionally heated. In order to evaluate the effect of the oxygen concentration on RS working gas mixtures with 10 and 7% oxygen were used as well as gradual variation of the O_2 content in the range of 5 to 10% in the deposition process. Previously, it was established that at oxygen concentration above 10% the obtained tantalum oxide is stoichiometric [6]. So, the lower O_2 concentration in the ambient is suggested to result in a substoichiometric oxide with increased density of oxygen vacancies, thereby facilitating the resistive switching effect. During the TaO_x preparation part of the substrate area was shielded to prevent oxide deposition there and ensure access to the bottom electrode. MIM structures were defined by e-beam evaporation through a shadow mask of circular 30 nm thick Pt top electrodes capped with 30 nm thick Au layer (figure 1). The current voltage (I-V) characteristics were measured with Keithley 4200 SCS in a dark chamber at room

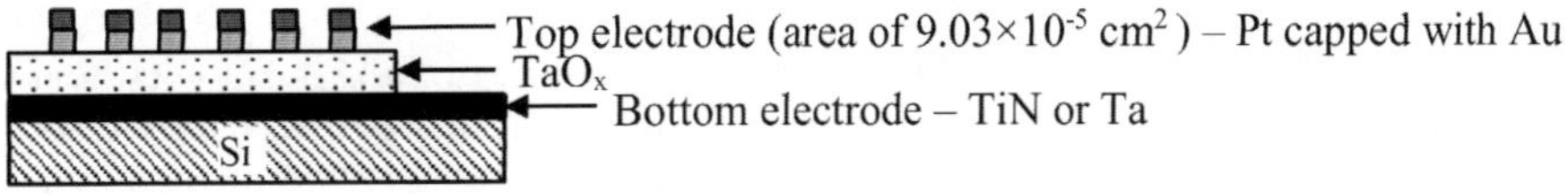

Figure 1. Schematic cross section of the test memristors.

IOP Conf. Series: Journal of Physics: Conf. Series **794** (2017) 012016 doi:10.1088/1742-6596/794/1/012016

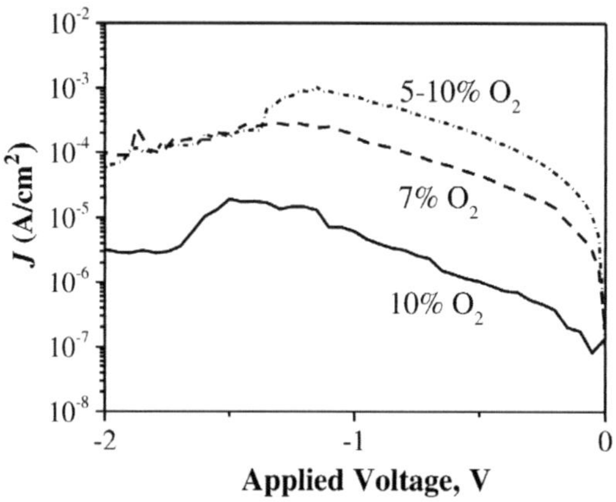

Figure 2. A comparison of the leakage currents of Au/Pt/TaO_x/(reactively sputtered TiN) capacitors deposited at a different oxygen concentration in the working gas.

temperature. The polarity of the applied voltage is defined in respect to the bottom electrode.

3. Results and discussion

Figure 2 shows the comparison of the density of initial leakage current J of capacitors with reactively sputtered TiN bottom electrodes and TaO_x films obtained at different O_2 concentrations. As seen the increase of the oxygen concentration in the working gas decreases J implying an improved stoichiometry of the layer. At applied voltage, V below about -1.7 V a slight reduction of J is observed which is further followed be a sharp increase at higher negative V (not shown here). The results on the resistive switching effect of these structures are presented in figure 3 for the used O_2 concentrations. As evidenced, bipolar resistive switching is observed for all investigated structures. The SET and RESET operations require different voltage polarity. SET occurs under negative V, while RESET appears at positive V. In order to obtain RS, however, a forming procedure has to be executed. At this procedure negative voltage linearly increasing from 0 to -3 V (in case of TaO_x obtained at 7% O_2 content in the ambient) or -5 V (for films deposited at Ar+10% and (5 to 10)% O_2 was applied to the structure with compliance function of the measurement unit turned on at certain level in order to prevent the permanent (hard) breakdown of the dielectric. The aim of the forming is to generate oxygen vacancies or conduction path, activating RS behaviour of the device. In tantalum oxide the main type of defects governing the leakage currents thorough the films are the oxygen vacancies, V_O^{n+} [5] which are positively charged. During the forming the existing and the generated by the electrical stress through breaking weak Ta-O bonds O vacancies move towards the inert cathode (Pt), while the negative oxygen ions O^{n-} drift towards the anode (TiN) where they are stored [7]. The generated vacancies and O ions are further used to switch the films resistance. As seen in figure 3 the voltage needed for the forming process is higher than corresponding SET voltage, V_{SET}, hence it is critical to limit J during the process to avoid device destruction. The need of a forming to achieve RS suggests that despite decreased O_2 concentrations in the working ambient in respect to the one for stoichiometric Ta_2O_5, the amount of initially embedded V_O^{n+} is not enough for visible resistance modulation, and Ta/O ratio of the dielectric layers obtained at the chosen conditions is probably close to the stoichiometric one. In case of layers obtained at 10 and 7% O_2 up to three forming cycles (see figure 3(b)) have to be performed to found RS effect. Note that for the samples prepared at 7% O_2 the maximum applied voltage for forming is lower than for the rest of the samples, indicating that probably the density of V_O^{n+} is higher and/or the rupture of Ta-O bonds occurs at lower electric fields because the bonds are weaker. For the memristors fabricated at 10% and (5 to 10%) O_2 resistive switching is achieved at one and the same maximum forming voltage. This might be related to the fact that part of the film in the latter case is deposited at 10% O_2. Then higher initial J of the samples

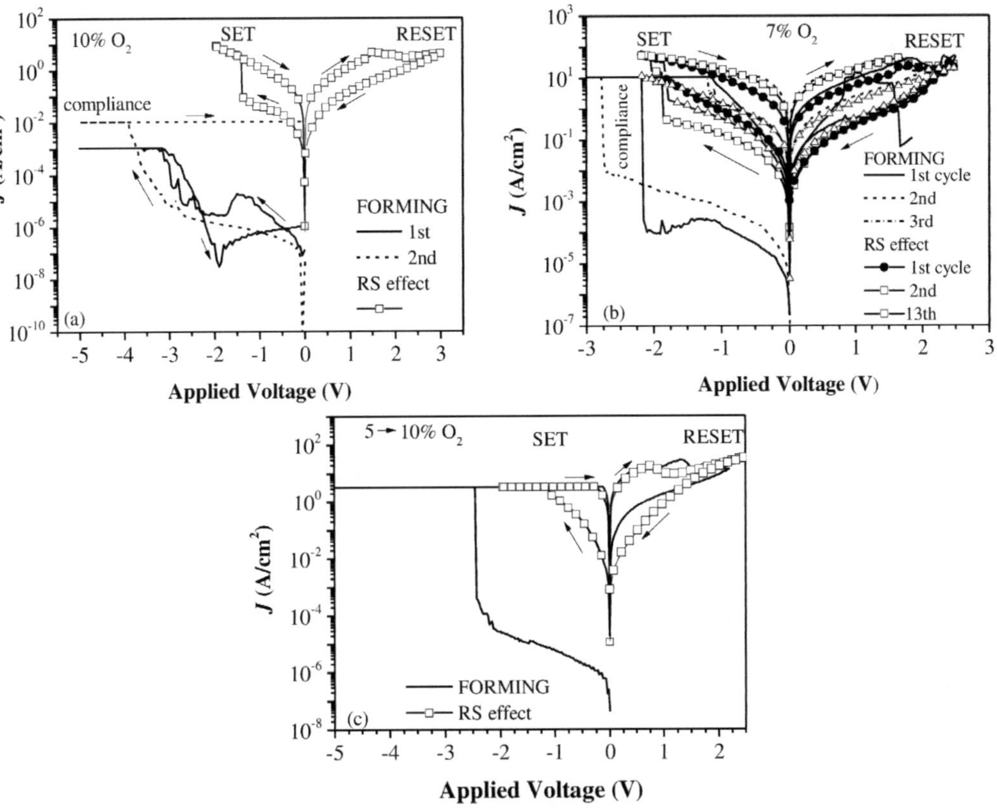

Figure 3. *J-V* characteristics of TaO_x memristors with reactively sputtered TiN bottom electrodes, fabricated at a different O_2 content of the working ambient: (a) 10% O_2; (b) 7% O_2; and (c) variable from 5 to 10% O_2.

produced with O_2 variation (figure 2) could be explained by the smaller thickness of the layers part corresponding to 10% oxygen. The lower compliance level used for the 10% O_2 samples reflects the tendency for hard electric breakdown, i.e. irreversible loss of the insulating properties for these layers. The obtained V_{SET} is between -1.1 and -1.9V. The lowest value (-1.1 V) correspond to the samples with gradient of the oxygen concentration and the highest to the structures deposited at 7% O_2. The transition from LRS to HRS occurs at about 2.5V and does not seem to depend on the TaO_x deposition conditions.

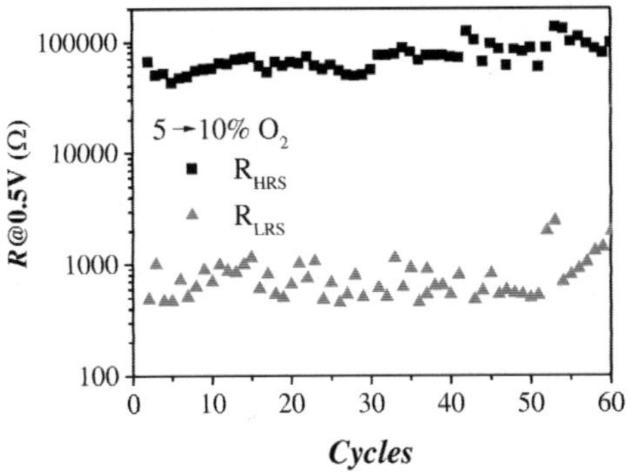

Figure 4. The resistance in high and low resistive state determined at 0.5 V as a function of the SET/RESET cycles of memristors with TaO_x fabricated with variable oxygen concentration of the working ambient and reactively sputtered TiN bottom electrodes. R_{HRS} and R_{LRS} are the resistance at high and low resistive state, respectively.

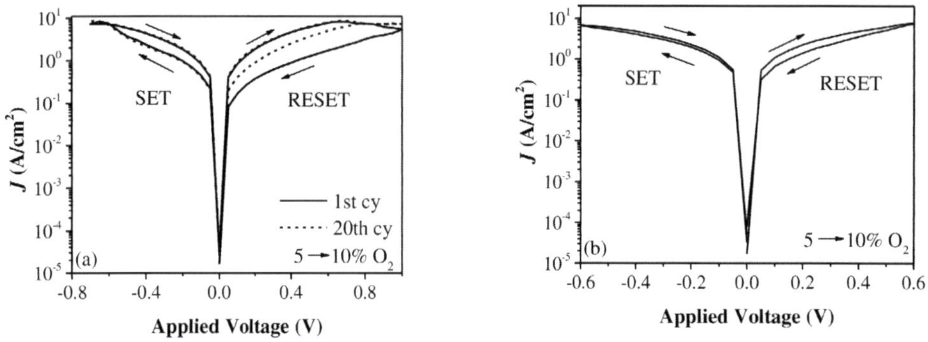

Figure 5. *J-V* characteristics of TaO_x memristors prepared using variable O_2 content of the working ambient with bottom electrodes of: (a) TiN deposited by sputtering of TiN target; and (b) Ta.

The variation of *J* between HRS and LRS (so called ON/OFF ratio) is an important parameter of the resistive switching memory devices. It was evaluated at 0.5 V for the first SET/RESET cycle. The lowest ON/OFF ratio (7.6) is obtained for the layers prepared at 10% O_2 and for the films deposited at 7 and 5-10% O_2 it is ~104 and 135, respectively. Although the RS has been observed for all types of TaO_x films the effect is not stable in case of 10 and 7% O_2. RS effect in the memristors with TaO_x deposited at 10 and 7% O_2 extinguishes after a few SET/RESET cycles, (figure 3(b) depicts *J-V* curves corresponding to first and second SET/RESET and13th cycle clearly indicating that the difference between HRS and LRS diminishes for higher number of cycles). The most stable resistive switching is obtained for thefilms deposited with gradual change of the ambient oxygen content, as demonstrated in figure 4. The comparison of figures 3(c) and 5 reveals the dependence of the RS effect on the type of the bottom electrode. The difference in HRS and LRS for the structures with TiN deposited by magnetron sputtering of TiN target in pure Ar (figure 5 (a)) is significantly lower than the corresponding one for reactively sputtered TiN (figure 3 (c)). The ON/OFF ratio at 0.5 V is 4.8. In case of Ta bottom electrodes RS is hardly observed. Note that presented data are obtained without applying a forming process, and *J-V* curves in SET direction before the ON state represent the initial current of the structures which is ~6 orders of magnitude higher than the current for reactively sputtered TiN. Most likely, because of the preferential sputtering the deposition of TiN from TiN target in pure Ar atmosphere results in Ti-rich layer which can further react with the tantalum oxide layer and thereby increasing the oxygen vacancies (hence conductivity) to a level at which RS is difficult to be obtained. The same can be also applied to the case of Ta bottom electrodes, as a consequence Ta partial oxidation by O from the overlying TaO_x occurs.

In order to acquire some understanding on the nature of the RS effect we briefly discuss the conduction mechanism of memristors with TaO_x deposited at 10 and 7% O_2 with reactively sputtered TiN electrodes in negative *V* region where SET operation occurs. Films obtained at variable O_2 are not considered as *J-V* curves were measured with applied compliance. For memristors produced at 10% O_2 the conduction in the low *V* region (V> -1V) is Ohmic in both HRS and LRS (slope log*J*-log*V* is close to 1.0), figure 6(a). At higher negative *V*, however, the conduction is governed by other mechanism (indicated by the change of the slope). The symmetry of the *J-V* curves in respect to 0V (figure 3(a)) suggests that it is bulk limited. The different slope of high voltage part of HRS and LRS curves generally infers either a change of the mechanism or modification of its parameters upon SET. Space charge limited current (SCLC) or Poole-Frenkel (PF) mechanism (inset of figure 6(a)) could be invoked in case of LRS. SCLC theory [8] predicts slopes of log*J*-log*V* above 2.0 in case of traps with levels distributed into the band gap or deep traps lying below the Fermi level. In case of PF conduction the obtained (with a compensation factor of 2) refractive index, *n* of the layers is ~5 which is higher than *n* for Ta_2O_5 (2.2). The presence of non-oxidized Ta in the layer, however, increases *n* and values as high as ~11 have been detected [9]. So, the data suggests that conductive filament between electrodes is probably not formed

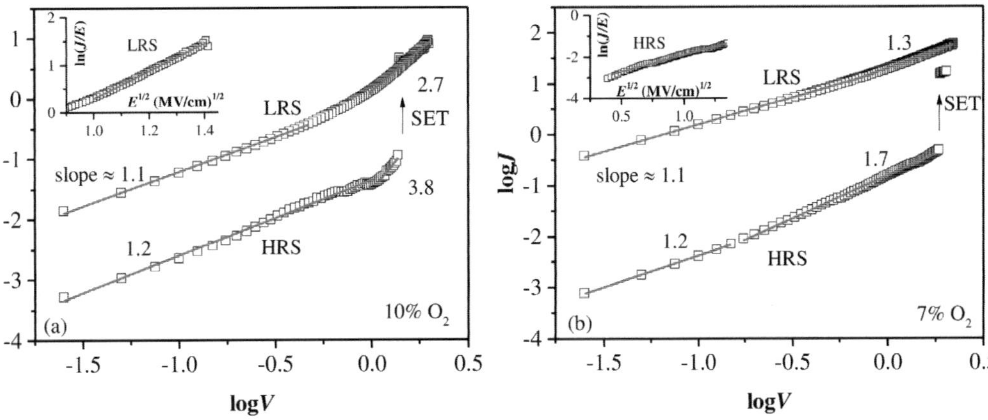

Figure 6. Log*J*-Log*V* plots of memristors with reactively sputtered TiN electrodes and TaO_x deposited at: (a) 10% O_2; and (b) 7% O_2. Insets Poole-Frenkel plots of the *J*-*V* curves at high negative *V*.

and the RS in these films is due to a modification of the defect states and/or increase of their density. For the layers obtained at 7% O_2 the conduction in HRS is also Ohmic at low negative *V*. At higher applied *V* it can be attributed to either SCLC (Child's law [9]) or PF effect (inset figure 6(b)). The obtained *n* in case of PF mechanism is 9.3 which correlates with the expected higher non-stoichiometry. In LRS, however, *J* has an almost Ohmic behaviour in the all voltage region, which is indicative for a formation of conductive filament through the film at the SET. Hence, the RS effect in the deposited at 7% O_2 films is most likely due to a creation and a rupture of a conductive filament.

Thus, the overall resistive switching mechanism in the investigated system could be described as follows. At negative applied voltage some positively charged oxygen vacancies are generated by breaking weak Ta-O bonds. The oxygen vacancies move to the Pt electrode where a vacancy rich region already exists as a result of forming process. The negative oxygen ions drift to the bottom electrode (TiN). When the applied voltage reaches a certain value (V_{SET}) there are enough vacancies to form a conductive path between the electrodes turning the memristor in LRS. The positive applied voltage causes the backward drift of oxygen ions to the Pt electrode. The oxygen ions interact with the vacancies reoxidizing Ta atoms. At V_{RESET} part of the conductive filament is cut off and the structure returns in HRS.

4. Conclusion

The possibility to obtain resistive switching with an ON/OFF ratio above 100 in tantalum oxide films was demonstrated. It was found that resistive switching effect is very sensitive to the oxygen concentration of the dielectric. In order to achieve stable resistive switching substoichiometric tantalum oxides with carefully optimized Ta/O ratio have to be used. This is further supported by the conduction analysis indicating that lower oxygen concentrations facilitate the creation of an Ohmic conductive filament. The best results are obtained for layers with a varying along the thickness oxygen concentration from almost stoichiometric to oxygen depleted. The type of the used electrodes as well as the means of their deposition has a significant impact on the resistive switching.

Acknowledgement

The present work was supported in the frame of contract BK05-14 by Institute of Solid State Physics.

References

[1] Zhao C, Zhao C, Taylor S and Chalker P 2014 *Materials* **7** 5117
[2] Pan F, Chen C, Wang Z, Yang Y, Yang J and Zeng F 2010 *Prog. Nat. Sci.: Mater. Int.* **20** 1-15
[3] Lanza M 2014 Materials 7 2155-2182

IOP Conf. Series: Journal of Physics: Conf. Series **794** (2017) 012016 doi:10.1088/1742-6596/794/1/012016

[4] Paskaleva A, Hudec B, Jančovič P, Fröhlich K and Spassov D 2014 *Facta Universitatis: Electronics and Energetics* **27** 621
[5] Atanassova E and Paskaleva A 2007 *Microelectron. Reliab.* **47** 913-923
[6] Atanassova E, Dimitrova T and Koprinarova J 1995 *Appl. Surf. Sci* **94** 193-202
[7] Xu N, et al. 2008 *Appl. Phys. Lett.* **92** 232112
[8] Demiryont H, Sites J and Geib K 1985 *Appl. Opt.* **24** 490-495
[9] Kao K 2004 *Dielectric phenomena in solids* (San Diego: Elsevier Accademic Press)

Electric, dielectric and optical properties of Ga_2O_3 grown by metal organic chemical vapour deposition

A Paskaleva, D Spassov and P Terziyska

Institute of Solid State Physics, Bulgarian Academy of Sciences, 72 Tzarigradsko Chaussee, Sofia 1784, Bulgaria

e-mail: paskaleva@issp.bas.bg

Abstract. Thin film (15-130 nm) of gallium oxide were grown by the industry relevant metal organic chemical vapour deposition (MOCVD) technique on p-type Si to check the possibility for integration of newly rediscovered wide bandgap material with the Si technology. Electric, dielectric and optical properties were studied and analyzed. To perform electrical characterization, Ga_2O_3 films were integrated into Al/Ga_2O_3/*p*-Si metal–oxide–semiconductor (MOS) capacitors. Relative dielectric permittivity, flat-band voltage shift and effective oxide charge density were obtained from C-V measurements. Spectroscopic ellipsometry measurements reveal that Ga_2O_3 deposited by MOCVD is a direct bandgap material with a large optical bandgap of about 5.1 eV. Both ellipsometrical and electrical results show formation of a thick interfacial SiO_2.

1. Introduction

Gallium oxide has attracted significant attention as material due to its very interesting physical and electrical properties. Depending on growth conditions, β-Ga_2O_3 can be an insulator or a semiconductor; growing under oxidizing conditions results in insulating β-Ga_2O_3, whereas that growing under reducing conditions is conductive [1]. Intentionally doped with IV-group elements β-Ga_2O_3 also attains semiconducting properties [2-4]. Ga_2O_3 is considered as a promising material for various applications such as: optoelectronic devices in deep UV range, gas sensing, transparent electronic devices, metal-oxide- semiconductor (MOS) capacitors, resistive switching memory devices, antireflection coatings due to its favorable properties - wide band gap(about 5 eV), medium dielectric constant, good dielectric strength and deep-UV (<300 nm) transparency [3, 5]. Since many of these devices have been and will likely continue to utilize silicon, it is important to investigate the properties of Ga_2O_3 integrated into a Si device. Although the properties of Ga_2O_3 as a wide bandgap semiconductor attracts more interest, its potential as a dielectric material has also been considered and studied [6-8]. For example, Ga_2O_3 metal-oxide semiconductor field effect transistors (MOSFETs) with high breakdown voltages, large ON/OFF ratios, and high temperature operation have been recently demonstrated [9, 10]. Variety of thin film deposition methods such as different physical vapor deposition techniques, sol-gel, pulsed laser deposition, atomic layer deposition and chemical vapour deposition (CVD) were implemented to achieve better quality of Ga_2O_3 thin films. CVD is the method of choice in this paper since it is applicable for growth of a large number of metals, elemental or compound semiconductors and dielectric thin films with structure ranging from amorphous to polycrystalline and single-crystalline [11,12]. It is an industry-

IOP Conf. Series: Journal of Physics: Conf. Series **794** (2017) 012017 doi:10.1088/1742-6596/794/1/012017

relevant technique combining high material quality with reasonable costs due to the ability to deposit various materials homogeneously on a large-scale.

In this work, dielectric, electrical and optical properties of MOS capacitor with Ga_2O_3 grown by MOCVD on Si substrate have been investigated.

2. Experimental procedure

p-type Si (100) wafers were used as substrates in the deposition process. They were chemically cleaned and etched just before loading in the reactor to control the thickness of the native SiO_2.Thin films of Ga_2O_3 were grown at 750 °C by MOCVD in a horizontal cold-wall reactor employing the metal-organic compound trimethylgallium (TMGa) and oxygen as precursors for Ga and oxygen, respectively. The TMGa bubler was kept at temperature of 10 °C. Argon was used as a transport gas. No special buffer layers or substrate pretreatments were employed on the Si substrates.

Ga_2O_3 films with different thickness in the range of 15-130 nm were deposited. Metal-Oxide-Si (MOS) capacitors with an area S=2.8x10^{-3} cm^2 are formed by evaporating Al through a shadow mask. The electrical properties were studied by capacitance–voltage (*C–V*) measurements. The measurements were carried out in a dark chamber employing Agilent 4980A LCR meter. Characterization by spectroscopic ellipsometry is also performed to obtain information about the optical constants and the thicknesses of layers. The ellipsometry measurements were performed using J.A. Woollam Co., Inc. M2000D rotating compensator spectroscopic ellipsometer with a CCD spectrometer with wavelength range from 193 to 1000 nm. Experimental data were acquired at angles of incidence of 65, 70 and 75 degrees and were modelled using the CompleteEASE Woollam Co., Inc. software.

3. Results and Discussion

The typical frequency dependent *C-V* curves for thinnest (15 nm) Ga_2O_3 sample are presented in figure 1 (a). Similar very strong dependence of capacitance on the frequency is observed also for thicker samples. In figure 1 (b) the capacitance in accumulation in dependence on frequency is presented for samples with different thicknesses. As is seen for thicker samples the dependence is weaker. The *C-V* curves of different samples measured at 1 kHz are compared in figure 1 (c). The parameters extracted from 1 kHz curves – effective dielectric constant, ε_{eff}, flat-band voltage, V_{fb}, effective oxide charge,Q_{eff}, are presented in table 1. Here, thicknesses are determined ellipsometrically using a single layer model. The effective oxide charge (Q_{eff}) is calculated from the *C-V* curves using the flat-band voltage shift and is found to be negative with a density in the order of 4x10^{12} cm^{-2}and is only slightly dependent on the thickness. Net negative oxide charge in Ga_2O_3 layers has been observed also in other works [8]. It should be pointed-out that most of the metal oxides used as high-k dielectrics (e.g. HfO_2, ZrO_2, Ta_2O_5, etc.) in nanoelectronic devices exhibit effective positive oxide charge. Therefore, Ga_2O_3 may be considered as a possible dopant of high-k dielectrics in applications where compensation of oxide charges, hence low net oxide charge is needed. As is seen in figure 1 (a) with decreasing frequency the curves shift to more positive voltages, which also reveals negative charge trapping. Therefore, the frequency behavior of *C-V* curves implies existence of electron traps with significant density. The shift of *C-V* curves and the hysteresis can be attributed to the combined effect of the slow and fast interface states. In this context, the lack of hysteresis and the steeper *C-V* curve measured for 15 nm Ga_2O_3 films implies the better interface quality of the thinnest samples. The decreased slope of the curves for thicker films is a clear indication of increased density of fast interface states. The thicker samples demonstrate also significant hysteresis, which reveals charge exchange between Si substrate and electrically active traps in the dielectric stack.

As is seen from table 1 the effective permittivity of layers depends on the thickness and decreases very strongly with decreasing thickness. This strong dependence is assigned to the growth of interfacial SiO_2 and formation of a double-layer structure. In this case the structure could be represented by two capacitors C_{SiO} and C_{GaO} with SiO_2 and Ga_2O_3 dielectrics, connected in series and the total capacitance

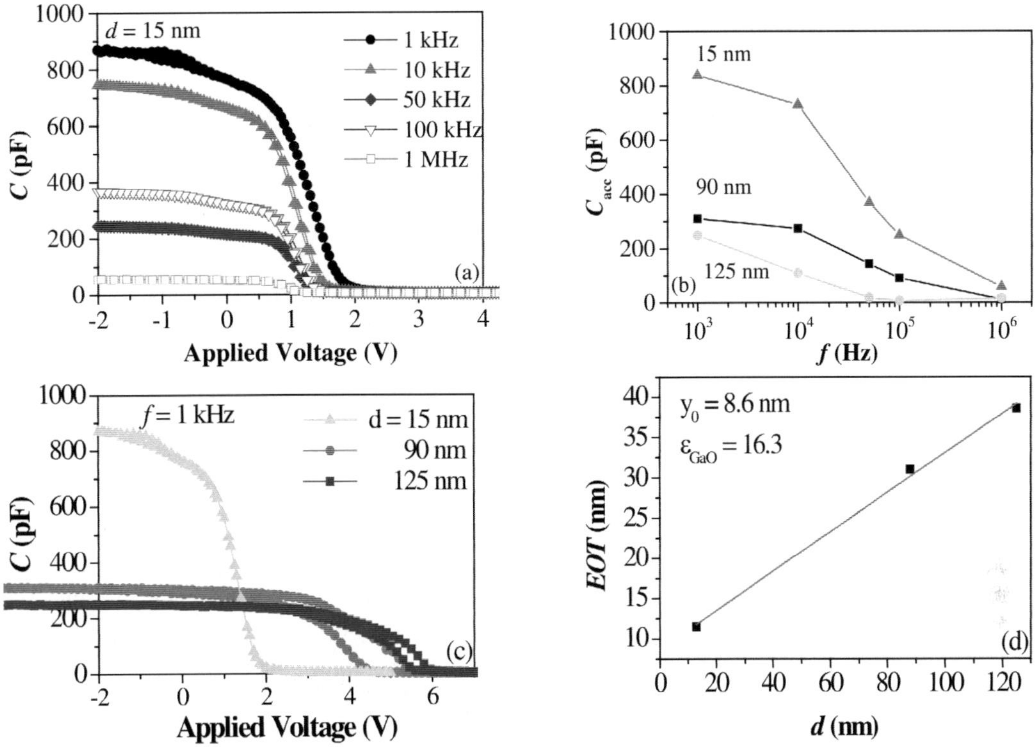

Figure 1. (a) C-V curves of 15 nm Ga_2O_3 layers measured at different frequencies; (b) frequency dependence of capacitance in accumulation for Ga_2O_3 layers with various thicknesses; (c) comparison of 1 kHz C-V curves of Ga_2O_3 layers with various thicknesses; (d) dependence of equivalent oxide thickness on physical thickness.

Table 1. Effective dielectric constant ε_{eff}, flat-band voltage shift V_{fb} and effective oxide charge Q_{eff} as obtained from 1 kHz C-V curves for Ga_2O_3 films with different thickness.

Thickness, nm	ε_{eff}	V_{fb} Volts	Q_{eff} cm^{-2}
15	4.4	2.1	-4.5×10^{12}
90	11.4	5.4	-4.0×10^{12}
125	12.6	6.2	-3.6×10^{12}

could be written as:

$$\frac{1}{C}=\frac{1}{C_{SiO}}+\frac{1}{C_{GaO}}, \tag{1}$$

which can be re-written also as:

$$\frac{EOT}{\varepsilon_{SiO}}=\frac{d_{SiO}}{\varepsilon_{SiO}}+\frac{d_{GaO}}{\varepsilon_{GaO}}, \tag{2}$$

where the equivalent oxide thickness of the stack EOT is evaluated from the capacitance in accumulation of *C–V* curves. d_{SiO} and d_{GaO} are the thicknesses of SiO_2 and Ga_2O_3 layers and ε_{SiO} and ε_{GaO} - their permittivities. In figure 1(d), the *EOT* vs Ga_2O_3 thickness (d_{GaO}) dependence is presented. As is seen from equation (2) this dependence should be a straight line and d_{SiO} could be found from its intercept with the y-axis, and the slope of the curve gives the permittivity of Ga_2O_3 dielectric. The fitting gives a value of SiO_2 thickness d_{SiO} = 8.6 nm. Therefore, electrical measurements reveal formation of very thick SiO_2 at the Ga_2O_3/Si interface. The permittivity of Ga_2O_3 is estimated to be about 16.3, which is slightly larger than the reported values of about 10-15 by other authors [8, 13].

The reasons for the strong frequency dispersion (figure 1 (a)) can be: i) underoxidized films in which oxygen deficiency leads to poor insulating properties [14]. In the case of existence of defective conductive regions interspersed in insulating material a dramatic increase of dielectric constant with decreasing frequency is observed and the obtained permittivity at low frequencies is unrealistically large [14]. In our case although the same behaviour of capacitance with frequency is observed, the values of ε_{GaO} at low frequency are consistent with values typically reported. Also, as will be seen below, the results from spectroscopic ellipsometry reveal well oxidized layers; ii) Maxwell-Wagner polarization which arise from charge accumulation at the interface between two materials which have different conductivities [15]. Having in mind the results obtained, this is the most probable reason for the large frequency dispersion in the samples under consideration. The binary metal oxides are inherently thermally unstable in contact with Si and formation of thin interfacial SiO_2 layer is unavoidable. In most of the cases, however, the thickness of this interfacial layer is less than 3 nm and the obtained here thickness of about 8 nm is very surprising. Therefore, we performed spectroscopic ellipsometry measurements to determine the thicknesses of Ga_2O_3 and interfacial SiO_2.

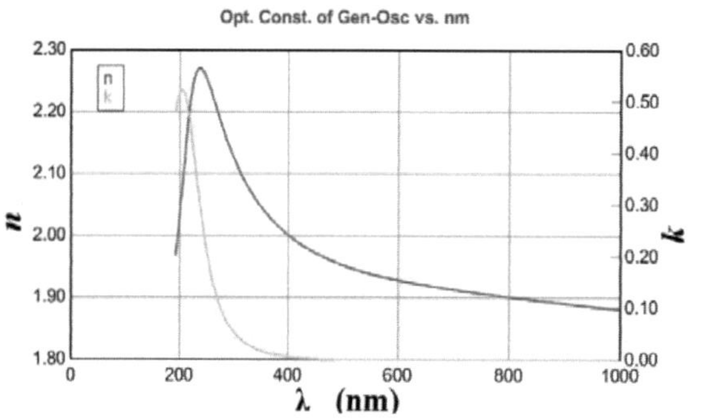

Figure 2. Refractive index *n* and the extinction coefficient *k* as a function of wavelength for Ga_2O_3 layer with a thickness of about 130 nm.

For ellipsometric modelling a three-layer model was used to represent the sample, consisting of a silicon substrate, a silicon oxide layer (as a 1st layer), a Ga_2O_3 layer as a second layer and a roughness layer as a third layer. For the silicon substrate and the silicon oxide layer we used the models from the CompleteEASE software database. The experimentaly obtained parameters Ψ and Δ were then fitted in the transparent area from 500 to 1000 nm using the Cauchy model. The Ga_2O_3 layer was then parametrised using first the b-spline model and then the General oscillator model with one Cody-Lorentz oscillator. The thicknesses of the silicon native oxide layer and the Ga_2O_3 layer were determined in the transparent area from 500 to 1000 nm and fixed. The dispersion of the dielectric functions was then determined by fitting Ψ and Δ in the entire wave length (λ) range from 193 to 1000 nm. The General oscillator parametrization ensures Kramers-Kroning consistent optical properties [16, 17]. The roughness layer was modeled by Bruggeman's EMA (Effective Medium Approximation) of 50% voids and 50% bulk material [16]. The values for the thickness as determined from the above described model are: 128.8 nm for the Ga_2O_3 layer and 7.7 nm for the silicon oxide layer. The roughness of the Ga_2O_3

layer as determined from the model is 1.67 nm. Therefore, the spectroscopic ellipsometry results are in a very good agreement with electrical results and also confirm formation of very thick interfacial SiO_2. This implies a strong thermodynamic instability of Ga_2O_3 in contact with Si. The SiO_2 formation is obviously stimulated by a high deposition temperature (750 °C) of MOCVD process.

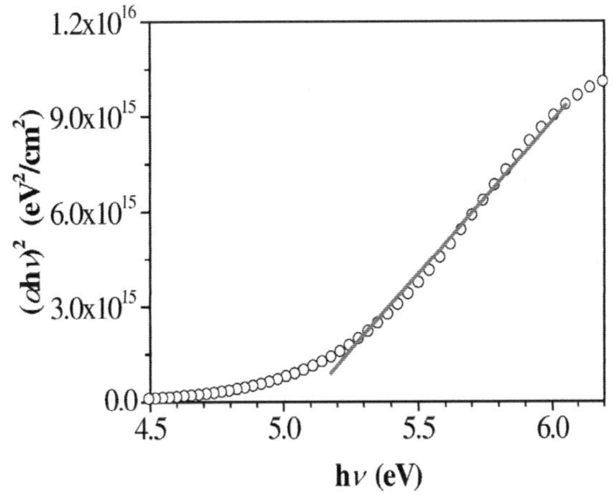

Figure 3. Tauc plot for Ga_2O_3 layer with a thickness of about 130 nm.

The dispersions of the refractive index *n* and the extinction coefficient *k* are presented in figure 2. The values of n, especially its maximum value (~ 2.27), are larger than the reported by other authors [18]. As the higher n indicates more dense films without pores, this means that the MOCVD technique results in denser Ga_2O_3 layers. This can explain also the higher dielectric constant as compared to films deposited by other techniques.

The optical bandgap can be derived from the relation: $\alpha h\nu \sim (h\nu - Eg)^h$, where $\alpha = 2\pi k/\lambda$ is the absorption coefficient at photon energy $h\nu$ and h depends on the type of transition (h=1/2 for direct bandgap and h=2 – for indirect) [19]. Figure 3 shows that linear fit is obtained for h=1/2, which reveals a direct bandgap. The extrapolation of the linear part of the curve to zero yields $E_g = 5.1$ eV. The bandgap of Ga_2O_3 layers is reported to be in the range 4.3 – 5.1 eV depending on its thickness and crystalline status [18]. Also, the better oxidation results in increase of E_g and better insulating properties of Ga_2O_3 [18]. Therefore, the large $E_g = 5.1$ eV obtained for our films reveals good oxidation, i.e. the strong frequency dispersion of capacitance is not due to poor oxidation.

4. Conclusion

The following conclusions could be drawn from the results obtained. Ga_2O_3 layers deposited on Si by MOCVD exhibit better dielectric and optical properties – larger relative dielectric constant of ~16, larger optical bandgap of ~5.1 eV and larger refractive index (revealing more dense films) as compared to the reported values for Ga_2O_3 layers deposited by other techniques. However, the high deposition temperature (750 °C) of MOCVD process stimulates formation of a very thick SiO_2 interfacial layer which could be a serious problem to integrate Ga_2O_3 on Si. The different conductivities of Ga_2O_3 and SiO_2 layers resulting in an accumulation of charge at the Ga_2O_3/SiO_2 interface is the most probable reason for the observed strong frequency dispersion of capacitance. Technological approaches to eliminate or suppress formation of SiO_2 should be found in order to successfully integrate Ga_2O_3 in Si technology.

References

[1] Nakano Y and Jimbo T 2003 *Appl. Phys. Lett.* **82** 218–220

IOP Conf. Series: Journal of Physics: Conf. Series **794** (2017) 012017 doi:10.1088/1742-6596/794/1/012017

[2] Gogova D, Schmidbauer M and Kwasniewski A 2015 *CrystEngComm.***17** 6744-6752
[3] Higashiwaki M, Sasaki K, Kuramata A, Masui T and Yamakoshi S 2012 *Appl. Phys. Lett.* **100** 013504.
[4] Gogova D, Wagner G, Baldini M, Schmidbauer M, Irmscher K, Schewski R, Galazka Z, Albrecht M and Fornari R 2014 *J. Crystal Growth* **401** 665–669.
[5] Wager J F, 2003 *Science* **300(5623**) 1245–1246.
[6] Passlack M, Schubert E F, Hobson W S, et al 1995 *J. Appl. Phys* **77** 686-693
[7] Dakhel A A and Alnaser W E 2013 *Microel. Reliab.* **53** 676
[8] Altuntas H, Donmez I, Ozgit-Akgun C and Biyikli N 2014 *J Vac. Sci. Technol. A* **32** 041504
[9] Higashiwaki M, Sasaki K, Kamimura T, Wong M H, Krishnamurthy D, Kuramata A, Masui T, and Yamakoshi S 2013 *Appl. Phys. Lett.* **103** 123511
[10] Hwang W S, Verma A, Peelaers H, Protasenko V, et al 2014 *Appl. Phys. Lett.* **104** 203111
[11] Pierson H O, 1999 *Handbook of Chemical Vapor Deposition (CVD): Principles, Technology, and Applications* (Norwick, New York: Noyes Publ.)
[12] Gogova D, Siche D, Fornari R, Monemar B, Gibart P, Dobos L, Pécz B, Tuomisto F, Bayazitov R and Zollo G 2006 *Semicond. Sci. Technol.* **21** 702-706
[13] Fu D and Kang T W 2002 *Jpn. J. Appl. Phys.* **41** L1437
[14] Ramanathan S, Park C-M and McIntyre P C 2002 *J. Appl. Phys.* **91** 4521
[15] Jameson J R, Griffin P B, Plummer J D and Nishi Y 2006 *IEEE Trans. Electr. Dev.* **53** 1858
[16] *CompleteEASE® Software Manual, J.A.Woollam Co. Inc*
[17] Hilfiker J N, Singh N, Tiwald T, Convey D, Smith S M, Baker J H and Tompkins H G 2008 *Thin Solid Films* **516** 7979-7989
[18] Al-Kuhaili M F, Durrani S M A and Khawaja E E 2003 *Appl. Phys. Lett.* **83** 4533
[19] Pankove J I, 1971 *Optical Processes in Semiconductors (*New York: Dover)

IOP Conf. Series: Journal of Physics: Conf. Series **794** (2017) 012018 doi:10.1088/1742-6596/794/1/012018

Resistive switching behavior of SiO_x layers with Si nanoparticles

D Nesheva[1]*, B Pantchev[1], N Nedev[2], B Valdez[2], E Manolov[1], V Dzhurkov and R Nedev[3]

[1]Institute of Solid State Physics, Bulgarian Academy of Sciences,72 Tzarigradsko Chaussee Blvd,1784 Sofia, Bulgaria

[2] Universidad Autónoma de Baja California, Instituto de Ingeniería, Benito Juárez Blvd., s/n, C.P. 21280, Mexicali, Baja California, México

[3]Universidad Politécnica de Baja California, Calle de la Claridad s/n, C.P. 21376 Mexicali, Baja California, México

E-mail: nesheva@issp.bas.bg

Abstract: First results on resistive switching in SiO_x film containing crystalline silicon nanoparticles are reported. SiO_x layers (x = 1.15) with thickness of 50 nm were deposited on n-Si crystalline substrates and annealed for 60 min at 1000°C to grow crystalline nanoparticles. Part of the samples were annealed in an inert atmosphere, while the rest were subjected to a two-step (O_2+N_2/N_2) annealing process. Current-voltage (I-V) characteristics were by applying positive or negative voltage to the top contact. For both types of samples the I-V characteristics were asymmetric with lower currents measured at negative voltage, especially in the case of two-step annealed samples. In most of the N_2 annealed structures switching behavior high-low/low-high resistance state was observed in both polarities at voltages with amplitudes in the range (2 - 4) V. Uncontrolled switching low/high resistance was also seen, more frequently at positive voltages. In contrast, the two-step annealed samples showed stable behavior. The transition high-low resistance state was achieved by negative voltages in the (-2, -5) V range leading to an increase of the current by more than three orders of magnitude. The structures were reset to the high resistive state, by positive voltage in the range (3 - 4) V. Uncontrolled switching was not observed in the two-step annealed samples for both polarities and they showed higher reliability regarding the number of switching cycles.

1. Introduction

Resistive random access memory (RRAM) has drawn significant interest as a promising candidate for next generation nonvolatile memory due to its potential scalability beyond 10 nm feature size, fast switching speed, low operating power, and good reliability as compared with flash memory [1,2]. Large number of metal-oxide-metal (MIM) structures with CMOS (complementary metal-oxide-semiconductor) compatible metal oxides have been studied for memory switching properties (Al_2O_3, MoO_x, Ta_2O_5, NiO_x, TiO_x, WO_x, HfO_x, ZrO_x etc.). Most of these MIM structures show bipolar switching and their behavior depends on both the metal oxide composition and the metal contacts used.

Resistive switching and related nonlinear conduction have also been observed in SiO_x, $x \approx 2.0$ layers sandwiched between two poly-Si layers [3]. SiO_x resistive switching memory is attractive for its fully CMOS compatible material composition and processing [4]. Moreover SiO_2 is a conventional material with high availability and ease of integration with current and future technology platforms. Most

RRAM devices using SiO_x as the active switching medium have unusual electrical characteristics where resistive switching only occurs in non-oxidizing ambient and a unique unipolar operation. Two types of resistive switching memory have been demonstrated based on SiO_x ($x \approx 2.0$). The first type relies on the formation and rupture of metal filament injected from the electrode and it features bipolar switching behavior. The second type normally shows unipolar current-voltage (I-V) characteristics with the reset voltage larger than the set value. This type of resistive switching is an intrinsic property of SiO_x ($x \approx 2.0$) and it is electrode-independent [5]. The intrinsic resistive switching in metal-free silicon oxide offers a simpler route to integration with CMOS technology.

The mechanism of the intrinsic resistive switching in SiO_x is characterized as filament conducting. In electroforming stage, voltage bias application causes formation of conductive filament(s). The real-time formation and evolution of filament in a silicon oxide resistive switch has been imaged through in situ transmission electron microscopy [6]. It has been revealed that the electroforming process involves the local enrichment of silicon from the silicon oxide matrix. Semi-metallic silicon nanocrystals have been observed, which can account for the conduction in the filament. Joule heating at higher currents disrupts the filaments, returning the system to the high resistance state. The switching between high resistance state (HRS) and low resistance state (LRS) occurs at an electroforming voltage (V_{ef}) which is an important parameter for SiO_x RRAM application. Lower V_{ef} is demanded since a large electroforming voltage will increase the consumed power and will decrease the overall memory density. The underlying cause for the resistive switching in the SiO_x ($x \approx 2$) layer can be induced by unintended factors, such as defects in SiO_x surface, pores, material-assisted local electric-field enhancement, current local heating, etc. [3,7].

In the RRAM devices using SiO_x ($x \approx 2$) as the active switching medium the resistive switching only occurs in non-oxidizing ambient [3,4 8-10]. The operation of such a device is possible only under vacuum due to oxidation of silicon conductive pathways on the device surface under ambient conditions. A study of resistive unipolar switching in a silicon-based ITO/n-poly-Si-SiO_x-p-Si device in which the active layer is silicon-rich silica (SiO_x, $x < 2.0$) has demonstrated that in contrast to other work in the literature, switching in that device occurred in ambient conditions, and was not limited to the surface of the active material [11]. A switching mechanism driven by competing field-driven formation and current-driven destruction of filamentary conductive pathways has been proposed. It has been assumed that conduction is dominated by trap assisted tunneling through noncontinuous conduction paths consisting of silicon nanoinclusions in a highly nonstoichiometric suboxide phase. It has been found that the reset currents were higher in devices annealed at 950°C. In those samples a high density of silicon nanoinclusions can exist within the SiO_x film as a result of phase separation allowing highly conductive pathways to occur.

In this paper initial results on resistive switching in SiO_x film containing crystalline silicon nanoparticles are reported. SiO_x layers ($x = 1.15$) with thickness of 50 nm were deposited on n-Si crystalline substrates by thermal evaporation of SiO in vacuum and then annealed in an inert atmosphere for 60 min at 1000°C in order to grow Si nanocrystals. The switching behavior has been studied by measuring DC current-voltage (I-V) characteristics metal/oxide (SiNCs-SiO_x)/n-Si (MOS) structures. In contrast to [11], bipolar switching behavior has been observed.

2. Experimental details

Silicon-rich SiO_x layers with initial composition $x = 1.15$ and thickness of 50 nm were deposited on n-type (100) crystalline silicon (c-Si) substrates ((4-6) Ω×cm) maintained at room temperature. Thermal evaporation of SiO at a vacuum of 1×10^{-3} Pa was used for the film preparation [12].The deposition rate and film thickness were monitored by a calibrated quartz microbalance system. The silicon wafers were cleaned chemically prior to film deposition by applying a standard procedure for the microelectronics industry. All as-prepared layers were annealed at 250°C for 30 min in an Ar atmosphere to ensure good film stability at ambient conditions. This annealing does not lead to phase separation and Si nanoparticle formation [12]. Silicon nanocrystals (NCs) were grown in a SiO_x matrix by furnace annealing at 1000°C in a nitrogen or nitrogen + oxygen/nitrogen atmosphere for 60 min.

Our previous results have shown that annealing of $SiO_{1.1}$ under these conditions leads to phase separation and formation of silicon nanocrystals with diameter of ~ 5-6 nm [13,14] in a SiO_2 matrix [15,16]. The thickness of the oxide formed between the nanocrystals and the adjacent films is ≥ 3 nm. It has also been observed that the N_2 annealed samples the Si nanocrystals grown in the region close to the top surface (~ 5-6 nm) get transformed into SiO_2 due to native oxide formation after exposure to air [16]. The two-step thermal treatment, first in an O_2+N_2 (10 min), followed by N_2 annealing leads to formation of ~ 15 nm SiO_2 region close to the top surface [17].

After the furnace annealing metallization was carried out through a mask and top electrodes (referred from now on as gates) of the MOS capacitors were formed with area of ~ 2×10^{-3} cm^2. Aluminum and gold were used as top-contact metals for the N_2 and O_2+N_2/N_2 annealed samples, respectively. Aluminum electrode was also prepared on the bottom side of the silicon substrate. DC current-voltage characteristics were measured by using Semiconductor Characterization System Keithley 4200-SCS with preamplifiers. The hold and delay times of all measurements were 5 and 1 s, respectively. The polarity of the applied voltage given below refers to the top electrode.

3. Results and Discussion

Typical current-voltage (I-V) characteristics of a N_2 annealed Al/c-Si/SiO_2-Si/Al structures are shown in figure 1. The numbers in the figure denote the sequence of the measurements. There were no longtime pauses between the measurements of the consecutive curves. As seen from the figure switching from low conductive to high conductive state is observed both on applying positive and negative voltage on the top electrode. A large number (more than 15) of samples were studied and it has been found that the voltage at which the switching takes place is not constant; it varies in the range (2 - 5) V. The current increase is normally about 3 orders of magnitude. It has to be noticed that the switching process is better controlled on applying negative voltage. The number of the observed switching cycles from HRS to LRS and back is small (less than 10) and finally the structures remain in the LRS.

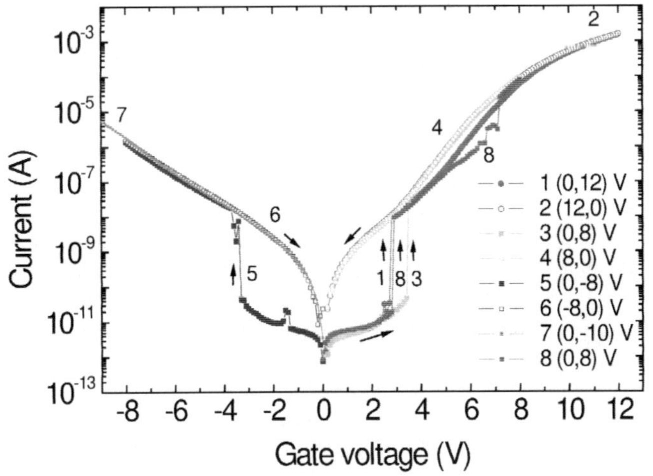

Figure 1. Current-voltage characteristics of an Al/c-Si/SiO_2-Si/Al structure prepared by annealing in nitrogen at 1000°C for 60 min. The numbers in the figure denote the sequence of the measurements.

Figure 2 shows three curves measured first on increasing the applied negative voltage from 0 V to (-6) V (curve 1) and then going back from (-6) V to 0 V (curves 2, 3). It illustrates that uncontrolled switching from LRS to HRS may take place when a negative voltage is applied. During the voltage scan corresponding to curve 2 the MOS structures remains at LRS, while at one of the next scans it spontaneously switches from LRS to HRS (curve 3). No such spontaneous switching has been found on applying positive voltage but again uncontrolled transfer from LRS to HRS was observed in many samples; for example curves 2, 3 and 4, 5 in figure 1. Hence, the experiments on this group of samples have shown two drawbacks, which make these structures not prospective for memory applications: (i) small number of switching cycles and (ii) uncontrolled return to the low conductive state.

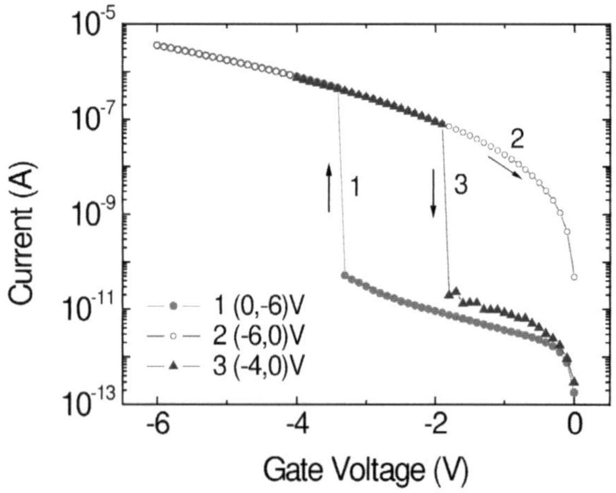

Figure 2. Current-voltage characteristics measured on an Al/c-Si/SiO_2-Si/Al structure showing a controlled switching from high resistive to low resistive state (curve 1) and uncontrolled switching from low resistive to high resistive state (curve 3).

It is seen from figure 1 that the I-V characteristics in the high resistive state are nearly symmetric while in the low resistive state they are asymmetric - on applying the same voltage the current is significantly higher when the gate voltage is positive. Assuming that the current through the oxide layer is due to electrons injected from the c-Si substrate at positive voltage and from the top Al contact at negative voltage, the observed symmetry for the low conductive state can be understood keeping in mind that the energy barriers at both interfaces are nearly equal in the case of stoichiometric SiO_2 (~3.1 eV for the Si/SiO_2 interface and ~3.2 eV for the Al/SiO_2 one) and supposing that they remain close to each other for SiO_2-Si NCs composite layers. As a possible explanation of the I-V characteristics asymmetry in the high conductive state one can consider the following idea, though other reasons cannot be excluded. According to the accepted mechanism of the intrinsic resistive switching in SiO_x, in electroforming stage voltage bias application causes formation of filament (or filaments) which include highly conductive silicon nanocrystals. The band gap of silicon nanocrystals, due to the quantum size effect, is larger than that of bulk crystalline silicon but much smaller than the band gap of SiO_2. Hence, the barrier height at both interface c-Si/SiNCs and Al/SiNCs may be quite different (this point will be explored in a further work).

Based on the accepted switching mechanism in SiO_x and keeping in mind the experimental results in figure 1, one can assume that the Al/c-Si/SiO_2-Si NCs/Al structures, if further optimized by varying the size and density of Si NCs and/or the contact type to avoid the uncontrolled return to the HRS, could be used as memory elements by applying an appropriate negative voltage on the gate electrode to switch from HRS to LRS and a positive voltage for the LRS to HRS switching.

In order to reduce to some extent the current at negative applied voltages and thus to avoid the uncontrolled filament "burning", a SiO_2 layer was formed by the two-step annealing process described in the Experimental details section. This layer can limit the electron injection from the top electrode and one can expect elimination of the spontaneous return to the HRS. Also the Al top contact was replaced by Au ones. These changes lead to another type of metal/insulator interface compared to the Al/SiO_2-Si NCs one. Figure 3 shows current-voltage characteristics of an Al/c-Si/SiO_2/SiO_2-Si/SiO_2/Au structure prepared by a two-step annealing at 1000°C. As seen from the figure the first switching at negative voltages (forming process) occurred at (-5) V. The following switching processes, as can be expected, take place at lower voltage, though some variations in the switching voltage still exist. They could be related to charge trapping in the top SiO_2 layer produced by 10 min annealing of $SiO_{1.15}$ at 1000°C in an O_2+N_2 atmosphere. For producing of top dioxide layer with a higher quality, optimization of the annealing procedure should be carried out or separate SiO_2 deposition on top of the SiO_2-Si NCs layer may be applied (for example by chemical vapor deposition or sputtering).

It is important that the current increase with the voltage is significantly slower than that in figure 1 and no uncontrolled switching was observed at negative applied voltages in all samples studied. Based on the results obtained on a large number of samples one can conclude that a voltage of (-5) V can ensure a controllable switching between HRS and LRS. It is demonstrated in figure 3 that the structures can be reset to the initial HRS in a controllable way by applying positive voltages higher than 3 V. The currents measured in the HRS were in the range (10^{-12} – 10^{-11}) A, while the current measured in the LRS was above 10^{-8} A at all switching cycles. No uncontrolled switching was observed for the positive polarities, as well, and the samples showed higher reliability regarding the number of switching cycles. These results indicate that the O_2+N_2/N_2 structures with a gold top contact are promising for resistive memory applications.

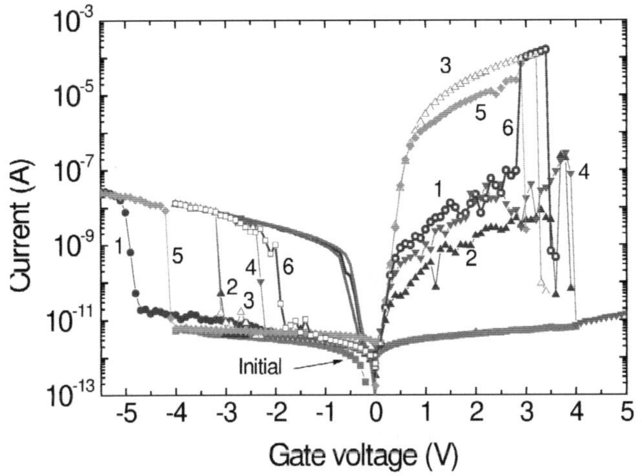

Figure 3. Typical current-voltage characteristics of an Al/c-Si/SiO_2/SiO_2-Si/SiO_2/Au structure prepared by a two-step annealing at 1000°C. The numbers in the figure denote the consequence of curve measurement. For curves with the same numbers first a scan at negative voltages was applied and then at positive ones.

Figure 4 shows a comparison between two HRS/LRS transitions performed with scans from 0 to (-3) and (-4) V, curves 1 and 5, respectively. The LRS curve measured after the (-4) V voltage scan (curve 6) has the typical shape of a low resistance characteristic with currents between (10^{-9} – 10^{-8}) A at voltages in the (-0.5)–(-2) V range. In contrast, the (-3) V scan leads to smaller currents in the 0 - (-1.5) V range (curve 2) resulting in a not well defined 'current window' (the currents measured in the 0 - (-2) V range). The next scan (curve 3), performed in the low resistive state, up to (-4) V stabilizes the LRS and the next curve 4 coincides with curve 6. These results demonstrate that voltages higher than (-4) V are required in order to realize reliable switching and high quality information storage.

4. Conclusions

Initial results have been reported on resistive switching in Al/c-Si/SiO_2-Si/Al and Al/c-Si/SiO_2/SiO_2-Si/SiO_2/Au structures containing crystalline silicon nanoparticles. Current-voltage characteristics taken at room temperature in air were asymmetric with lower currents measured at negative voltage, especially in the case of the samples from the second group (two-step annealed). Switching from high resistive state to low resistive state was observed in both groups of structures when applying both positive and negative voltage with amplitudes in the range (2 - 5) V to the top contact. In structures from the first group, annealed in N_2, uncontrolled switching from LRS to HRS was seen, more frequently at positive voltages indicating that this type of structures have to be further optimized to be suitable for resistive memory applications.

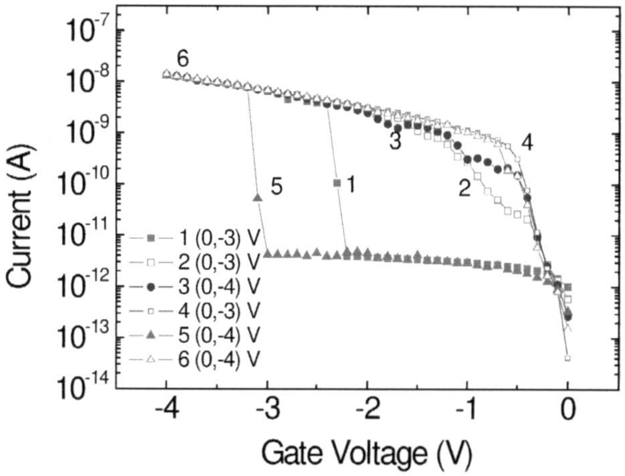

Figure 4. Current-voltage characteristics of an Al/c-Si/SiO_2/SiO_2-Si/SiO_2/Au structure measured at negative applied voltages. All curves were taken starting from 0 V and the numbers in the figure denote the sequence of the measurements.

In the two-step annealed samples the transition from high to low resistance state was achieved by applying negative voltages and a current increase of more than three orders of magnitude was observed. The structures were repeatedly reset to the initial, highly resistive state, by applying positive voltage in the range (3 – 4) V. Uncontrolled switching was not observed for both polarities. The stable behavior of the 'O_2+N_2/N_2' samples i.e. the higher reliability regarding the number of switching cycles and the absence of uncontrolled switching makes these structures promising for memory applications.

Acknowledgements
Financial support from the Project INERA under the contract REGPOT 316309 for the paper presentation at the 19th International School on Condensed Matter Physics "Advances in Nanostructured Condensed Matter: Research and Innovations", Varna, 2016 is gratefully acknowledged. This work was supported in part by CONACYT, Mexico under Contract No. 169831.

References

[1] Waser R and Aono M 2007 *Nat.Mater.* **6** 833
[2] Sawa A 2008 *Materials Today* **11** 28
[3] Yao J, Zhong L, Natelson D and Tour J M 2011 *J. Am. Chem. Soc.* **133** 941
[4] Yao J, Zhong L, Natelson D and Tour J M 2011 *Appl. Phys.* A **102** 835
[5] Yao J, Sun Z, Zhong L, Natelson D and Tour J M 2010 *Nano Lett.* **10** 4105
[6] Yao J, Zhong L, Natelson D and Tour J M 2012 *Scientific Reports* **2** 242
[7] Cavallini M, Hemmatian Z, Riminucci A, Prezioso M, Morandi V and Murgia M 2012 *Advanced Materials* **24** 1197
[8] Chang Y F, Chen P Y, Fowler B, Chen Y T, Xue F, Wang Y, Zhou F and Lee J C 2012 *J. Appl. Phys.* **112** 123702
[9] Chang Y F, Ji L, Wu Z J, Zhou F, Wang Y, Xue F, Fowler B, Yu E T, Ho P S and Lee J C 2013 *Appl. Phys. Lett.* **103** 033521
[10] Chang Y-F, Ji L, Wang Y, Chen P-Yu, Zhou F, Xue F, Fowler B, Yu E T and Lee J C 2013 *Appl. Phys. Lett.* **103** 193508
[11] Mehonic A, Cueff S, Wojdak M, Hudziak S, Jambois O, Labbe Ch, Garrido B, Rizk R and Kenyon A J 2012 *J. Appl. Phys.* **111** 074507
[12] Nesheva D, Bineva I, Levi Z, Aneva Z, Merdzhanova T and Pivin J C 2003 *Vacuum* **68** 1
[13] Nesheva D, Nedev N, Manolov E, Bineva I and Hofmeister H 2007 *J. Phys. Chem. Solids* **68** 725

[14] Curiel M, Petrov I, Nedev N, Nesheva D, Sardela M, Murata Y, Valdez B, Manolov E and Bineva I 2010 *Materials Science Forum* **644** 101
[15] Nesheva D, Nedev N, Levi Z, Bruggemann R, Manolov E, Kirilov K and Meier S 2008 *Semicond. Sci. Technol.* **23** 045015
[16] Curiel M A, Nedev N, Nesheva D, Soares J, Haasch R, Sardela M, Valdez B, Sankaran B, Manolov E, Bineva I and Petrov I 2010 *Mat. Sci. Eng.* B **174** 132
[17] Herrera R, Curiel M, Arias A, Nesheva D, Nedev N, Manolov E, Dzhurkov V, Perez O, Valdez B, Mateos D, Bineva I, delaCruz W and Contreras O 2015 *Mat. Sci. Sem. Pross.* **37** 229

IOP Conf. Series: Journal of Physics: Conf. Series **794** (2017) 012019 doi:10.1088/1742-6596/794/1/012019

Structural and morphological characterization of ternary nanocrystalline Cu-In-S thin films prepared by laser ablation

I Bineva[1*], B Pejova[2,3], V Mihailov[1], A Dinescu[4], M Danila[4] and S Karatodorov[1]

[1]Institute of Solid State Physics, Bulgarian Academy of Sciences, 72 Tzarigradsko Chaussee Blvd., 1784 Sofia, Bulgaria
[2]Institute of Chemistry, Faculty of Natural Sciences and Mathematics, Ss. Cyril and Methodius University, POB 162, 1000 Skopje, Macedonia
[3]Research Center for Environment and Materials, Macedonian Academy of Sciences and Arts, Krste Misirkov 2, 1000 Skopje, Macedonia
[4]National Institute for R&D in Microtechnologies – IMT Bucharest, 126 Erou Iancu Nicolae Str., 077190 Bucharest, Romania

E-mail: irka@issp.bas.bg

Abstract. Thin nanocrystalline Cu-In-S films are prepared by pulsed laser ablation in vacuum with Nd:YAG laser operating at 1064 nm on Si (100) substrates. As a target, a $CuInS_2$ quantum dots powder, synthesized by sonochemical template-free colloidal approach and pressed to form a pellet is used. The structure and morphology of the deposited films were studied employing X-ray diffraction, scanning electron microscopy, and atomic force microscopy techniques. The analysis of the deposited thin films revealed predominant cubic $Cu_{1.7}In_{0.05}S$ phase. It has been found that deposited films were nanocrystalline. The effect of laser pulse energy and deposition time on the structure and morphology of as-prepared films is studied. Slight mean diameter increase from 8 nm to 13 nm with deposition time and applied power was observed, as the time dependence is more pronounced. On the contrary, six fold increase from 7.4 to 44.3 nm of the surface root mean square roughness was estimated with augmentation of the applied laser pulse energy.

1. Introduction

Members of the ternary Cu-In-S semiconductor system with various compositions and structures play a major role in thin film photovoltaic technology [1]. The ternary compounds from the mentioned system are very similar to the II-IV semiconductors from structural viewpoint, which reflects in the similarity of optoelectronic properties [2], with the advantage of being environmentally friendly. These properties can be controllably tuned if the semiconductors are prepared in low-dimensional form [2-4]. Recently, $CuInS_2$ nanocrystals attracted an ever-increasing attention due to their application in quantum dots sensitized solar cells [5, 6].

So far, many physical and chemical techniques have been employed for preparing Cu-In-S thin films. Among them, the pulsed laser deposition (PLD) is a simple, but versatile, experimental method that finds use and in extensive areas of thin film deposition and multi-layer research [7, 8].

IOP Conf. Series: Journal of Physics: Conf. Series **794** (2017) 012019 doi:10.1088/1742-6596/794/1/012019

This technique offers the possibility to transfer material to a random substrate with no preliminary treatment (absence of seed layer or molecular linker). Literature data concerning laser ablation as a technique for deposition of thin films composed by QDs are, however, scarce [5, 9].

In the context of our efforts towards the development of controllable methods for deposition of low-dimensional semiconductor films, in the present study, we employ this technique as a direct method to deposit thin films from the ternary Cu-In-S semiconductor system starting from a highly nanostructured target material, initially deposited by a template-free, sonochemically-assisted colloidal chemical route. We pay a special emphasis on the structure and surface morphology properties of the synthesized material in thin film form as a function of the laser pulse density and the deposition time.

2. Experimental details

2.1. Sample preparation

As a target for subsequent thin film deposition by the laser ablation technique, powder sample composed of $CuInS_2$ quantum dots (QDs) pressed in a form of pellet has been used. The QD solid powder was synthesized by our recently developed sonochemically-assisted template-free colloidal chemical technique [10]. Synthesized target material was identified by the X-ray diffraction technique and it was concluded that it was composed by highly nanostructured $CuInS_2$ with cubic crystal structure (zincblende structural type) [10].

Q-Switched Nd:YAG laser source (model - Quanta Ray GCR:3 from Spectra-Physics, Inc.) operating at fundamental wavelength λ = 1064 nm, pulse duration τ = 7 ns was used for target ablation. The depositions were conducted in a stainless steel vacuum chamber which was evacuated down to residual pressure of 10^{-4} Torr. The laser beam was incident at 45° on the target surface and the irradiation spot was of 1 mm^2 area. The expulsed material was deposited on the Si (100) substrates placed parallel to the target at a separation distance of 7 cm. Laser power is altered starting from the ablation threshold at fixed distance between the target and substrate, chosen to maximize the surface quality. The laser pulse energies used were 60 mJ, 100 mJ, 300 mJ, and 500 mJ at repetition rate 10 Hz. The deposition time was varied between 10 and 40 min.

2.2. Material characterization

The X-ray diffraction method for polycrystalline samples is employed to identify the chemical composition and structural types of the deposited films and also to get much more in-depth insights into their structural characteristics. Grazing Incidence XRD (GIXRD) experiments were performed on a Rigaku SmartLab X-ray thin film diffraction system (Rigaku Corporation, Japan) with $CuK\alpha_{1,2}$ irradiation (λ=1.5418 Å). The GIXRD patterns realized at an angle of incidence α = 0.35° were registered via continuous method applying 0.01° steps (2θ, θ is the Bragg angle) and a counting speed of 12.000 deg/min. The scattered X-ray radiation was detected with SC-70 detector. All XRD patterns were taken within an angular interval 2θ from 20° to 70° in which the main diffraction peaks of CIS are located. SEM measurements were utilized to complement the XRD studies, e.g. in the course of confirmation of the nanostructured character of the deposited films, as well as to estimate the size distribution of the nanoparticles constituting the films. SEM observations have been made using e-Line EBL equipment in SEM mode (Raith GmbH, Germany), working with accelerating voltage of 10 kV. The Cu-In-S films were observed in plane and on cross-section with magnification from 20 to 150 thousand times. Cross-section SEM was used to determine the film thickness. The grain size was estimated with SPIP^{tm} image processing software version 6.0.9 (Image metrology A/S).

AFM measurements were performed on scanning probe microscope Multimode V (Bruker, ex. Veeco, Santa Barbara, CA). The images were taken in tapping mode, as each sample has been investigated on multiple points of the surface. Measurements in scale of 5 µm have been performed with scanning rate in the interval 0.7-1.0 Hz and images resolution of 512 lines per scan direction. Aluminium coated silicon cantilevers TAP150-Al-G (Budget Sensors Innovative Solutions Bulgaria Ltd., Bulgaria) have been used. All images were just flattened and no further processing was carried out. Roughness measurements were performed using the Nanoscope 7.30 programme.

3. Results and discussion

3.1. *Structural investigations by X-ray diffraction*

XRD patterns of Cu-In-S films obtained at each of the applied laser pulse energies (500, 300, 100, and 60 mJ) are shown in figure 1. The spectra shown are with low intensity; their constituting diffraction bands are obviously broadened, indicating a nanocrystalline character. It can be seen that the spectra intensity differs, as it is quite low for the film, obtained in the vicinity of the ablation threshold even for 40 min deposition time, probably due to ultra-small film thickness. The sharp peak around 50° which appears in some of the patterns is reckoned to belong to c-Si (100) substrate. Four main bands are observed, with intensity increasing with the applied power, and location at the diffraction angle of 27.8, 32.3, 46.5, and 54.9°. The peaks are indicated as {111}, {200}, {220} and {311} crystal planes of cubic structure. Comparison with the literature data revealed that the observed diffraction peaks well match to cubic $Cu_{1.7}In_{0.05}S$ JCPDS file No 45-1304. Although for Cu-In-S crystals the typical stable structure is chalcopyrite $CuInS_2$, a loss of sulphur during the ablation is possible. Deviation from the nominal stoichiometry in terms of S deficiency is reported [8, 11] as a consequence of evaporation loss during deposition or annealing in vacuum. The positions of all the peaks in the spectra are slightly shifted to higher 2θ in comparison to referred data, which is likely caused by compressive lattice strain (smaller d_{hkl} values in the case of nanostructured material as compared to the bulk one, as follows from Bragg's law). Some less prominent peaks as well as shoulders of the main peaks are noticed in the spectra. Their reliable indication is embarrassed by the low intensity and quite high noise of the patterns and for trustful results, additional measurements of thicker films are needed. Although the presence of small quantities of Cu_xS and In_2S_3 cannot be fully excluded at this stage, more probably there is an additional phase of chalcopyrite stoichiometric $CuInS_2$ (JPDS file 27-0159). The diffraction bands, typical for In are not found-out and, hence, In phase, reported by the others [9] is not revealed.

The most prominent diffraction peak in the patterns, located at 2θ~47 was least-squares fitted with two bands using Voight function. A typical fit is shown in figure 2. The values for θ and full width at

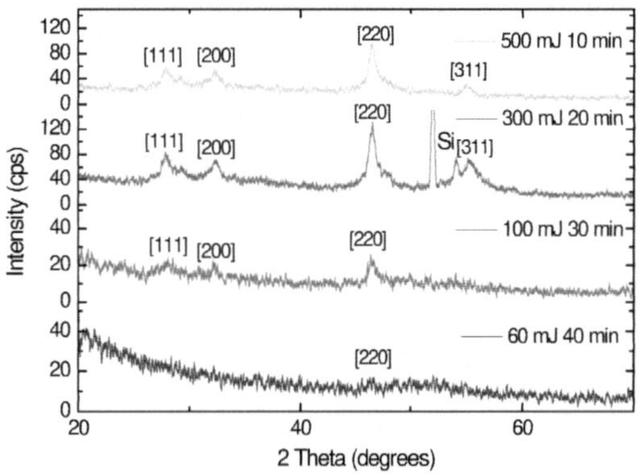

Figure 1. XRD patterns of Cu-In-S thin films.

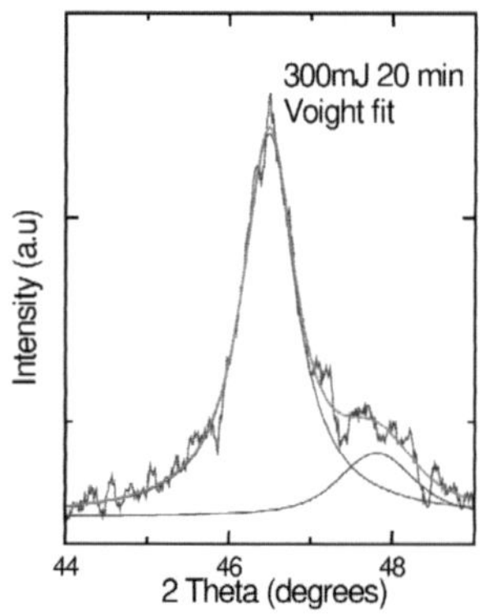

Figure 2. A typical fit of experimental XRD peaks.

half-maximum intensity of the main peak β obtained from the fits were then applied to calculate the average crystal size, using the Scherrer equation and the results are presented in table 1. An increase of quantum dots size with the laser energy was expected, since more $CuInS_2$ plasma should be generated from the target when employed higher laser energy and the $CuInS_2$ plasma continues to grow on the substrate to form quantum dots with larger diameter [5, 11, 12]. The results listed in table 1 confirmed to some extent our expectation. When compared calculated diameters for the film obtained with

Table 1. Angular position, full width at half maximum, calculated size, interplanar spacing, lattice constant, and uniform strain in $Cu_{1.7}In_{0.05}S$ thin films determined by Voight fit of the {220} peak in the XRD spectra.

Sample	*2 Theta (deg)*	*FWHM (deg)*	*Size (nm)*	*Interplanar spacing d*	*Lattice constant a (Å)*	*Uniform Strain*
500 mJ 10 min	46.46	0.76	11.90	1.955	5.528	0.011
300 mJ 10 min	46.54	1.01	8.95	1.951	5.519	0.012
300 mJ 20 min	46.48	0.79	11.40	1.954	5.526	0.011
100 mJ 30 min	46.43	0.75	12.05	1.956	5.532	0.010
60 mJ 40 min	46.35	0.66	13.69	1.959	5.541	0.008

different applied energy for the same time, one can see a relatively small increase of the crystallites size with the applied energy. However, it can be noticed, that changes in the diameter with the ablation time are more prominent. XRD spectra of the films deposited with the same laser pulse energy for 10 and 20 min are shown in figure 3. With the deposition time, a strong increase in the diffraction peaksintensity and reduction of the FWHM, resulting from the increase of the nanocrystallites size can be seen. In fact, looking at table 1, it is evident that the biggest calculated crystallites mean diameter is for the films deposited with the weakest power – at the ablation threshold, but for the longest time. The bigger grain size in the as-deposited films, despite the assumed from the XRD intensity ultra-small film thickness, implies a low density of nuclei on the substrate surface which may be due to the applied low laser power [13] and hence, the very low deposition rate. Perhaps, the embryo density is power dependent, while the time is crucial for the subsequent growth of the grains. The combination of very low average deposition rate and long deposition time might result in high mobility of ad-atoms on the substrate surface, and hence, in bigger grain size.

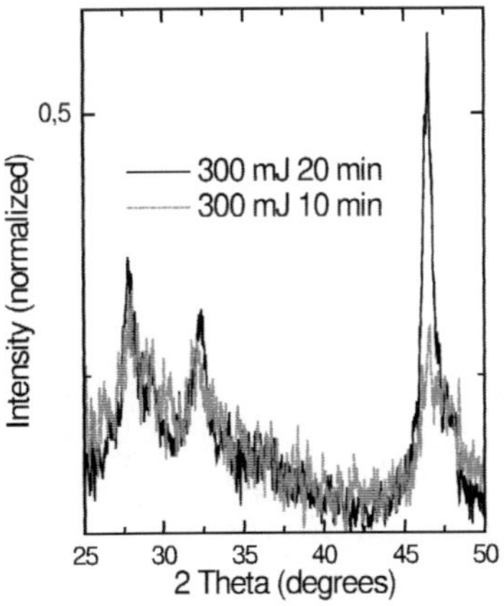

Figure 3. XRD pattern of Cu-In-S thin films, deposited at 300 mJ for 10 min and 20 min. The spectra are normalized.

Taking into account the fact that nanometer-sized crystals could exhibit an intrinsic lattice strain, caused by a high fraction of surface atoms, the experimental XRD data were also exploited for refinement of unit cell parameters and for estimation of lattice strain. The interplanar distance d_{hkl} was determined using the Bragg equation $d_{\{hkl\}} = \lambda/2\sin\theta$. The unit cell parameter a in the case of cubic crystal system is calculated using the relation Eq. (1):

$$a = d_{\{hkl\}}(h^2+k^2+l^2)^{1/2} \tag{1}$$

where (h, k, l are the lattice planes; $d_{\{hkl\}}$ is the corresponding interplanar distance; for the {220} peak $a = \sqrt{8}d_{\{220\}}$. Further on, the average uniform strain is computed, following the equation Eq. (2):

IOP Conf. Series: Journal of Physics: Conf. Series **794** (2017) 012019 doi:10.1088/1742-6596/794/1/012019

$$\langle \varepsilon \rangle = \frac{\left| a_{\text{nanocrystal}} - a_{\text{bulk}} \right|}{a_{\text{bulk}}} = \frac{|\Delta a|}{a_{\text{bulk}}} \tag{2}$$

where a_{bulk} is the lattice constant value of 5.587 (JCPDS file No 45-1304).
From the results listed in the table 1, it can be seen that the calculated lattice constants are quite smaller than the referred one, with the value of ~ 5.53 closer to the parameter a -5.52 of chalcopyrite structure. The high internal stress implies a lattice distortion, which, apart from the expected strain, caused by the size reduction, could also originate from the $CuInS_2 \rightarrow Cu_{1.7}In_{0.05}S$ phase transformation during deposition, combined with the low substrate temperature. Moreover, since $CuInS_2$ compound is slightly non-stoichiometric [8, 14] as well as $Cu_{1.7}In_{0.05}S$ is considered [14] as a solid solution based on Cu_2S and In_2S_5, some deviations from the reported bulk units parameters values are not surprising. As can be seen from the table 1, however, the observed trends in lattice strain can be actually related [15] to the crystal size—the smaller size of nanocrystals constituting the films induces larger lattice strain regardless of the film thickness.

3.2. Morphology

Information about the grain size has also been obtained from the SEM investigation. SEM surface images of samples deposited with 300 mJ pulse energy are depicted in figure 4(a). It is seen that SEM measurements confirm the nanocrystalline character of the films and that most of the grains have close to spherical shape. The SPIP[tm] image processing was used to get the statistical information and the results confirm the conclusion based on the XRD data.

SEM images in cross section were used to estimate the film thicknesses. A typical image is shown in figure 4 (b). The results have shown film thicknesses up to 100 nm, as in the case with the sample, processed close to the ablation threshold, at 60 mJ, formation of discontinuous layer is not observed.

THE GRAPHIC ON THIS PAGE WAS NOT AVAILABLE ON THE E-MEDIA VERSION OF THIS PAPER.

(c)

Figure 4. Grain size analysis, film thickness and morphology of Cu-In-S thin film, deposited at 300 mJ for 20 min: (a) High resolution SEM image. Part of the crystal grains are outlined to make the size and shape clearer; (b) cross section SEM. The film thickness is marked with white line; (c) 3D AFM surface image.

The control and optimization of surface topography are essential for the successful application of Cu-In-S thin films in related device configurations. Quantitative information for the film surface can be obtained with AFM. On figure 4(c) is shown 3D AFM image, revealing quite rough surface of the same film in 5 μm scan area. Upon increasing the applied laser energy is observed an increase of the surface roughness. The root mean square roughness, estimated from AFM images, strongly increases with applied pulse energy from 7.4 nm for 60 mJ, and 22 nm for 100 mJ to 44.3 nm for 300 mJ. The obtained results indicate the suitability of the applied method for deposition of nanostructured CIS films on a random substrate with controllable surface roughness.

4. Conclusions

Thin nanocrystalline Cu-In-S films are prepared by pulsed laser ablation of $CuInS_2$ quantum dots powder, synthesized by the sonochemical template-free colloidal approach. X-ray diffraction and SEM measurements revealed that deposited films are nanocrystalline with predominant cubic phase and a small grain size (~ 10 nm). Deviation from the nominal stoichiometry in terms of the S deficiency has been noticed. These observations have been related to loss of sulphur during the ablation.

On the basis of analysis of data obtained by XRD and SEM measurements, the evolution of crystal size upon laser pulse energy and time of deposition was followed. Slight mean diameter increase from 8 nm to 13 nm with deposition time and applied power was observed, as the time dependence is more pronounced. It was found, that the embryo density is power dependent, while the time is crucial for the subsequent growth of the grains.

AFM investigations revealed a strong anticipated increase of the surface roughness from 7.4 to 44.3 nm with increasing applied laser pulse energy.

Acknowledgements

This work has been supported under the bilateral agreement between the Bulgarian Academy of Sciences and Macedonian Academy of Sciences and Arts.

References

[1] Nishitani M, Kohara N, Negami T, Wada T, Igari S and Shimokawa R 1998 *Sol. Energ. Mat. Sol. C.* **50** 63-70
[2] Leach A D P and Macdonald J E 2016 *Phys. Chem. Lett.* **7** 572-85.
[3] Tan Z et al. 2011 *Adv. Mat.* **23** 3553-58
[4] Uehara M, Watanabe K, Tajiri Y, Nakamura H and Maeda H 2008 *J. Chem. Phys.* **129** 134709.
[5] Han M, Chen W, Guo H, Yu L, Li B and Jia J 2016 *J. Power Sources* **318** 121-27
[6] Jara D H, Yoon S J, Stamplecoskie K G and Kamat P V 2014 *Chem. Mater.* **26**(24) 7221-28
[7] Hiramatsu H, Yanagi H, Kamiya T, Hirano M, Matsunami N, Shimizu K-I and Hosono H 2008 *Phys. Status Solidi A* **205**(8) 2007–12
[8] Zhang L, Fang J, Li M, Shang X, He Y and Kang T W 2014 *J. Korean Phys. Soc.* **64**(3) 410-14.
[9] Bensch W, Quiroga-González E, Kienle L, Duppel V, Lee D-K and Janek J 2010 *Solid State Sci.* **12** 1953-59
[10] Sherif E, Pejova B and Bineva I "Achieving the "impossible" with nanochemistry: A sonochemically-assisted colloidal chemical route to 3D assemblies of close packed cubic $CuInS_2$ quantum dots with controlled structure and tunable band gap energy" *submitted.*
[11] Vanalakar S A et al. 2015 *J. All. Comp.* **619** 109-21
[12] Girtan M, Vlad A, Mallet R, Bodea M A, Pedarnig J D, Stanculescu A, Mardare D, Leontie L and Antohe S 2013 *Appl. Surf. Sci.* **274** 306-13
[13] Guo R, Nishimura J, Matsumoto M, Higashihata M, Nakamura D and Okada T 2008 *Jpn. J. Appl. Phys.* **47** 741-45
[14] Tomashik V. (2006). Copper-Indium-Sulphur. In G. Effenberg S I (Ed.), *Non-Ferrous Metal Systems. Part 1.* Germany: Materials Science International Services GmbH
[15] Pejova B and Bineva I 2016 *J Mater Sci: Mater Electron* 1-16

IOP Conf. Series: Journal of Physics: Conf. Series **794**(2017) 012020 doi:10.1088/1742-6596/794/1/01202(

Dielectric and electrical behaviours of polymeric (PEO/PVP):$NaIO_4$ composite for solid electrolytes

Y G Marinov[1*], G B Hadjichristov[1], A G Petrov[1], H K Koduru[2], L Marino[2] and N Scaramuzza[2]

[1] Georgi Nadjakov Institute of Solid State Physics, Bulgarian Academy of Sciences, 72 Tzarigradsko Chaussee, Sofia 1784, Bulgaria.
[2] Dipartimento di Fisica, Università della Calabria, Via Pietro Bucci, Cubo 31C, 87036 Rende (CS), Italy.
* E-mail address: ymarinov@issp.bas.bg

Abstract. Composite material prepared from polyethylene oxide (PEO) and polyvinylpyrrolidone (PVP) doped with Sodium (meta)periodate ($NaIO_4$) salt was studied by complex impedance spectroscopy at room temperature. The polymers PEO and PVP were mixed in a weight ratio 70:30 %, and the concentration of the embedded $NaIO_4$ compound was 7.5 wt.%. The effect from $NaIO_4$ filler on the dielectric permittivity of the three-component mixed system was analyzed in the frequency range 0.1 Hz – 1 MHz. As compared with the two-component polymer host PEO/PVP, a distinctly enhanced electrical and dielectrical response and an increase of the value of dielectric constant of the polymeric (PEO/PVP):$NaIO_4$ composite were present. This suggests the potential of this material for soft electronics and applications such as solid electrolytes.

1. Introduction

Solid polymer electrolytes (SPEs) for solvent-free metal-ion rechargeable batteries are attractive due to their promising advantages, including safety, ease of packing, excellent flexibility and low level of environmental pollution [1,2]. At present, the SPEs become the most appropriate materials for compact design of solid-state ion-conducting electrochromic devices, and in the fabrication of solid-state high energy density rechargeable batteries to run portable electronic equipment for long duration [3,4]. Sodium (Na) based rechargeable batteries (SRBs) have recently captured much attention because they are environmentally friendly, non-toxic, low cost and earth abundant materials [5,6]. Substantial research, development, and demonstration efforts are currently in progress to replicate the performance of well-established Li-ion batteries with sodium ion batteries. SPEs for SRBs simply consist of a sodium salt dissolved in a polymer matrix.

Polyethylene oxide (PEO) based SPEs have been employed extensively for battery applications since Armand et al. [7] have demonstrated the feasibility of using this polymer as an ion-conductive electrolyte. PEO is one of the widely investigated host polymer for synthesizing SPEs for alkali metal ion-conducting batteries due to its high electrochemical stability, good solvation power, complexation and ion dissociation abilities [8].

Several researchers have reported on various sodium salts complexed PEO based polymer electrolytes for SRBs applications [9,10]. Below the melting temperature ($\approx 330^{\circ}$ K), the pure PEO electrolytes consist of both semi-crystalline and amorphous regions, as well as intermediate regions at the crystalline/amorphous interphase. The semi-crystalline nature of PEO leads to both amorphous and crystalline phases at room temperature, which subsequently limits its ionic conductivity [11]. The most straightforward approach to overcome this problem is to modify the PEO matrix in order to decrease its degree of crystallinity. The suppression in the degree of crystallinity of the polymer chains improves their mobility, thereby leading to better ionic conduction. Therefore, ion hopping is believed to be predominant in amorphous rich PEO based SPEs [12].

One of the most promising alternative choice of enhancing the amorphous phase in PEO based SPE systems is blending of PEO with suitable amorphous polymer. Polyvinyl pyrrolidone (PVP) has been identified as a compatible partner to PEO which exhibits high-degree amorphosity. PVP demonstrates a high glass transition temperature, mechanical and thermal stabilities, the carbonyl group (C=O) in PVP enables to form different complexes of alkali metal ion salts at a wide range of concentrations [13]. In the work presented here, thin films of pure PEO/PVP and sodium periodite ($NaIO_4$) salt complexed PEO/PVP blends were prepared by solution casting technique. Their electric and dielectric properties are investigated by applying impedance spectroscopy and related to their microstructural properties.

2. Experimental

Poly(ethylene oxide) (PEO) with an average molecular weight 5×10^6 and Polyvinyl pyrrolidone (PVP) of molecular weight 3.6×10^5 were obtained from Aldrich and were dried in a vacuum oven at 50°C for 24 hours before use. Sodium periodate salt ($NaIO_4$) was procured from Aldrich and was dried under vacuum at 80°C for 24 hours before use. Methanol (from Aldrich) was used as a solvent without any further purification. Pure and 7.5 wt% $NaIO_4$ salt complexed PEO/PVP blend electrolyte films of thickness approximately 110 μm were prepared by conventional solution cast technique.

The structural properties of blend electrolytes were analyzed by recording X-ray diffraction (XRD) patterns in the 2θ range of 10° – 80° with a constant step 0.02°, counting time 35 s/step on a Bruker D8 Advance diffractometer with Cu Kα radiation and Lynx Eye detector. To understand complexation of additives with the host polymer blend, Fourier-transform infra-red (FTIR) spectra of the films in nitrogen atmosphere were recorded in transmittance mode in the range 400 – 4000 cm^{-1} with a spectral resolution of 2 cm^{-1} using Bruker, model Vertex 70' spectrometer. Impedance measurements were performed by Potentiostat/Galvanostat SP-200 (Bio-Logic) set in two electrode configuration. Mechanically pressed two copper plates were placed in contact to the film surfaces forming sandwich structure. The electrode surface was fixed at 1 cm^2. Each sandwich structure was measured from 0.1 Hz to 1 MHz at 20 mV perturbation voltage. The measurements were carried out at room temperature (during the impedance experiments, the temperature of the samples was 20 $^\circ$C).

3. Results and Discussion

3.1. *Microstructural properties*

Figure 1(a) represents XRD patterns of PEO/PVP and $NaIO_4$ salt complexed PEO/PVP blend electrolytes. The diffraction pattern of PEO/PVP blend displays characteristic diffraction peaks at 2θ = 19.30° and 2θ = 23.40° corresponding to characteristic crystalline peaks of PEO, which originates from the ordering of polyether side chains and strong intermolecular interaction between PEO chains through hydrogen bonding [14]. The decrease in intensity and increase of broadness of the diffraction peaks due to the complexation of $NaIO_4$ salt (in our case at a concentration of 7.5 wt.%) reveal the decline in crystalline nature of the blend electrolytes. This could be due to strong interaction of dissolved $NaIO_4$ salt with the host polymer matrix and intermolecular interaction among the polymer chains of the considered polymer blend electrolytes, which lead to increase of their amorphous phase. Further, FTIR spectra for PEO/PVP and (PEO/PVP): $NaIO_4$ (figure 1 b) were recorded to understand the complexation properties of these blend electrolytes, such as ion–polymer interactions, intermolecular interactions, and their miscibility.

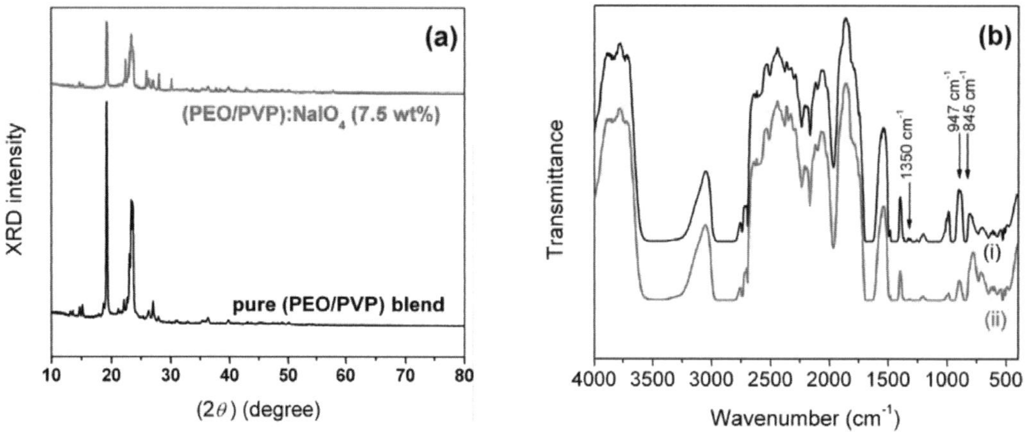

Figure 1. (a) XRD patterns of pure and $NaIO_4$ salt complexed PEO/PVP blend films obtained under identical experimental conditions; (b) FTIR spectra for polymer electrolyte films: pure PEO/PVP blend (i) and complexed (PEO/PVP):$NaIO_4$ (ii).

The low-wavenumber region 800 cm^{-1} – 1000 cm^{-1} of the FTIR spectra provides information about modifications in local structure of the host polymer due to the interaction of the salt with polymer backbone. The vibrational modes present at around 845 cm^{-1} and 947 cm^{-1} (indicated in figure 1 b with arrows) are accountable for CH_2 rocking motion with a little C–O stretching motion and C–O stretching motion with some contribution from CH_2 rocking motions, respectively. The intensities of these two characteristic bands are associated with helical structure of PEO [15,16]. The relatively weak band at 1350 cm^{-1} signifies the amorphous portion in PEO matrix, whose intensity increased with the salt concentration. It evidences for increase of amorphous portion in the matrix of PEO polymer. In the FTIR spectra, the presence of all allowed vibrational modes and co-existence of well resolved respective band positions corresponding to ether oxygen groups (C–O–C) of PEO and carbonyl groups (C=O) of PVP are in congruent with previous reports in the literature and support the presence of PEO, PVP and their miscibility in resultant blend electrolyte films [17,18].

3.2. Dielectric properties

Dielectric materials are recognized as media that have an ability to store electrical energy. This property can be estimated by measuring the permittivity of dielectric materials. In alternating electric fields, the relative permittivity exhibits complex behaviour and is defined by $\varepsilon^* = \varepsilon' - j\varepsilon''$. The real ($Z_r$) and imaginary ($Z_i$) parts of complex impedance (Z^*) can be used for evaluation of real (ε') and imaginary (ε'') parts of dielectric permittivity by the following equations [19]:

$$\varepsilon' = \frac{-Z_i}{\omega C_0(Z_r^2 + Z_i^2)} \tag{1}$$

and

$$\varepsilon'' = \frac{Z_r}{\omega C_0(Z_r^2 + Z_i^2)} \tag{2}$$

Here, C_0 is the vacuum capacitance of the sample given by $\varepsilon_o A/d$, where $\varepsilon_o = 8.85 \times 10^{-12}$ Fm^{-1} is the permittivity of free space, A and d are the active area and the thickness of the sample, respectively. The angular frequency is $\omega = 2\pi f$, where f is the frequency of the applied electric field. ε' has the same significance as that of the ordinary dielectric constant of the material. It measures the energy stored in the material during each charging cycle, to be returned to the electric field at the end of the cycle.

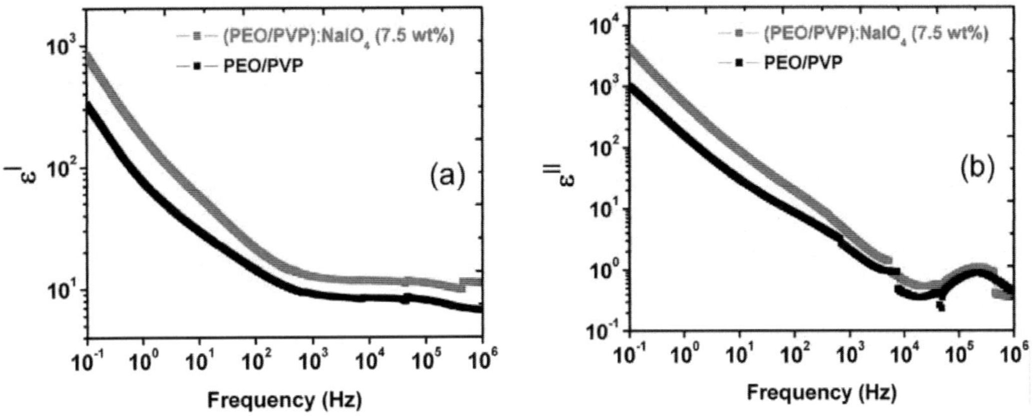

Figure 2. Variation of real (a) and imaginary (b) parts of dielectric constant as a function of frequency for PEO/PVP blend (dark lines) and salt complexed (PEO/PVP):$NaIO_4$ electrolyte (red lines).

Figure 2(a) shows the frequency dependence of dielectric constants of pure and salt complexed blend films studied here. The high values of the dielectric constant in the low-frequency region can be attributed to a build-up of space charge near electrode/electrolyte interface, which partially blocks the charge transport [20]. The dielectric constant decreases with the increasing frequency, more steeply at lower frequencies and less at higher frequencies. It reveals that, at very high frequencies (e.g., ~ 1 MHz) a periodic reversal of the field takes place so rapidly that the charge carriers will hardly able to orient themselves in the field direction resulting in the observed decrease of dielectric constant [21].

As seen from figure 2(a), the dielectric constant of (PEO/PVP):$NaIO_4$ polymeric complex is greater than that of pure PEO/PVP. The increase in the dielectric constant can be attributed to fractional increase in the number of charges due to the addition of salt, resulting also in a higher conductivity. The enhanced dielectric and electric response suggest that the studied (PEO/PVP):NaIO4 composite is a promising material for soft electronics and applications such as SPEs.

3.2. Electric modulus

Further information about the electrical properties and the conductivity relaxation behaviours of the considered SPEs can be obtained from the complex electric modulus M^*. The real (M') and imaginary (M'') parts of M^* can be calculated from the impedance data using the following relations [22,23]:

$$M' = \frac{\varepsilon'}{\varepsilon'^2 + \varepsilon''^2} \tag{3}$$

and

$$M'' = \frac{\varepsilon''}{\varepsilon'^2 + \varepsilon''^2} \tag{4}$$

where ε' and ε'' are the real and imaginary parts of the complex dielectric function ε^* that depends on frequency, temperature and structure. At low frequency, both ε' and ε'' are very high due to electrode polarization effect. Namely to minimize the effect of electrode polarization, Macedo et al have established the electric modulus formalism [23]. Generally, in SPEs the movement of ions from one site to another will perturb the electric potential of the surroundings. Motion of the other ions in this region will be affected by perturb potential. Such a cooperative motion of ions will lead to non-exponential decay, or a conduction processes with distribution of relaxation time.

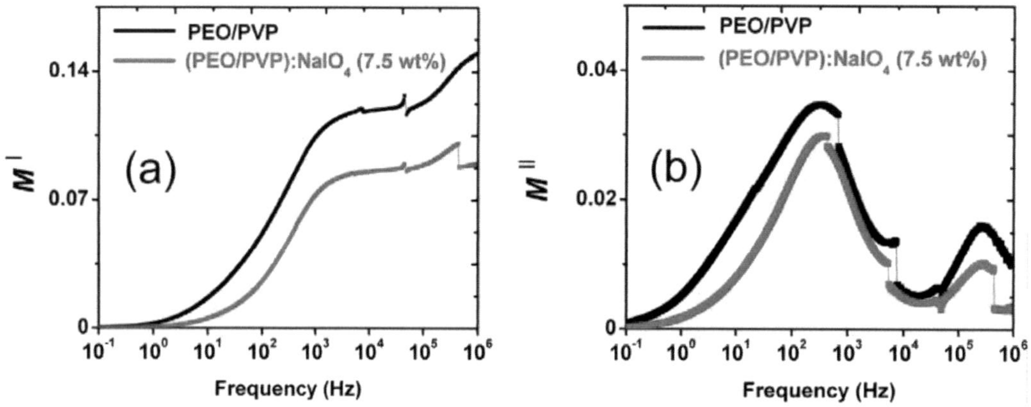

Figure 3. Frequency dependence of real (a) and imaginary (b) parts of electric modulus for pure and $NaIO_4$ salt complexed PEO/PVP electrolytes.

The frequency dependence of the electric modulus of pure and $NaIO_4$ salt complexed PEO/PVP blend electrolytes is shown in figure 3. Evidently, the real part (M') of the electric modulus values of the samples increase with an increase frequency and saturate at higher frequency region (figure 3 a). The spectra of the imaginary parts (M'') of the electric modulus of the studied electrolyte samples displayed distinct relaxation peaks (figure 3 b), which are related to the conductivity processes, whereas no corresponding feature can be seen in the dielectric loss spectra as depicted in figure 2(b). It evidences, that ionic and polymer segmental motions are strongly coupled. The presence of relaxation peaks in M'' spectra indicate that the considered materials are potential ionic conductors. The low-frequency side of the peaks represents the range of frequencies in which charge carriers are mobile over long distances. The high-frequency side of the M'' peaks represents the range of frequencies in which the ions are spatially confined to their potential wells and the ions can make only short-range motion with in the wells. The frequency range where the peak occurs is indicative of the transition from long-range to short-range mobility by increasing frequency of the applied electric field [24]. The low value of M'' in the low-frequency region suggests that the electrode polarization makes negligible contribution to electric modulus. The broad nature of the M'' peaks can be interpreted as being the consequence of distribution of relaxation time. Thus, the electric modulus studies could be used to detect the percolation threshold phenomena [25].

4. Conclusions

Pure and $NaIO_4$ (7.5 wt.%) salt complexed PEO/PVP blend electrolyte films prepared by solution cast technique were studied by means of impedance spectroscopy. Measured under identical experimental conditions, thin films of (PEO/PVP):$NaIO_4$ exhibit an enhanced dielectric response as compared to pure PEO/PVP. The distinct peaks present in the spectra of imaginary parts of electric modulus can be attributed to a strong coupling between ionic and polymer segmental motions. The clear effect from the inclusion of the $NaIO_4$ salt resulting in the increasing value of dielectric constant of the studied (PEO/PVP):$NaIO_4$ blends, indicates their usefulness as a material for SPEs. A further advantage of this flexible polymeric composite is the possibility to be relatively easily optimized (compositionally and structurally) in order to enhance its SPE performance, e.g., by increase of the concentration of $NaIO_4$ salt complexed in the PEO/PVP, thereby increasing the amorphous portion in the (PEO/PVP):$NaIO_4$ SPE.

Acknowledgements: The authors acknowledge the INERA (EU project Research Potential (FP7-316309-REGPOT-2012-2013-1) support for their participation at 19th ISCMP "Advances in Nanostructured Condensed Matter: Research and Innovations", Aug 28th – Sept 2nd, 2016, Varna, Bulgaria. Work supported by research project DFNI-TO2/18 "Nanostructured liquid crystals for tuneable photonic devices" by the Ministry of Education and Science, National Science Fund of Bulgaria.

References

[1] Tarascon J M and Armand M 2001 *Nature* **414** 359
[2] Zhang J J, Zhao J H, Yue L P, Wang Q F, Chai J C, Liu Z H, Zhou X H, Li H, Guo Y G, Cui G L and Chen L Q 2015 *Adv. Energy Mater.* **5** 1501082
[3] Liu W, Liu N, Sun J, Hsu P -C, Li Y, Lee H –W and Cui Y 2015 *Nano Lett.* **15** 2740
[4] Xu J, Dou S, Liu H and Dai L 2013 *Nano Energy* **2** 439
[5] Palomares V, Serras P, Villaluenga I, Hueso K B, Carretero-Gonzalez J and Rojo T 2012 *Energy & Environ. Sci.* **5** 5884
[6] Wang L, Lu Y, Liu J, Xu M, Cheng J, Zhang D and Goodenough J B 2013 *Angew. Chem.* **52** 1964
[7] Armand M 1994 *Solid State Ionics* **69** 309
[8] Fenton D F, Parker J M and Wright P V 1973 *Polymer* **14** 589
[9] Patel M, Chandrappa K G and Bhattacharyya A J 2010 *Solid State Ionics* **181** 844
[10] Mohan V M, Raja V, Bhargav P B, Sharma A K and Rao V V R N 2007 J. *Polym. Res.* **14** 283
[11] Kiran K K, Ravi M, Pavani Y, Bhavani S, Sharma A K and Narasimha Rao V V R 2011 *Phys. B* **406** 1706
[12] Zardalidis G, Ioannou E, Pispas S and Floudas G 2013 *Macromolecules* **46** 2705
[13] Kumar K N, Saijyothi K, Kang M, Ratnakaram Y C, Krishna K H, Jin D and Lee Y M 2016 *Appl. Phys. A* **122** 698
[14] Agrawal R C and Hashmi S A 2007 *Ionics* **13** 295
[15] Wright P V 1998 *Electrochim. Acta* **43** 1137
[16] Noor S A M, Ahmad A, Talib I A and Rahman M Y A 2010 *Ionics* **16** 161
[17] Zain N M and Arof A K 1998 *Mater. Sci. Engineering B* **52** 40
[18] Simon G P, Shen Z and Cheng Y B 2003 *Europ. Polymer J.* **39** 1917
[19] Aziz S B, Abidin Z H Z and Arof A K 2010 *Express Polym. Lett.* **4** 300
[20] Bhaskaran R, Selvasekarapandian S, Hirankumar G and Bhuvaneswari M 2004 *Ionics* **10** 129
[21] Raga H M 2011 *Physica B* **406** 3759
[22] Kumar K K, Ravi M, Pavani Y, Bhavani S, Sharma A K and Narasimha Rao V V R 2012 *J. Non-Cryst. Solids* **358** 3205
[23] Okutan M and Sentürk E 2008 *J. Non-Cryst. Solids* **354** 1526
[24] Nobre M A L and Langfredi S J 1999 *Phys. Chem. Solids* **62** 20
[25] Nadkarni G S and Simmons J G 1970 *J. Appl. Phys.* **41** 545

New cyclometalated Iridium(III) beta-dicetone complex as phosphorescent dopant in Organic light emitting devices

P Ivanov*[1], P Petrova[1], S Stanimirov[2] and R Tomova[1]
[1] Institute of Optical Materials and Technolologies "Acad. J. Malinowski", Bulgarian Academy of Sciences, Acad. G. Bonchev str. bl. 109, 1113 Sofia, Bulgaria
[2] Laboratory of Organic Photochemistry, Faculty of Chemistry and Pharmacy, University of Sofia, 1 James Baurchier blvd., 1164 Sofia

E-mail: ivanov.petar@mail.bg

Abstract. A new Bis[4-(benzothiazolato-N,C2′-2-yl)-N,N-dimethylaniline]Iridium(III) acetylacetonate $(Me_2N\text{-bt})_2Ir(acac)$ was synthesized and identified by ^1H NMR and elemental analysis. The application of the new compound as a dopant in the hole transporting layer (HTL) of Organic light emitting diode (OLED) structure: HTL/EL/ETL, where HTL was N,N'-bis(3-methylphenyl)-N,N'-diphenylbenzidine (TPD), incorporated in Poly(N-vinylcarbazole) (PVK) matrix, EL - electroluminescent layer of Bis(8-hydroxy-2-methylquinoline)-(4-phenylphenoxy)aluminum (BAlq) and ETL - electron-transporting layer of Tris-(8-hydroxyquinoline) aluminum (Alq_3) or Bis[2-(2-benzothiazoly) phenolato]zinc ($Zn(btz)_2$). We established that the electroluminescent spectra of OLEDs at different concentrations of the dopant were basically the sum of the greenish-blue emission of BAlq and yellowish-green emission of Ir complex. It was found that with increasing of the dopant concentration the relative electroluminescent intensity of Iridium complex emission increased and this of BAlq decreased and as a result the fine tuning of OLED color was observed.

1. Introduction

During the past decade, intense research efforts have been devoted to the design of new emitting materials for organic light emitting devices (OLEDs). Phosphorescent OLEDs (PhOLEDs) offer an intriguing future for the next generation of flat-panel displays and lighting techniques because of their high quantum efficiencies relative to fluorescent OLEDs [1-2]. Among all emitters that have been investigated to date, phosphorescent complexes have attracted more attention than the fluorescent ones due to their capability of achieving 100% internal quantum efficiency [2-4]. Extensive investigations of this materials have focused on heavy metal complexes of Os(II), Pt(II) and Ir(III) [3, 4]. Mixing of the singlet and triplet excited states of these d^6 complexes via spin-orbit coupling, removes the spin-forbidden nature of the radiative relaxation of the triplet state, leading to high phosphorescence efficiencies [5]. The choice of iridium metal is of special interest because its stable and accessible oxidation and reduction states and quasi-octahedral geometry permitted introducing of specific ligands in a controlled manner. Varying the respective set of ligands on the iridium(III) center permits a variety of efficient emission wavelengths, such as blue, green, yellow or red, which can be used in color tuning and full-color applications [6-8].

2. Experimental

2.1 Instruments and measurements
^{1}H-NMR (600 MHz) spectra were recorded on a Bruker Avance II+ 600 at room temperature. The ^{1}H-NMR spectra were referenced to internal standard tetramethylsilane (TMS). The elemental analyses were made on Elementar Vario EL III. The UV and fluorescent spectra of the complexes in solutions were recorded on a Thermo Spectronic Unicam 500 spectrophotometer and a Varian Cary Eclipse fluorescence spectrophotometer, respectively. All measurements of OLED structures were performed at room temperature and under ambient atmosphere, without any encapsulation. The electroluminescent spectra (EL) and CIE (Commission internationale de l'éclairage) coordinates were obtained by Ocean Optics HR2000+ spectrometer. The current-voltage (I-V) curves were measured by programmable with Labview power supply. The luminescence (L) was determined in DC (direct current) mode and the light output was detected using a calibrated Hamamatsu silicon photodiode S2281-01. The electroluminescent efficiency (η_{EL}) were calculated by equations (1) and used for quantifying the properties of the OLEDs.

$$\eta_{EL} = L/I,\ cd/A \tag{1}$$

where *L* is the luminescence (in cd/m^2), I is current density (in A/m^2).

*2.2. Synthesis of Iridium complex*Ligand 1 (figure 1) was obtained according procedure reported in [9]. The other initial reagents and solvents were purchased from Merck and used without any purification. *Synthesis of μ-chloro-bridged iridium dimmer (2).*
Cyclometalated Ir(III) μ-chloro-bridged dimmer $(Me_2N\text{-}bt)_2Ir(\mu\text{-}Cl)_2Ir(Me_2N\text{-}bt)_2$ was synthesized according a method reported by Nonoyama [10]. 2-(4- dimethylaniline)benzo[d]thiazole (1) (2.2 equiv), iridium trichloride hydrate (1 equiv), dissolved in 2-ethoxyethanol: water = 3:1 were stirred for 20 h at 110 °C under argon. After cooling to the room temperature, the reaction mixture was filtrated, and the orange-yellow precipitate was obtained. Then the precipitate was washed with ethanol and hexane several times, and dried at room temperature.

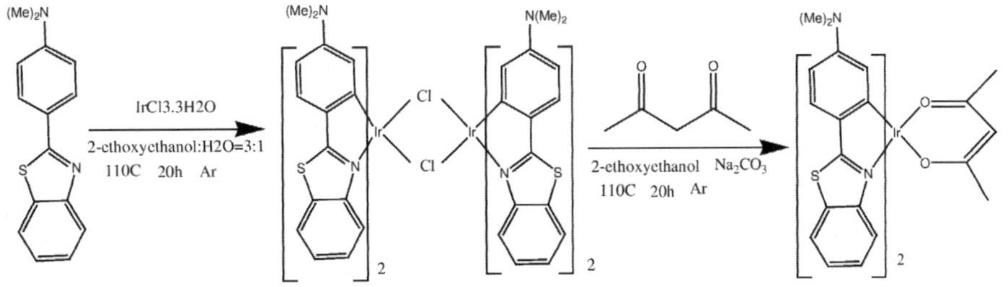

1 2 3

Figure 1. Synthesis of iridium acetylacetonate complex $(Me_2N\text{-}bt)_2Ir(acac)$.

2.2.2. Synthesis of iridium acetylacetonate complex (3), $(Me_2N\text{-}bt)_2Ir(acac)$.
Dimmer (2), 2,4-pentanedione (2.5 equiv) and Na_2CO_3 (5 equiv) were stirred for 20 h at 110 ^{0}C in 2-ethoxyethanol under argon. After cooling to the room temperature, the obtained product was filtrated giving an orange-yellow precipitate. Then the precipitate was washed with water, ethanol, and hexane several times. Finally, the crude product was purified using liquid chromatography (silica gel, dichloromethane) and dried.

Iridium(III)bis[4-(benzothiazolato-N,C$^{2'}$-2-yl)-N,N-dimethylaniline]acetylacetonate, $(Me_2N\text{-}bt)_2Ir(acac)$: Yield: 70 %.^{1}H NMR (600 MHz, $CDCl_3$) δ[ppm]: 1.71 (s, 6H), 2.51 (s,12H), 5.08

(s, 1H), 5.58 (d, J=2.4 Hz, 2H), 6.15 (dd, $J_{1=}$ 2.4 Hz, $J_{2=}$ 8.4 Hz, 2H), 7.20–7.22 (m, 2H), 7.26–7.28 (m, 2H), 7.39 (d, J=9 Hz, 2H), 7.69 (dd, J_1=0.6 Hz, J_2=7.8 Hz, 2H), 7.96 (dd, J_1=0.6 Hz, J_2=8.4 Hz, 2H); $C_{35}H_{33}N_4O_2S_2Ir$. Elemental analysis: Calc. C 52.68 %, H 4.17 %, N 7.02 %. Found: C 52.51 %, H 4.10 %, N 6.92 %.

2.3. OLED fabrication

Devices with area 1cm^2 were prepared on commercial polyethylene terephtalate (PET) substrate, precoated with ITO (40Ω/sq). The (30nm) composite films of PVK:$TPD_{10\ wt\%}$: $(Me_2N\text{-}bt)_2Ir(acac)x_{\ wt\%}$ (relatively to PVK) were formed by spin-coating from 0.75% PVK solution in CH_2Cl_2 at 2000 rpm. The films of BAlq (40nm), $Zn(btz)_2$ (35nm) and Al cathode (120nm), were deposited in vacuum, better than 10^{-4} Pa, by thermal evaporation at rates 2-5 Å/s, controlled with a quartz crystal microbalance.

3. Results and discussion

3.1. UV-VIS absorption and photoluminescence

In figure 2 the normalized UV-VIS absorption spectra of complexes $(bt)_2Ir(acac)$ and $(Me_2N\text{-}bt)_2Ir(acac)$ and photoluminescent spectra (PL) of: $(bt)_2Ir(acac)$, $(Me_2N\text{-}bt)_2Ir(acac)$, PVK and TPD dissolved in CH_2Cl_2 are presented. The absorption spectra of both complexes (curves 1 and 2) had almost identical shapes with peaks at 327 and 332nm (table 1), typical for that kind of complexes.

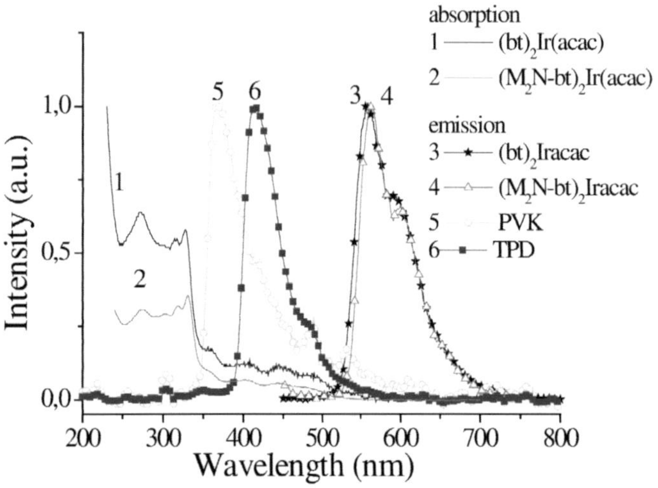

Figure 2. Normalized UV-VIS absorption spectra of: $(bt)_2Ir(acac)$ and $(Me_2N\text{-}bt)_2Ir(acac)$ and PL spectra of dissolved in CH_2Cl_2: $(bt)_2Ir(acac)$; $(Me_2N\text{-}bt)_2Ir(acac)$; PVK and TPD.

The spectra can be considered as composed of two parts. The region below 360 nm contains bands for electron transitions, which leads to excitations in the bt-ligand. Most probably, the band around 330 nm is due to the transition to lowest singlet ligand centered (^{1}LC) excited state. This statement is in accordance with published before articles discussing similar complexes. The shoulders above 360 nm are assigned to singlet and triplet metal-to-ligand charge transfer bands (^{1}MLCT, ^{3}MLCT), respectively. Effective spin-orbit coupling (SOC) in this complex leads to mixing of the singlet and triplet states making the band for resonant transition $^1S^0 \rightarrow {}^3$MLCT detectable in absorption spectrum (~480 nm). The band for ^{1}MLCT transition is centered at 440 nm. The both complexes showed emission in yellowish-green region (table 1) with peaks at 558 nm for $(bt)_2Ir(acac)$ and 563 nm for

IOP Conf. Series: Journal of Physics: Conf. Series **794** (2017) 012021 doi:10.1088/1742-6596/794/1/012021

$(Me_2N\text{-}bt)_2Ir(acac)$ and shoulders respectively at 594 and 604 nm (curves 3 and 4). It is seen that the PL spectrum of $(Me_2N\text{-}bt)_2Ir(acac)$ was red shifted with 5 nm with respect to the PL spectrum of the unsubstituted $(bt)_2Ir(acac)$, that is a typical for derivatives containing electron donating substituent [11, 12]. The PVK (curve 5) and TPD (curve 6) emitted a blue light with peaks at 367, 385 and 414 nm respectively. That's why it could be expected that energy transfer from PVK and TPD to both complexes would be possible since ^{1}MLCT and ^{3}MLCT absorption spectra of the dopants had a broad spectral overlap with fluorescent spectra of PVK and TPD. The PL spectra of thermal evaporated neat $Zn(btz)_2$ and BAlq films shown in figure 2 (curves 5 and 6) were very close to each other, with comparatively broad peaks centered at ≈ 490 nm.

Table 1. Photophysical data of $(bt)_2Ir(acac)$ and $(Me_2N\text{-}bt)_2Ir(acac)$ in Ar saturated CH_2Cl_2

Complex	*Absorption λ[nm],(lgε)**	*Emission λ_{max} [nm]*	*QY*
$(bt)_2Ir(acac)$	272(3.93), 315(3.86), 327(3.88), 407(3.21), 442(3.19), 489(3.03)	558	0.50
$(Me_2N\text{-}bt)_2Ir(acac)$	408 (4.41), 446 (4.08)	563	0.16

* ε is the molar absorption coefficient (molar absorptivity)

3.2. Electroluminescence

The electroluminescent spectra (EL) of various OLEDs viewed in the normal direction at a benchmark luminance of 18 V DC with structure doped $HTL_{(PVK:TPD)}$/BAlq/Alq_3 with different concentrations of $(Me_2N\text{-}bt)_2Ir(acac)$ are shown in figure 3. For comparison the normalized EL spectra of devices with undoped HTL and EL of BAlq, $Zn(btz)_2$ and BAlq/$Zn(btz)_2$ are presented in figure 4.

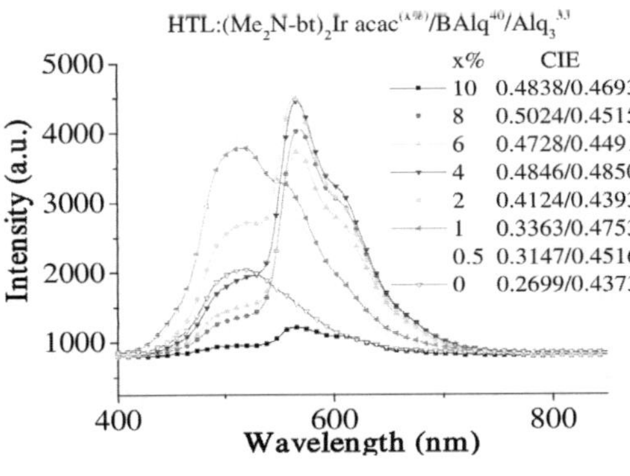

Figure 3. EL spectra of devices with structure: doped $HTL_{(PVK:TPD)}$/BAlq/Alq_3 with different concentrations of $(Me_2N\text{-}bt)_2Ir(acac)$, viewed in the normal direction at a benchmark of 18 V DC.

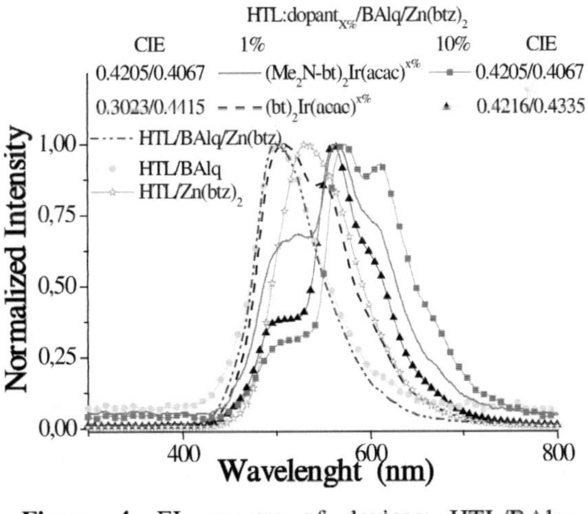

Figure 4. EL spectra of devices: HTL/BAlq; HTL/$Zn(bt)_2$ and doped $HTL_{(PVK:TPD)}$/BAlq/$Zn(btz)_2$ with 1 and 10 wt% of $(Me_2N\text{-}bt)_2Ir(acac)$ and $(bt)_2Ir(acac)$ devices, viewed in the normal direction at a benchmark of 18 V DC.

It is seen that device with EL of BAlq irradiated at λ_{max} 496 nm and with $Zn(btz)_2$ – at 524 nm (figures 4). The shape of EL of OLED with consequent evaporated layers of BAlq/$Zn(btz)_2$ was just the same as that of OLED with EL of BAlq. In our opinion, it is due to the fact that in the both cases the recombination zone is near to the HTL/BAlq interface. As the $HOMO_{Zn(btz)2}$ level (5,05 eV) was lower than $HOMO_{BAlq}$ (5,90 eV) level, $Zn(btz)_2$ didn't stop the holes, which flowed unimpeded from $Zn(btz)_2$ layer. By this reason in BAlq/$Zn(btz)_2$ structure $Zn(btz)_2$, acted the role only of ETL. The EL spectra of doped devices did not include any $Zn(btz)_2$, PVK and TPD emissions and were basically the sum of emissions of $(Me_2N\text{-}bt)_2Ir(acac)$ and BAlq. The absence of any emission peaks of PVK and TPD pointed out the effective charge trapping on the Ir(III) complexes and/or effective energy transfer to the Ir(III) complexes.

Increasing of the $(Me_2N\text{-}bt)_2Ir(acac)$ concentration caused decreasing of the relative intensity of the greenish-blue emission (496 nm) and increasing of the yellow emission (559 nm) of devices that suggests that the exciton recombination zone migrates from HTL/BAlq interface deeper in the doped HTL layer. The Commission Internationale de L'Eclairage CIE (x/y) coordinates of OLEDs shown in figure 3 were red shifted from yellow green 0,2699/0,4373 for undoped device structure: doped HTL/BAlq/Alq to orange - 0,3562/0,4224 at 10 wt% of $(Me_2N\text{-}bt)_2Ir(acac)$.

As can be seen in figure 4 at one and the same concentrations of the dopants the Full width at half maximum (FWHM) of EL spectra of devices with $(Me_2N\text{-}bt)_2Ir(acac)$ were broader with 42 nm in comparison with FWHM of devices doped with $(bt)_2Ir(acac)$ due to more intensive and widely red shifted with 35 nm shoulder. Obviously the replacement of an H atom in the 2-phenylbenzothiazole ligand with a dimethylamine radical leads to a strong red shifting of the emitted light which is typical for electron donating groups.

The influence of the dopant concentration on electroluminescent performance of the OLED structure HTL: $(Me_2N\text{-}bt)_2Ir(acac)$ are presented in figure 5. We established that with increasing of the dopant concentration at one and the same voltage the current density decreased while the luminescence initially increased and then decreased (figure 5 a). This shows that a small amount of dopant improves the current efficiency of device due to enhancing of the carrier mobility as can be seen in figure 5 b. As the dopant concentration goes beyond a certain value, the dopant tend to aggregate and the aggregations degrade the device performance [2]. The best performance had device doped with 1 wt% $(Me_2N\text{-}bt)_2Ir(acac)$ with its 533 Cd/m^2 luminescence; 95 A/m^2 current density and 5.69 Cd/A current efficiency demonstrated at 16 V DC in comparison with undoped device with its 253 Cd/m^2;70 A/m^2 and 3.51 Cd/A at the same voltage.

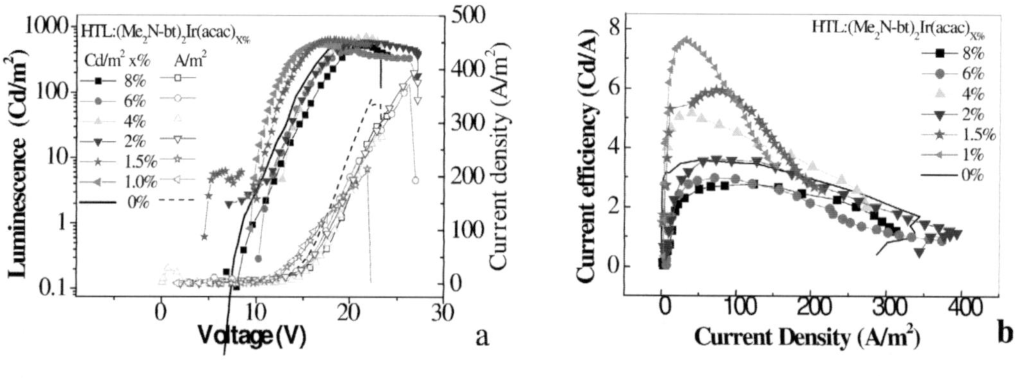

Figure 5. a) L-V-I and **b)** η_{EL} –I characteristics of devices with structure: doped HTL/BAlq/Alq_3 at different concentrations of $(Me_2N\text{-}bt)_2Ir(acac)$, viewed in the normal direction.

The electroluminescence in PhOLEDs comes either from the energy transfer from the host to the guest molecules, as described previously, or from the charge trapping by the guest molecules, where the dopant acts as a trap site which can capture a charged carrier. As was shown by Diouf at al [13], four main elements - exciton confinement, charge trapping, dopant concentration and molecular structure directly affect the performance of PhOLEDs. These factors are dependent each other and individually predominant depending on the situation. In our case weakly increasing of OLED's efficiency with increasing of dopant concentration are consistent with the charge trapping of the Ir complex, rather than the energy transfer, as the dominant mechanism in investigated OLEDs.

4. Conclusion

The new iridium complex $(Me_2N\text{-}bt)_2Ir(acac)$ irradiating in yellow region has been synthesized and characterised as a dopant in hole transporting layer of OLED. It was established that the fine tuning of the emitted light of OLED in yellow orange region can be obtained depending on the dopant concentration. It was found that the lightly-doped (around 1.5 wt%) samples exhibit better luminescence efficiency than the highly-doped samples at all applied voltages and exhibit peak luminescence efficiency at relatively low applied voltage in comparison with the undoped device.

References

[1] Wang Y, Herron N, Grushin V, LeCloux D and Petrov V 2001 *Appl.Phys. Lett.* **79** 151
[2] Feng Sh-W, Shih M-Ch, Huang C and Chung Ch-T 2009 *Thin Solid Films* **517** 2719
[3] Kappaun S, Slugovc Ch and List E J 2008 *Int. J. Mol. Sci.* **9** 1527
[4] You Y., An Cheng-Guo , Lee D, Kim J and Park S 2006 *J. Mater. Chem.* **16** 4706
[5] Baldo M, Lamansky S, Burrows P, Thompson M and Forrest S 1999 *Appl. Phys. Lett.* **75** 4
[6] Thompson M, Lamansky S, Djurovich P, Murphy D, Abdel-Razzaq F, Kwong R, Forrest S, Baldo M and Burrows P US Patent 20020034656A1
[7] Grushin V, Herron N, Lecloux D, Marshall W, Petrov V and Wang Y 2001 *Chem. Commun. (Cambridge)* 1494
[8] Chang W, Hu A, Duan J, Rayabarapu D and Cheng C 2004 *Organomet J. Chem.* **689** 4882
[9] Deligeorgiev T, Kaloyanova S, Vasilev A and Vaquero J 2010 *Phosphorus Sulfur* **185** (11) 2292
[10] Nonoyama M 1974 *Bull. Chem. Soc. Jpn.* **47** 767
[11] Chen C and Shi, J 1998 *Coord. Chem. Rev.* **171** 161
[12] Wang. R, Deng L, Zhang T and Li J 201 *Dalotn Trans.* **41** 6833
[13] Diouf B, Jeon W, Pode R and Kwon J 2012 *Hindawi Publishing Corporation Advances in Materials Science and Engineering* doi:10.1155/2012/794674

IOP Conf. Series: Journal of Physics: Conf. Series (2017) 012022 doi:10.1088/1742-6596/794/1/01

Thin films from hydrophilic poly(N,N-dimethyl acrylamide) copolymers as optical indicators for humidity

K Lazarova[1], L Todorova[2], D Christova[2], M Vasileva[1], R Georgiev[1], V Madjarova[1] and T Babeva[1]

[1]Institute of Optical Materials and Technologies ''Acad. J. Malinowski'', Bulgarian Academy of Sciences, Acad. G. Bonchev str., bl. 109, 1113 Sofia, Bulgaria.
[2]Institute of Polymers , Bulgarian Academy of Sciences, Acad. G. Bonchev Str., bl. 103-A, 1113 Sofia, Bulgaria.

babeva@iomt.bas.bg

Abstract. In the present paper we study thin films from poly(N,N-dimethyl acrylamide)-poly(ethylene oxide) (PDMAA/PEO) copolymers of different composition and structure in order to implement them as sensitive media for optical indicators for humidity. PDMAA/PEO di- and triblock copolymers were synthesized via redox polymerization in aqueous media. Thin films were deposited on silicon substrates by spin coating method using polymers solutions with appropriate concentrations. Refractive index, extinction coefficient and thickness of the films are calculated from reflectance spectra of the films deposited on silicon substrates using non-linear curve fitting method. Sensing properties of the films were tested by films exposure to different humidity levels followed by *in-situ* monitoring of the changes in the optical properties. The influence of the polymer structure and postdeposition annealing on the optical and sensing properties of the films was investigated. The potential application of selected polymers for optical sensing of humidity were demonstrated and discussed.

1. Introduction

The quantity of water vapors found in the air can vary drastically from almost zero to the point of saturation. Low or high humidity, or changes in-between, can harm sensitive materials and objects, affect many manufacturing processes in industries or disturb human comfort. Traditional humidity sensors are generally resistive and capacitive one based on the electrical measurement. Despite some recent progress in room temperature sensing [1-4], generally electric sensors suffer from high working temperature and lack of selectivity. Besides, most of the traditional sensors have relatively low accuracy due to the cross-temperature feature. In recent years an optical sensing of humidity emerges as an alternative approach. It offers simplicity of detection, room temperature detection, high accuracy and sensitivity, and immunity to electromagnetic interference or explosion danger. Various measuring scheme have been developed so far that operate on the basis of fibre-optic technology [5-10], surface plasmon resonance [11], microring resonator and whispering modes [12-14], Fabry-Perot resonances [15, 16], holographic grating [17, 18], photonic crystals [19, 20], etc. Great variety of materials that change their refractive index, extinction coefficient or thicknesses are implemented as sensitive media: tungsten disulfide [6], SiO_2 nanoparticles [7], porous SiO_2 [11], undoped and dye doped polyvinyl alcohol [8, 15], polyimide [9], agarose gel and hydroxyethylcellulose / polyvinylidenefluoride [10], elastic polymer SU-8 [12], poly(acrylic acid) / polyethyleneimine polymers [13], polyacrylamide microfiber rings [14], chitosan [16], undoped and zeolite doped self-developed polyacrylamide / poly(vinyl alcohol) photopolymer [17, 18], alternating thin films of titania and poly(2-hydroxyethyl methacrylate-co-glycidyl methacrylate) (PHEMA-co-PGMA) [19], PHEMA and epoxy based resist [20], self-assembled poly(styrenesulfonate-methylbutylene) (PStS-b-PMB) block copolymers with tailored hygroscopic properties [21], graphene oxide [4]. Among them, polymers are receiving considerable attention because of the tailored functionality, easy deposition and relatively low cost. When they are prepared in thin film form they offer additionally fast response due to the short diffusion path length.

In this study we have chosen poly(N,N-dimethyl acrylamide) (PDMAA) because it is a hydrophilic polymer well known for its biocompatibility and good mechanical and film-forming properties [22]. PDMAA is less polar than polyacrylamide and hence highly soluble in water as well as in large number of organic solvents. Therefore strong swelling can be expected when exposed to water vapors. We expected that the combination of PDMAA with another hydrophilic polymer such as poly(ethylene oxide) will extend further their ability for water sensing.

In this paper we propose and demonstrate experimentally the optical sensing ability of hydrophilic poly(N,N-dimethyl acrylamide) / poly(ethylene oxide) block copolymers. The optical and sensing properties of spin-coated films are studied as a function of postdeposition annealing temperature (60 °C, 120 °C and 180 °C) and relative humidity (5% RH and 95 % RH). The implementation of films as active media for color indicators for humidity is demonstrated and discussed.

2. Experimental part

2.1. Synthesis of PDMAA and PEO-PDMAA copolymers

The monomer N,N-dimethyacrylamide (DMAA) and the initiator diammonium cerium(IV) nitrate were purchased from Sigma-Aldrich (Germany). Poly(ethylene oxide) methyl ether (PEO-ME) and poly(ethylene oxide) (PEO) both of molar mass 2000 g/mol were supplied from Fluka (Switzerland). The solvents and other reagents used were of standard laboratory reagent grade.

Double hydrophilic di- and triblock copolymers were synthesized by means of redox polymerization of DMAA in deionized water when using 3.3 g/L PEO-ME or respectively PEO as initiating moiety. The reaction mixture was purged with nitrogen for 30 min at ambient temperature and the initiator diammonium cerium (IV) nitrate (0.01 equiv. in respect to the PEO precursor) was added. The polymerization was carried out at 35 °C under nitrogen atmosphere and the obtained copolymers were purified by dialysis against water and finally isolated by freeze-drying. The reaction schemes of the synthetic procedures are shown in figure 1.

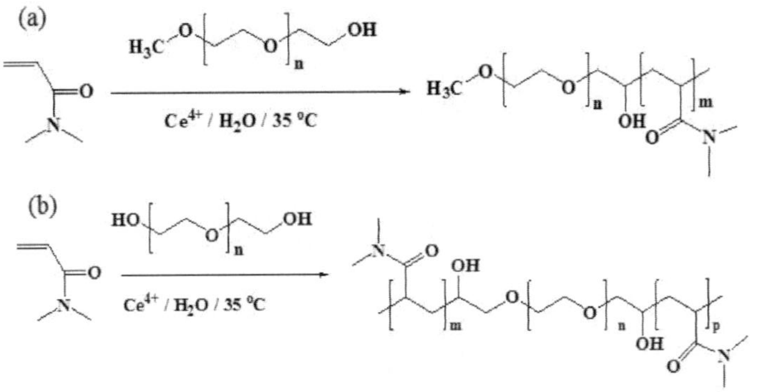

Figure 1. Reaction schemes of the synthesis of double hydrophilic block copolymers: (a) PEO-*b*-PDMAA; and (b) PDMAA-*b*-PEO-*b*-PDMAA.

PDMAA homopolymer was synthesized by aqueous free radical polymerization of DMAA initiated by 4,4-azobis (4-cyanovaleric acid) (ABCVA). The reaction was carried out at 60 °C under nitrogen atmosphere for 6 hours. The reaction mixture was evaporated and the obtained polymer was purified by precipitation from dichloromethane in cold diethyl ether.

The structure and composition of the prepared copolymers were analyzed by Nuclear Magnetic Resonance (NMR). The molar masses were calculated based on the integrated proton spectra when comparing the integrals of the signals of PEO and PDMAA repeating units. The compositions and molar masses (M_n) of diblock and triblock copolymers were PEO_{45}-*b*-$PDMAA_{37}$, M_n = 5700 g/mol and $PDMAA_{28}$-*b*-PEO_{45}-*b*-$PDMAA_{28}$, M_n = 7500 g/mol, respectively.

2.2. Thin film deposition and characterization

Polymer solutions of concentration 0.5 wt. % were prepared. PDMAA and PEO-*b*-PDMAA were dissolved in methanol-water mixture (80:20 volume ratio), while chloroform was used as a solvent for PDMAA-*b*-PEO-*b*-PDMAA. Thin films were deposited by the method of spin coating by dropping 0.2 ml of the polymer solutions onto preliminarily cleaned Si-substrates and spinning at a rate of 3000 min^{-1} for 30 s. Prior to characterization all films were subjected to annealing at temperatures of 60 °C, 120 °C and 180 °C for 30 min.

Optical constants (refractive index, *n* and extinction coefficient, *k*) along with film thickness, *d* were determined through non-linear curve fitting of reflectance spectra measured at normal light incidence by UV-VIS-NIR spectrophotometer (Cary 5E, Varian) [23]. The experimental errors for the three parameters, *n*, *k* and *d* were 0.005, 0.003 and 2 nm, respectively.

The humidity sensing performance was evaluated by measuring the reflectance spectra of the films at relative humidity (RH) of 5% and 95% using homemade bubbler system for generation of vapors from liquids [24]. The predetermined humidity levels were established in the measuring cell through flow of dry argon in the case of 5 % RH and flow of argon saturated with water vapors for reaching 95% RH. The measured reflectance spectra at both humidity levels were further used for determination of *n*, *k* and *d* of the films.

3. Results and discussion

The dispersion curves of the refractive index of the studied polymer films for three different postdeposition annealing temperatures (60°C, 120°C and 180 °C) are calculated from the measured reflectance spectra of the films [23] and are presented in figure 2. All films exhibit normal dispersion, i.e refractive index decreases with wavelength, that can be expected considering that films are transparent in the studied spectral range. The values of the extinction coefficient (*k*) vary in the ranges *k*=0.007 – 0.015. For all studied polymers there is no change of film thickness at 60 °C, while for 120 °C and 180 °C a decrease is observed for both di- and triblock polymers and no change for PDMAA films. For (PEO-*b*-PDMAA) the thickness changes from 61 nm (60 °C) to 49 nm at 120 °C and reaches value of 46 nm at 180 °C, while for PDMAA-*b*-PEO-*b*-PDMAA thin films *d* decreases from 90 nm at 60 °C to 87 nm and 77 nm for 120 °C and 180 °C, respectively.

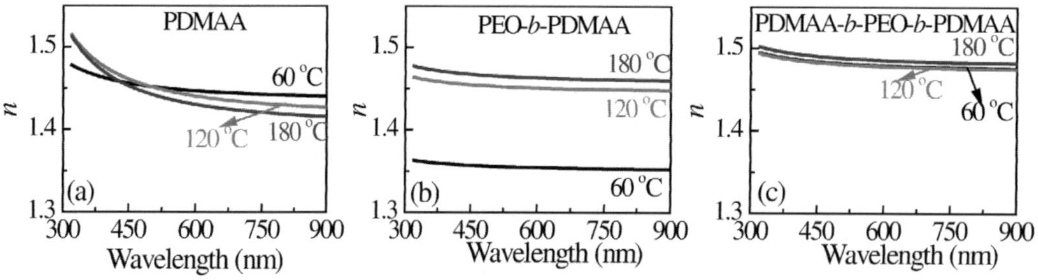

Figure 2. Refractive index of PDMAA (a), PEO-*b*-PDMAA (b) and PDMAA-*b*-PEO-*b*-PDMAA (c) thin films as a function of wavelength at dented postdeposition annealing temperatures.

The dependences of n and k (at wavelength of 600 nm) on the annealing temperature are displayed in figure 3. It is seen that the post deposition annealing influences in a different way the optical properties of polymer films. A decrease in n for PDMMA films is observed starting from 1.457 (60 °C), going through 1.440 (120 °C) and finally reaching 1.430 (180 °C), while for PEO-*b*-PDMAA and PDMAA-*b*-PEO-*b*-PDMAA films n increases substantially in the first case – starting from 1.355 going through 1.451 and reaching 1.463, and slightly for the second one: from 1.479, through 1.478 to 1.485 (all values are for wavelength of 600 nm).

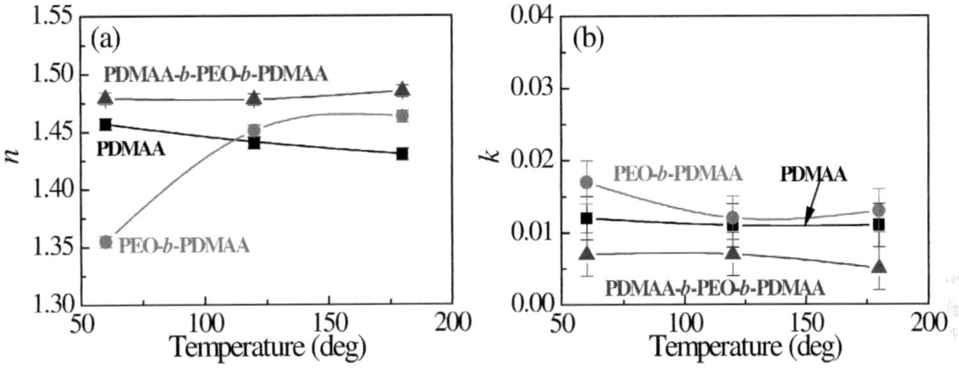

Figure 3. Refractive index (a) and extinction coefficient (b) at wavelength of 600 nm with respective errors as a function of post deposition annealing temperature.

Usually the postdeposition annealing leads to densification of the films due to the removal of residual solvent and consequent contraction of the film that is consistent with the reduction of film thickness. This is the possible reason for optical behavior of PEO-*b*-PDMAA and PDMAA-*b*-PEO-*b*-PDMAA films where n increases at higher annealing temperatures due to the increase of film density. It is seen from figure 3(a) that the change is mostly pronounced for diblock copolymer films that have been prepared using methanol:water mixture as solvent as compared to triblock copolymer films prepared using the more volatile compound (chloroform). The observed decrease in n for PDMAA thin films indicates no contraction of the film with temperature that is confirmed with the obtained similar values for the thickness at different temperaures. Thus the removal of residual solvent accompanied with no contraction lead to an increase of free volume of the film and decrease of its effective refractive index.

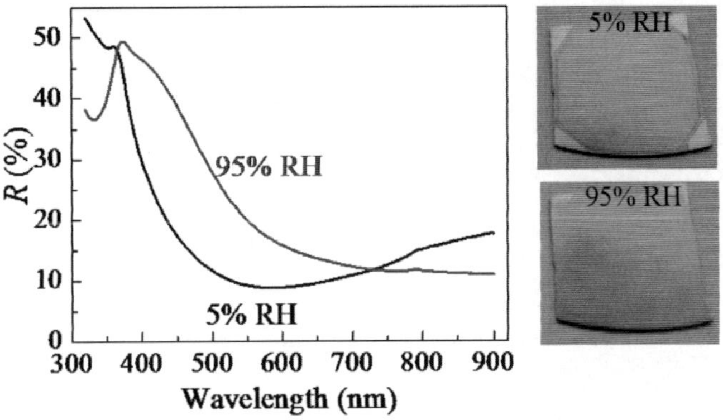

Figure 4. Reflectance spectra of PDMAA-*b*-PEO-*b*-PDMAA thin film measured at relative humidity of 5% and 95% and pictures of respective sample at denoted humidity.

According to the extinction coefficient, there is no dependence on the annealing temperature, or on the polymers structure. The k-values vary from 0.007 for the triblock copolymer films to 0.011 and 0.012 for PDMAA and diblock copolymer, respectively and are almost the same in the framework of the experimental errors (Fig. 3(b)).

Figure 4 presents the typical example for the sensing experiments. The reflectance spectra of the film (PDMAA-*b*-PEO-*b*-PDMAA is chosen for illustration) are measured at two different levels of relative humidity, 5% RH and 95 % RH, supplied by the bubbler system explained elsewhere [24]. At high humidity there is a significant red shift of the spectra that leads to a considerable change of the color of the film from light brown to light blue. The change of the color is instant and reversible. This means that there is no condensation of water inside the film; otherwise the film will be destroyed from the local dissolution of the polymer. Therefore no increase of refractive index can be expected due to replacement of air inside the film with water. Then the swelling of polymer is the only reason for the observed red shift.

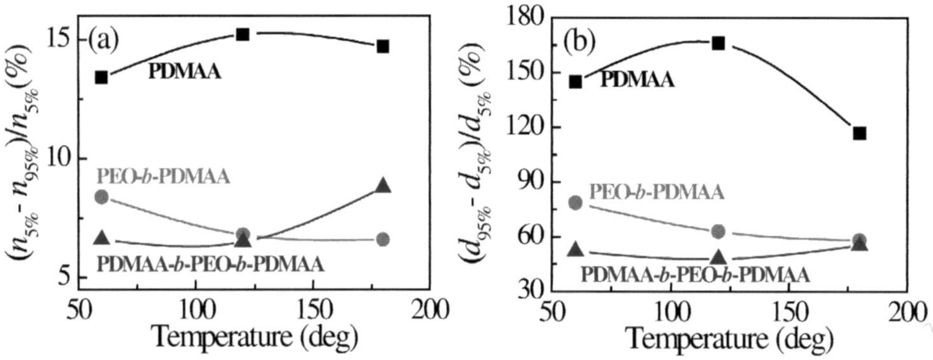

Figure 5. Relative change in refractive index (a) and thickness (b) of denoted thin polymer films due to exposure to relative humidity from 5 % RH to 95 % RH as a function of temperature of postdeposition annealing.

Reflectance spectra of all films are measured and the change of refractive index and thickness are calculated using already developed programs [23]. The results are shown in figure 5 as a function of postdeposition treatment. For all films there is a decrease in n and increase in d when relative humidity changes from 5% to 95 % RH. The strongest humidity response is obtained for PDMAA films annealed at 120 °C: n and d change with 15 % and 166 %. However compared to the homopolymer, the block copolymers offer possibility to improve the film properties particularly by optimizing macromolecular characteristics and block copolymer architecture (linear, branched, graft, etc.) which will be explored in the following studies of team. For PEO-*b*-PDMAA films the strongest response of n (8.4%) and d (79 %) are obtained at 60 °C, while for PDMAA-*b*-PEO-*b*-PDMAA the preferred treatment is annealing at 180 °C where the changes in n and d are 8.8 % and 55.1%, respectively. As expected significant swelling of all polymers is obtained that is accompanied with decrease of n due to the density reduction. The increase of relative humidity leads to an increase of extinction coefficient mostly pronounced for PDMAA films, where k changes almost twice - from 0.011 to 0.021.

It is also worthwhile to note that PEO homopolymer films have also been investigated. However the thin films are unstable and even small amounts of humidity present in the air dissolved them.

Conclusion

The successful application of thin films of poly(N,N-dimethyl acrylamide) copolymers as optical indicators for humidity is demonstrated. The change in humidity from 5 % to 95 % RH leads to color change of films that can be detected by naked eye. The color change is due to significant film swelling resulting in red shift of the reflectance spectra minimum. The swelling is also accompanied with decrease of refractive index owning to the reduction of film density. It is revealed that the structure and annealing temperature influence in different ways the optical response of polymer films. The highest changes are obtained for PDMAA films annealed at 120 °C where the thickness increases from 59 nm to 157 nm while the refractive index decreases from 1.47 to 1.25 (the values are for wavelength of 600 nm). However, the copolymers' performance can be improved considerably by optimization of the molar mass, chemical composition and structure, etc. The easy preparation, good sensitivity and fast response of hydrophilic poly(N,N-dimethyl acrylamide) copolymers provide promising opportunity for their application as optical indicators for humidity.

Acknowledgements: The financial support of the project DFNP-199/14.05.2016 of the Program for career development of young scientists, BAS is highly appreciated.

References

[1] Georgieva B, Podolesheva I, Spasov G and Pirov J 2014 *Sensors* **14** 8950-60
[2] Georgieva B, Podolesheva I, Zhivkov I, Strijkova V and Platikanova V 2007 *Journal of Optoelectronics and Advanced Materials* **9** 498-500
[3] Zhang D, Sun Y, Li P and Zhang Y 2016 *ACS Appl. Materials and Interfaces* **8** 14142-49
[4] Zhang D, Chang H, Li P, Liu R and Xue Q 2016 *Sensors and Actuators B: Chemical* **225** 233-40
[5] Yeo T L, Sun T and Grattan K T V 2008 *Sensors and Actuators* A **144** 280–95
[6] Luo Y, Chen C, Xia K, Peng S, Guan H, Tang J, Lu H, Yu J, Zhang J, Xiao Y and Chen Z 2016 *Optics Express* **24** 8956-66
[7] Gomez D, Morgan S, Hayes Gill B and Korposh S 2016 *Proc. SPIE* 9916 Art.№ 991623
[8] Sun H, Yang Z, Zhou L, Liu N, Gang T, Qiao X and Hu M 2015 *Opt. Comm.* **356** 556-9
[9] Alwis L, Sun T and Grattan K 2013 *IEEE Sensors Journal* **13** 767–71.
[10] Batumalay M, Harun S, Irawati N, Ahmad H and Arof H 2015 *IEEE Sensors Journal* **15** 1945-50
[11] Li Z, Meng X, Piao R, Zhao J, Tong K, Gu E and Li W, 2015 *Guangzi Xuebao/Acta Photonica Sinica* **44** Art.№ 0624001
[12] Eryürek M, Karadag Y, Anand S, Kilinç N and Kiraz A 2015 *Progress in Electromagnetics Research Symposium* **2015** 892-5
[13] Zhang J, Zhong J, Fang Y, Wang J, Huang G, Cui X and Mei Y 2014 *Nanoscale* **6** 13646-50
[14] Wang P, Gu F, Zhang L and Tong L 2011 *Applied Optics* **50** G7-G10
[15] Kumari M, Ding B and Blaikie R 2015 *Proc. SPIE* 9668 Art.№ 96680T
[16] Chen L, Li T, Chan C, Menon R, Balamurali P, Shaillender M, Neu B, Ang X, Zu P, Wong W and Leong K 2012 *Sensors and Actuators* B **169** 167–72.
[17] Naydenova I, Jallapuram R, Toal V and Martin S 2008 *Appl Phys Lett.* **92** 031109
[18] Naydenova I, Grand J, Mikulchyk T, Martin S, Toal V, Georgieva V, Thomas S and Mintova S 2015 *Chemistry of Materials* **27** 6097-6101
[19] Wang Z, Zhang J, Xie J, Li C, Li Y, Liang S, Tian Z, Wang T, Zhang H and Li H 2010 *Adv. Funct. Mater* **20** 3784–90
[20] Georgaki M, Botsialas A, Argitis P, Papanikolaou N, Oikonomou P, Raptis I, Rysz J, Budkowski A and Chatzichristidi M 2014 *Microelectronic Engineering* **115** 55–60
[21] Kim E, Kim S, Jo G, Kim S and Park M 2012 *ACS Appl. Mater. Interfaces* **4** 5179–87
[22] Huang S-Y, Lipp D W and Farinato R S 2001 *Acrylamide Polymers In: Encyclopedia of Polymer Science and Technology* New York: John Wiley & Sons Inc. 2001 vol 1 pp 41-79
[23] Lazarova K, Vasileva M, Marinov G and Babeva T 2014 *Optics & Laser Technology* **58** 114–8
[24] Lazarova K, Awala H, Thomas S, Vasileva M, Mintova S and Babeva T 2014 *Sensors* **14** 12207-18

Novel ammonia sensor based on polyaniline/polylactic acid composite films

S Sotirov, I Bodurov and M Marudova*

Plovdiv University "Paissi Hilendarski", 24 Tzar Assen str., 4000 Plovdiv, Bulgaria

*E-mail: margo@uni-plovdiv.bg

Abstract. We propose a new type of ammonia sensor based on composite film between polyaniline (emeraldine base) dissolved in dimethylformamide, and poly(DL-lactic) acid dissolved in chloroform. The two solutions were mixed in weight ratio of the components 1:1 and cast on Al_2O_3 substrate, on which silver electrodes were deposited previously. The active layer structure and morphology were examined by atomic force microscopy. The sensor resistance at constant humidity and different ammonia concentrations was measured. It was found that an increase in the ammonia concentration leads to resistance increase. This result is explained in the terms of ionic interactions between the polyaniline and the ammonia, which change the permittivity of the sensor active media. A response between 2% and 590% was shown depending on the ammonia concentration. The sensor is reversible and possesses response time of typically 100 s. Based on the changes of the sensor resistance, ammonia concentration from 10 ppm to 1000 ppm could be detected.

1. Introduction

Ammonia (NH_3) is one of the most widely produced inorganic chemicals, which is extensively used in many areas such as fertilizer production, clinical diagnosis and food processing [1]. However, ammonia is also a poisonous gas, which can cause both long term and immediate human health problems. For instance, even short-term exposure to 1000 ppm or above would lead to fatal respiratory system and lung disorder [2]. That is why there is an increasing interest in recent years in the development of ammonia detecting sensors.

There are several analytical techniques for gas detection - gas/liquid chromatography coupled with different detectors [3], ion mobility spectrometry [4], photoacoustic spectroscopy [5], surface-enhanced Raman scattering [6], and capillary electrophoresis [7], which usually use expensive, bulky and quite sophisticated equipment. Rapid real-time detection of the electronic sensors is a promising alternative to time-consuming laboratory techniques. The principle of operation involves conversion of chemical changes in sensing layer into electrically measurable response signal [8]. In this regard the application of conducting polymers, which reversibly change their characteristics depending on atmosphere composition, is a promising direction in development of sensing materials [9]. Among the family of conducting polymers, polyaniline (PANI) has attracted much attention due to its easy synthesis methods, good environmental stability, and because its surface charge characteristics can be easily tailored by changing the dopant during synthesis [10].

The PANI polymers exist in two forms – PANI salt and PANI base. Most of the ammonia sensors are produced using the salt form. In this case the contact of acid doped PANI with basic substances leads

IOP Conf. Series: Journal of Physics: Conf. Series **794** (2017) 012023 doi:10.1088/1742-6596/794/1/012023

to its dedoping by deprotonation and can be used for ammonia/amines concentration determination [11]. Only a few attempts were done in the development of ammonia sensor working with PANI based form. The mechanism of the ammonia – PANI base interaction is proposed in figure 1 [12].

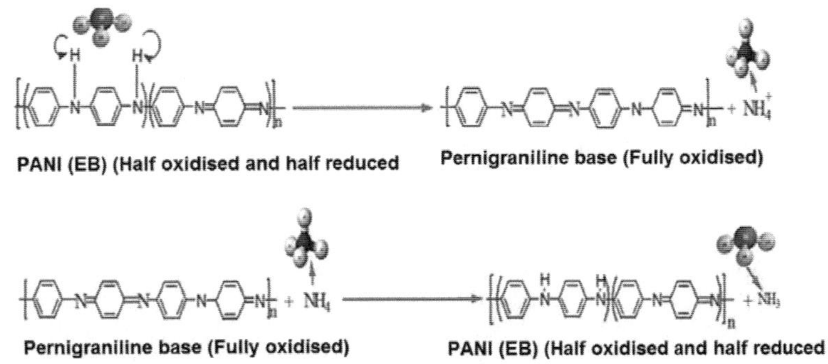

Figure 1. Proposed mechanism of ammonia PANI base interaction.

Polyaniline is a p-type material and when it interacts with NH_3, there is reduction in charge carrier density. Since the majority carriers (holes) density gets decreased due to electron donating nature of NH_3 gas. This results in decreasing the conductivity of material and film resistance increases [13].

A useful approach for the improvement of processability of conducting polymers involves blending them with soluble matrix polymers [14]. The blend films could potentially provide the following advantages: good mechanical stability; potential low cost; ease of measurement; operation at or near room temperature.

To the best of our knowledge and experience in this field, sensors based on polyaniline and polylactic acid have not been yet formulated and studied for detection of NH_3 gas. Following this idea the aim of the present work is developing of PANI base– polylactic acid blend films and investigating their sensing properties to ammonia.

2. Experimental

2.1. Materials

Polyaniline (emeraldine base) (PANI) with average molecular weight 65000 g/mol was purchased from Sigma-Aldrich and used without further purification and characterization. Poly(DL-lactic acid) (PDLA) with inherent viscosity 0.55-0.75 dL/g, delivered from Lactel Absorbable Polymers (USA), was used for the preparation of polymer matrix. All other chemicals were analytical grade purity.

2.2. Sensor fabrication

PANI/PDLA sensor was designed on Al_2O_3 substrate, on which silver electrodes were firstly deposited – figure 2. The dimensions of the substrate are 11 mm X 35mm, thickness 1mm. The electrodes are incorporated into two groups of 27 pieces in each; the distance between the electrodes is 0.2 mm. 2 %w/v of PANI was dispersed in dimethylformamide using a Compact Ultrasonic Laboratory Device UP100H (Hielscher Ultrasonics GmbH, Germany) for 30 min at room temperature. Then the dispersion was filtered through MF-Millipore™ Membrane Filters, pore size 0.45 μm. 2 %w/v of PDLA was dissolved in chloroform. The PANI dispersion and the PDLA solution were mixed by magnetic stirrer in weight ratio of the components 1:1 and cast on the Al_2O_3 substrate. In this way a sensing film with 1 μm thickness was constructed. Schematic structure of the ammonia sensor is shown in figure 3.

Figure 2. Al_2O_3 substrate with silver electrodes.

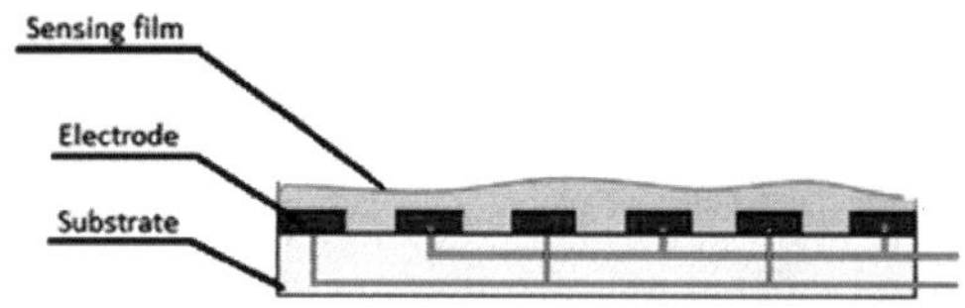

Figure 3. Schematic structure of the ammonia sensor.

2.3. Characterization of PANI sensor
The sensor structure and morphology were investigated by AFM Nanosurf EasyScan 2.

The gas sensing capabilities were carried out by High Precision LCR Meter GW INSTEK LCR-821. The changes in the resistance of the PANI/PDLA sensor was measured in the presence of air atmosphere and different ammonia concentrations. The humidity was maintained constant (RH 54 %) by use of desiccator. A known concentration of gas was injected in the desiccator by using a syringe. All measurements were done at room temperature. The response of the sensor was calculating using the following relation:

$$response(\%) = \frac{(R_g - R_a)}{R_a} \cdot 100, \quad (1)$$

where R_g is the sensor resistance when it is exposed to ammonia and R_a is the sensor's own resistance in air atmosphere at room temperature and RH 54 %.

3. Results and discussion

3.1. AFM analysis
Surface morphology of the PANI/PDLA sensor was obtained using AFM – figure 4. A homogeneous distribution of PANI beads with average diameter of 500 nm in the PDLA matrix was observed. The roughness of the surface was calculated as 12.8 nm. This type of morphology insured large surface of the active PANI compound and facilitated the adsorption of bigger amount of gas molecules. In this way the response of the sensor increased.

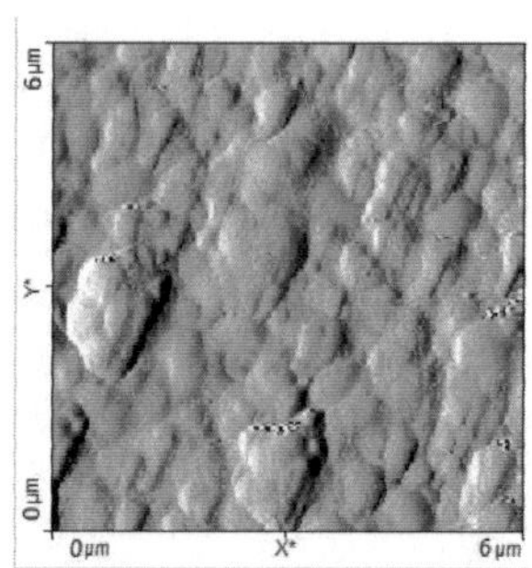

Figure 4. AFM image of PANI/PDLA sensor surface.

3.2. Ammonia sensing study
The PANI/PDLA sensor was subjected to successive injection of ammonia gas in the range 10 ppm – 1000 ppm. Typical changes of the sensor resistance at different ammonia concentrations are presented in figure 5. The resistance increased rapidly by exposing the sensor to ammonia and recovered towards

the original value typically 15 min after reverting to the air (this part of the process is not presented on the graphs), indicating good reversibility of the active PANI/PDLA blend film. The change was faster and bigger at higher ammonia concentrations. At ammonia concentration of 1000 ppm the resistance values reached a plateau and then slightly decreased even though the sensor was still exposed to ammonia. This observation could be explained by the specific interactions between the reactive sites of PANI and the ammonia [12].

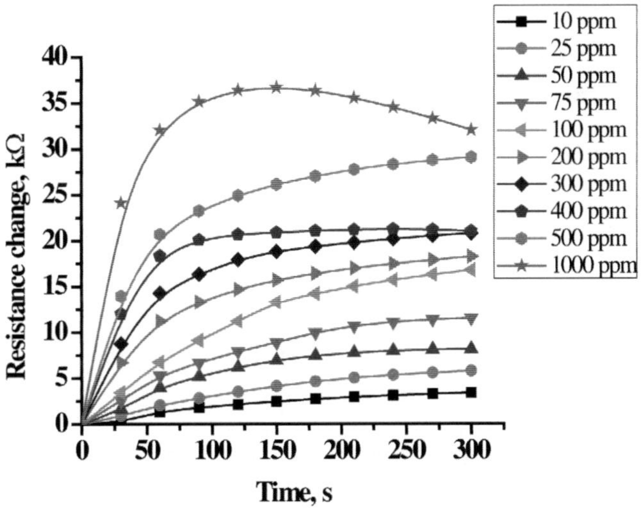

Figure 5. Change of resistance as a function of time for PANI/PDLA sensor, exposed to different ammonia concentrations.

PANI emeraldine base (EB) in half oxidized and half reduced state had been used for sensor construction. When PANI/PDLA sensor is exposed to NH_3 gas, which possessed electron donating property, the ammonia accepts H^+ ions from PANI (EB) by forming ammonium ion and converts it into pernigraniline base (fully oxidized) form. As a result the resistance of PANI/PDLA sensor increased. The pernigranilne base is unstable form of PANI. It is able to capture hydrogen and to convert to EB. The small reduction in the sensor resistance was due to reaching an equilibrium between the emeraldine base and the pernigraniline base at constant ammonia concentration. When the sensor was reverting to air, the pernigraniline base was converted back to emeraldin base and the sensor resistance recovered its initial value.

The response time of the sensor – the time for reaching plateau in the resistance/time dependence, depended on the ammonia concentration and decreased when the concentration increased. On the other hand the sensitivity of the signal, which could be expressed as the slope of the resistance/time dependence, was higher at the beginning of the measurement. Therefore it is convenient to calibrate the sensor using the response during the first 120 s of the measurement. Figure 6 presents the dependence of the response on the ammonia concentration 30 s, 60 s, 90 s and 120 s after the ammonia injection in the desiccator.

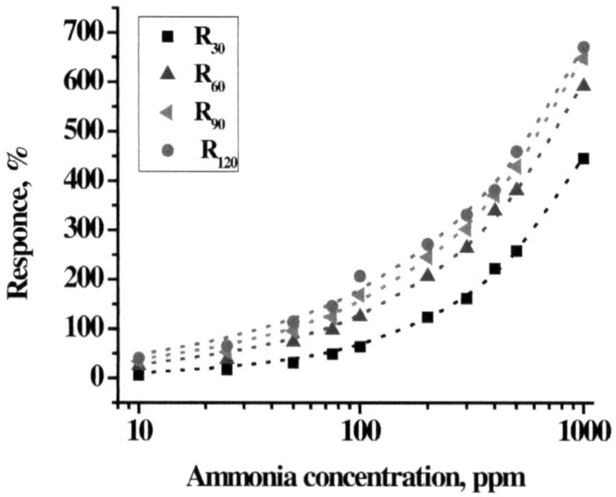

Figure 6. PANI/PDLA sensor response as a function of ammonia concentrations 30 s (R_{30}), 60 s (R_{60}), 90 s (R_{90}) and 120 s (R_{120}) after the ammonia injection.

All these dependences could be correlated to two-parameter power law:

$$response = a \cdot c^b, \quad (2)$$

where c is the ammonia concentration.

The values for the parameters a and b, as well as the coefficient of determination R^2, calculated for the correlation dependences at different times, are presented in table 1. Knowing the values of these parameters for the PANI/PDLA sensor, one can calculate an unknown ammonia concentration several times during the measurement and estimate the standard error of the measurement.

Table 1. Parameters a and b of the calibration power law dependences, constructed for the sensor response at different times after the ammonia injection.

Time after the ammonia injection (s)	a	b	R^2
30	1.5±0.17	0.82±0.018	0.998
60	5.7±0.50	0.67±0.014	0.998
90	8.8±0.63	0.62±0.011	0.998
120	12.5±1.34	0.58±0.017	0.996

The sensor reliability was strongly dependent on the reproducibility and stability exhibited by the sensor material. The reproducibility of the PANI/PLA blend film sensor was measured by repeating the response measurement a number of times. Results showed that the response of the material was almost constant confirming the reproducibility of the sensor material. Furthermore, the response of the sensor remained about 90 % of the initial response after 60 days, which indicated an acceptable long-term stability.

4. Conclusion

A novel ammonia sensor was successfully developed by casting of PANI base – poly(DL-lactic) acid bled film on Al_2O_3 substrate. The AFM images revealed the morphology of the composites and confirmed homogenous distribution of PANI particles with average diameter 500 nm in the PDLA matrix. The gas sensing property measurement showed that the PANI/PDLA nanocomposite film sensors exhibited fast response/recovery rate, high sensitivity, good reproducibility and acceptable long-term stability over a concentration range from 10 ppm to 1000 ppm. This outstanding sensing performance may be attributed to the large surface of the PANI nanoparticles immersed in the PDLA polymer matrix. Therefore, the proposed PANI/PDLA nanocomposite film sensors are strongly potential candidates for NH_3 detection.

Acknowledgements

This study was financial supported by Project MU-15-FFIT-003/23.04.2015 of Department of Scientific Research at the Plovdiv University.

References

[1] Zhihua L, Xucheng Z, Jiyong S, Xiaobo Z, Xiaowei H, Tahir HE, and Holmes M 2016 *Sensor Actuat. B-Chem.* **226** 553-62
[2] Xu DM, Guan MY, Xu QH, and Guo Y 2013 *J. Hazard. Mater.* **262** 64-70
[3] Chen G, Liu J, Liu M, Li G, Sun Z, Zhang S, Song C, Wang H, Suo Y and You J 2014 *J. Chromatogr.* A **1352** 8–19
[4] Jazan E and Mirzaei H 2014 *J. Pharm. Biomed. Anal.* **88** 315–20
[5] Wang J, Zhang W, Li L and Yu Q, 2011 *Appl. Phys.* B: *Lasers Opt.* **103** 263–69
[6] Jiang C, Liu R, Han G and Zhang Z 2013 *Chem. Commun.* **49** 6647–49
[7] Fakhari AR, Breadmore MC, Macka M and Haddad PR, 2006 *Anal. Chim. Acta* **580** 188–93
[8] Mikhaylov S, Ogurtsov N, Noskov Y, Redon N, Coddeville P, Wojkiewicz JL, and Pud A 2015 *RSC Advances* **5** 20218-26
[9] Pawar SG, Chougule MA, Patil SL, Raut BT, Godse PR, Sen S., and Patil VB 2011 *IEEE Sens. J.* **11** 3417–23
[10] Khuspe GD, Bandgar DK, Sen S and Patil VB 2012 *Synth. Met.* **162** 1822–27
[11] Chabukswar VV, Pethkar S, and Athawale AA 2001 *Sensor Actuat. B-Chem.* **77** 657-63
[12] Bandgar DK, Navale ST, Nalage SR, Mane RS, Stadler FJ, Aswal DK, and Patil VB 2015 *J. Mater. Chem.* **3** 9461-68
[13] Molapo KM, Ndangili PM, Ajayi RF, Mbambisa G, Mailu SM, Njomo N, and Iwuoha EI 2012 *Int. J. Electrochem. Sc.* **7** 11859-75
[14] Matsuguchi M, Io J, Sugiyama G, and Sakai Y 2002 *Synthetic Met.* **128** 15-19

Low pressure influence on the electrets stability of gamma irradiated PP and PET films

A Viraneva[*], I Bodurov and T Yovcheva

Department of Experimental Physics, Plovdiv University "Paisii Hilendarski",
24 Tzar Assen str., 4000 Plovdiv, Bulgaria

*E-mail: asia83@abv.bg

Abstract. The influence of low pressure on the surface potential decay of gamma irradiated polymer films of polypropylene (PP) and poly(ethylene terephtalate) (PET) were studied. Polymer film samples were subjected to integral irradiation doses (Eγ = 1.25 MeV, ^{60}Co source) of 5 kGy and 25 kGy accumulated in air at a dose rate of 0.26 Mrad/h. After irradiation, the samples were charged in a corona discharge by means of a corona triode system for 1 minute under room conditions. Positive or negative 5 kV voltages were applied to the corona electrode. Four different voltages of the same polarity as that of the corona electrode were applied to the grid. The electret surface potential V_0 was measured by the method of the vibrating electrode with compensation. After charging the electrets were placed into a vacuum chamber as the pressure was reduced step by step in the range of 1000 mbar to 0.1 mbar. At each step the samples were stored for 1 minute and the surface potential V was measured again. Then values of the normalized surface potential V/V_0 were calculated. Low pressure dependences of the normalized surface potential for positively and negatively charged PP and PET films were presented. It was established that the low pressure led to the surface potential decay of the electrets. The influence of low pressure was analyzed by the equation that describes processes of desorption from the electret surface accompanied with surface diffusion. The experimental results obtained show a significant change in the electret behaviour of the polimer films after gamma irradiation and storage at different low pressure. It was established that the surface potential decay depends on the corona polarity, gamma irradiation and the values of low pressure.

1. Introduction

Electret materials find broad application in various fields of science, engineering and technology because of their ability to create a permanent electric field long enough without additional power supply [1-3]. Over the years, the corona discharged has been one of the most commonly used methods of materials treatment as it is inexpensive and easy to realise from a technical point of view. In order to obtain stable electrets for numerous applications the influence of different factors (temperature, humidity, low pressure, gamma irradiation etc.) on charge decay has been studied [4, 5]. However, the gamma irradiation influence on the electret state of polymer electret films has been studied in relatively few articles, there are only a few publications about the influence of the charge decay. The gamma irradiation effect (up to 100 kGy) on the electret behavior of corona charged biobased polymer films of polylactic acid have been investigated in [6]. It was established that the degradation process is predominant. The values of the surface potential of the irradiated samples were higher than the non-irradiated ones, independently of the corona polarity and irradiation dose. The influence of low dose gamma irradiation on the stability of electret characteristics of corona charged PP and PET films was investigated in [7]. It was established that the significant changes in the electret behaviour of the polymer films induced by gamma irradiation are of complex origin. The PP electrets charged in a negative corona and irradiated by dose of 25 kGy achieved the highest surface potential. The possible surface potential decay mechanisms responsible for the above may be due to degradation, scissioning and crosslinking of the polymer chains with the increasing dose of gamma irradiation and formation of different charges trapping surface states. The surface potential decay of gamma irradiated low density polyethylene (LDPE) films has been examined in [8]. Samples with different absorbed doses between 5 kGy and 500 kGy were obtained by exposing samples to ^{60}Co for various times at room temperature. It was established that the surface potential decay is significantly affected by gamma irradiation with different doses. The effect of gamma irradiation on the surface conductivity of low density polyethylene (LDPE) was investigated in [9]. The samples had been previously irradiated at room temperature in air to 100 kGy and then to 1000 kGy with a dose rate of 10 kGy/h by using a ^{60}Co gamma-source. The results obtained show that the surface conductivity increases with the growth of the total dose. The influence of gamma irradiation on the structural properties of non-irradiated and gamma irradiated with different doses (0–2000 kGy) PET polymer films was analyzed in [10]. A significant increase of crystallinity of the investigated samples was established. For the formation of electrets different polymeric materials as polypropylene (PP) and poly (ethylene terephthalate) (PET) are widely used, because of their important commercial significance, structure and appropriate mechanical and electrical properties [11, 12]. In the literature, however, there is no data available for the influence of gamma irradiation on the charge decay of PP and PET electret films stored at low pressure. The aim of the present paper is to investigate the influence of low pressure on the electrets stability of corona charged PP and PET gamma irradiated films.

2. Experimental

2.1. Formation of gamma irradiated polymer films

40 µm thick PET films, (Hostaphan RNK, Mitsubishi Polyester Films GmbH, Germany) and 20 µm thick PP films (Asenova krepost Ltd, Bulgaria) were used. Initially, the films were cleaned in an ultrasonic bath with alcohol for 4 minutes, followed by rinsing with distilled water, and then were dried under room conditions. The 30 mm diameter samples were cut from the cleaned films and subjected to gamma irradiation treatment. The irradiations were performed in air by a ^{60}Co source with total doses of 5 kGy and 25 kGy accumulated in a single step at a dose rate of 0.25 kGy/h. It was checked by thermometric control that the sample temperatures during the irradiation did not appreciably increase above room temperature. This was expected because of the low dose rate of the irradiation.

2.2. Corona charging and measuring of the surface potential

The corona charging of the samples was carried out by the method of the corona discharge using a point-to-plane three-electrode corona discharge system [13] consisting of a grounded plate electrode, a corona electrode and a grid placed between them. The samples of the non-irradiated (0 kGy) and gamma irradiated (5 kGy and 25 kGy) PP and PET films were charged for 1 minute at room temperature. Positive or negative 5 kV voltage was applied to the corona electrode. Voltages of 500 V, 650 V, 800 V and 950 V of the same polarity as that of the corona electrode were applied to the grid. The electret surface potential was measured by the method of the vibrating electrode with compensation [1] and the estimated error was better than 5 %.

2.3. Low pressure measurement

After charging the samples, the initial electret surface potential V_0 was measured and then the electrets were placed into a vacuum chamber. The pressure in the vacuum chamber was reduced step by step in the range from 1000 mbar to 0.1 mbar. At each step the samples were kept at a respective constant pressure for 1 minute. After that the electrets were removed from the vacuum chamber, the surface potential V was measured again and the normalized surface potential V/V_0 was calculated.

3. Results and discussion

The dependences of the normalized surface potential V/V_0 on the low pressure for positively and negatively charged non-irradiated and gamma irradiated PP and PET films have been investigated. Dependences of normalized surface potential on normalized pressure for positively and negatively charged to different initial surface potentials (650 V and 800 V) non-irradiated and gamma irradiated PP films are presented in figures 1 and 2.

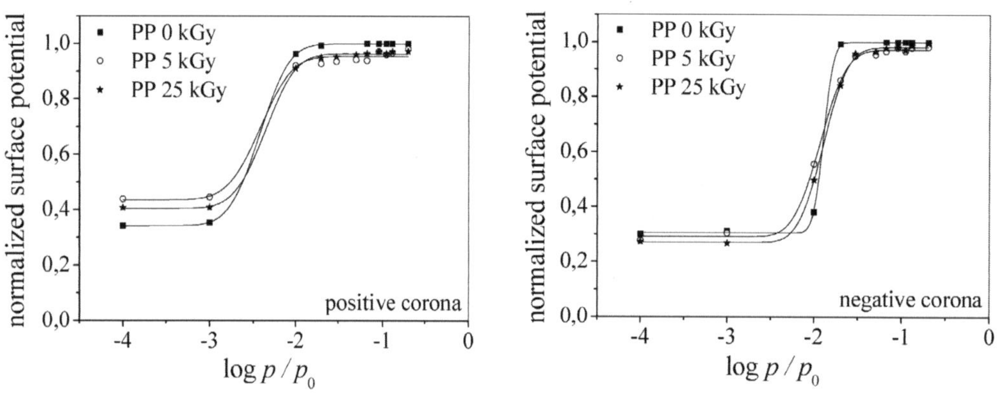

Figure 1. The normalized surface potential dependences on the normalized pressure for non-irradiated and gamma irradiated PP films positively and negatively charged to initial surface potential 650 V.

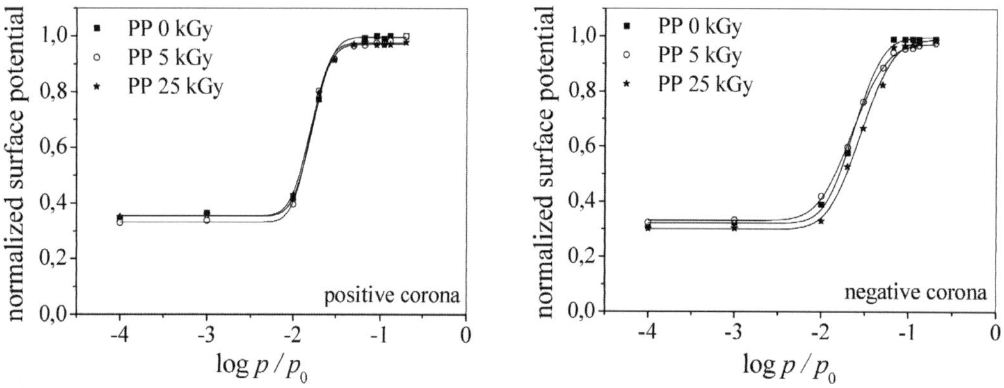

Figure 2. The normalized surface potential dependences on the normalized pressure for non-irradiated and gamma irradiated PP films positively and negatively charged to initial surface potential 800 V.

Dependences of normalized surface potential on normalized pressure for positively and negatively charged to different values of the initial surface potentials (650 V and 800 V) non-irradiated and gamma irradiated PET films are presented in figures 3 and 4.

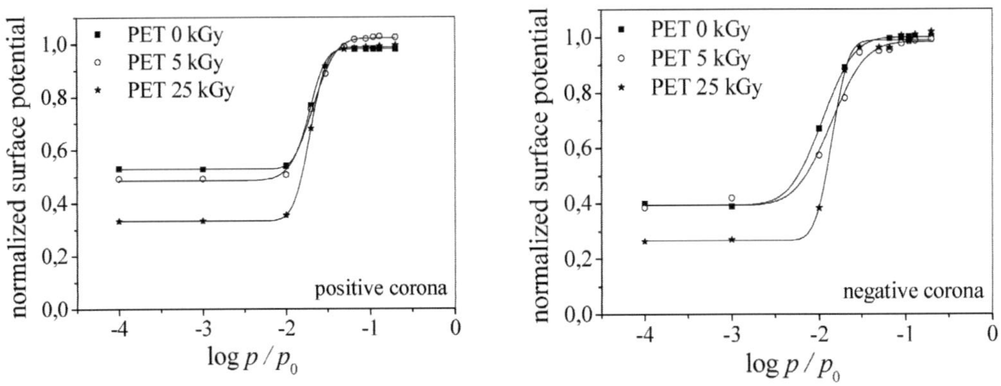

Figure 3. The normalized surface potential dependences on the normalized pressure for non-irradiated and gamma irradiated PET films positively and negatively charged to initial surface potential 650 V.

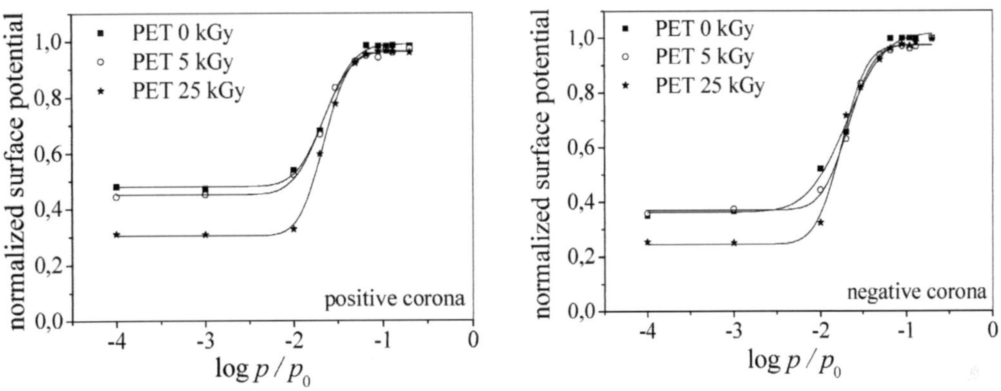

Figure 4. The normalized surface potential dependences on the normalized pressure for non-irradiated and gamma irradiated PET films positively and negatively charged to initial surface potential 800 V.

Analogues curves for positively and negatively charged to grid potentials 500 V and 950 V non-irradiated and gamma irradiated PP and PET films are obtained. In figures 1 – 4 the symbol V_0 marks the initial value of the surface potential measured just after electrets charging, and the symbol p_0 marks the atmospheric pressure. Each point in the figures is a mean value of 6 samples. The calculated standard deviation was 10 % better than the mean value with confidence level 90 %.

Table 1. The steady state values of the normalized surface potential at a pressure of 0.1 mbar for non-irradiated and gamma irradiated PP films.

V_0, V	V/V_0					
	PP positive corona			PP negative corona		
	0 kGy	5 kGy	25 kGy	0 kGy	5 kGy	25 kGy
500 V	0.47	0.45	0.43	0.42	0.40	0.39
650 V	0.36	0.42	0.41	0.30	0.28	0.27
800 V	0.34	0.33	0.35	0.32	0.31	0.30
950 V	0.32	0.34	0.33	0.25	0.22	0.22

Table 2. The steady state values of the normalized surface potential at a pressure of 0.1 mbar for non-irradiated and gamma irradiated PET films.

V_0, V	V/V_0					
	PET positive corona			PET negative corona		
	0 kGy	5 kGy	25 kGy	0 kGy	5 kGy	25 kGy
500 V	0.53	0.52	0.40	0.48	0.46	0.38
650 V	0.53	0.49	0.35	0.40	0.39	0.26
800 V	0.48	0.44	0.31	0.35	0.36	0.25
950 V	0.39	0.37	0.30	0.33	0.35	0.23

The results presented in figures 1-4 and tables 1 and 2 show that:

- Independently of the gamma irradiated doses in each curve three parts are observed. At higher pressures the surface potential is constant and equals initial surface potential. At lower pressures the surface potential is also constant but has different values for the different cases. For each curve a relatively narrow region of pressures exists where a sharp decay of surface potential occurs. A similar behavior has also been observed in [14].
- The steady state values of the normalized surface potential at the 0.1 mbar pressure for the samples charged in a positive corona are higher than those for the samples charged in a negative corona independently of the gamma irradiated doses. It was established in [15] that during the corona discharge in the air, at atmospheric pressure, different types of ions are deposited on a sample. In the case of a positive corona the ions are mainly $H^+(H_2O)_n$ and the ones for a negative corona - CO^{3-}. Those ions are bound in traps of various depths and are released from them depending on the surrounding conditions.
- The steady state values of the normalized surface potential at the 0.1 mbar pressure for PET samples are higher than those for PP samples independently of the corona polarity exept for samples irradiated by dose of 25 kGy. We assume that this is due to the different structures of the polymers, which leads to the formation of different localized surface states that capture charges.
- The surface potential values of the PET samples irradiated by a dose of 25 kGy were smaller in comparison with the non-irradiated samples and the irradiated ones by a dose of 5 kGy, independently of the corona polarity.
- The surface potential values of the gamma irradiated PP samples stored at low pressure do not depend on the irradiation dose.

Therefore, the low dose of 5 kGy does not cause changes of the electret's behavior at low pressure. The gamma irradiation by a greater dose of 25 kGy leads to creation of different localized surface states and the surface charge decrease [16]. The charge accumulation depends upon the density of localized surface states, which is varied by the radiation induced cross-linking and the degradation reactions of the molecule structure [17]. The results obtained (figures 1-4) were analyzed by the equation that describes the processes of desorption from the electrets surface accompanied with surface diffusion. This equation was used and described earlier in [18]:

$$\theta = a + \frac{1}{2}b\left(1 + erf\left(\frac{x-c}{\sqrt{2}d}\right)\right),$$

where $\theta = V/V_0$ is the normalized surface potential, $x = \log p/p_0$ and a, b, c and d are parameters. Values of the parameters *a, b, c* and *d* for all investigated samples are obtained by fitting.

The data obtained in table 3 for PP and PET electrets positively and negatively charged to initial surface potential 500 V could have practical benefits in its usage under conditions of reduction pressures. If you know the initial surface potential value V_0 of different materials, data obtained for parameters could be used to determine the pressure range in which there will be a potential sharp decay.

Table 3. Values of the parameters *a*, *b*, *c* and *d* obtained by fitting data to the equation.

Material type	Corona polarity	*a*	*b*	*c*	*d*
PP 0 kGy	+	0.43±0.01	0.54±0.01	-2.50±0.01	0.24±0.01
	-	0.45±0.01	0.51±0.02	-2.45±0.05	0.04±0.07
PET 0 kGy	+	0.55±0.01	0.39±0.01	-2.33±0.09	0.03±0.01
	-	0.51±0.01	0.45±0.01	-2.46±0.07	0.25±0.04
PP 5 kGy	+	0.48±0.01	0451±0.01	-2.41±0.07	0.28±0.05
	-	0.44±0.01	0.50±0.01	-2.44±0.06	0.24±0.03
PET 5 kGy	+	0.54±0.01	0.39±0.01	-2.46±0.09	0.28±0.05
	-	0.49±0.01	0.46±0.01	-2.39±0.17	0.24±0.1
PP 25 kGy	+	0.46±0.01	0.49±0.01	-2.53±0.07	0.23±0.03
	-	0.35±0.01	0.59±0.01	-2.34±2.27	0.18±1.25
PET 25 kGy	+	0.43±0.01	0.51±0.01	-2.16±0.01	0.07±0.01
	-	0.42±0.01	0.53±0.01	-2.44±0.13	0.22±0.07

4. Conclusion

The influence of low pressure on electret behavior of PP and PET non-irradiated and gamma irradiated films were studied. The results obtained show that charge decay depends on initial surface potentials, corona polarity and gamma irradiated dose. Our experimental data show that the low pressure at which PP and PET electrets have been stored should be the important factor of determining their charge decay. This is extremely important and should be taken into account in the study of the stability of electret sensors and devices that are used at room temperature and function under real conditions involving various low pressure levels.

References

[1] Sessler G and Gerhard-Multhaupt R 1999 *Electrets* 3rd Ed. (Laplacian: Press Morgan Hill California)
[2] Kestelman N 2000 *Electrets in engineering: Fundamentals and applications* (Springer)
[3] Goel M 2003 *Current science* **85** 443
[4] Gang-Jin C, Hui-ming X and Chun-feng Z 2004 *J. Zhejiang Univ. Scie.* **5** 923
[5] Aragoneses A, Tamayo I, Lebrato A, Cañadas J, Diego J, Arencón D and Belana J 2013 *J. Electrostat.* **71**, 611
[6] Yovcheva T, Marudova M, Viraneva A, Gencheva E, Balabanov N and Mekishev G 2013 *J. Applied Polym. Scie.* **128** 139
[7] Viraneva A, Yovcheva T, Krezhov K and Sotirov S 2015 *Bulg. Chem. Commun.* **47B** 121
[8] Chen G 2010 *J. Electrostat.* **68** 487
[9] Gao Y, Li Y and Du B 2013 *Diangong Jishu Xuebao Trans. China Electrotec. Soc.* **28** 125
[10] Siddhartha S, Dev Aarya K, Kumar Raghuvanshi S, Krishna J and Wahab M 2012 *Radiation Phys. and Chem.* **81** 458
[11] Li J, Zhou F, Min D, Li S and Xia R 2015 *IEEE Trans. on Diel. and Electr. Insul.* **22** 1723
[12] Herous L, Nemamcha M, Remadnia M, Dascalescu L 2009 *J. Electrostat.* **67** 198
[13] Yovcheva T 2010 *Corona charging of synthetic polymer films* (New York: Nova Science Publishers Inc.)
[14] Viraneva A, Yovcheva T and Mekishev G 2012 *Polypropylene* (InTech: Rijeka Croatia)
[15] Giacometti J, Fedosov S and Costa M 1999 *Brazilian J. Phys.* **29** 269
[16] Srivastava A and Virk H 2000 *Radiation phys. and chem.* **59** 31
[17] Du B and Gao Y 2009 Polymer Degradation and Stability **94** 139
[18] Viraneva A, Yovcheva T, Pisanova E, Gencheva E and Mekishev G 2010 *AIP Conf. Proc.* **1203** 382

Spectral matching and outdoor solar to electrical conversion efficiency in thin-film silicon multi-junction solar cells

S Reynolds[1], F Urbain[2,3] and V Smirnov[2]
[1] School of Science and Engineering, University of Dundee, Dundee DD1 4HN, Scotland.
[2] IEK-5 Photovoltaik, Forschungszentrum Jülich, D-52425 Jülich, Germany.
[3] Present address: IREC Catalonia Institute for Energy Research, Barcelona, Spain.

E-mail: s.z.reynolds@dundee.ac.uk

Abstract. Semi-empirical computer modelling is used to investigate spectral matching in tandem and triple-junction thin film silicon solar cells. In amorphous/microcrystalline silicon (a-Si:H/μc-Si:H) tandem cells, current mis-match is offset by an increase in fill-factor, resulting in a broad peak in efficiency versus average photon energy. For a-Si:H/a-Si:H tandem cells, photo-generated currents in both sub-cells increase with increasing average photon energy, and efficiency is predicted to increase monotonically over a wide spectral range. a-Si:H/a-Si:H/μc-Si:H triple cells exhibit spectral behaviour similar to a-Si:H/μc-Si:H tandem cells, but with a smaller fill-factor dependence. Variations in spectral quality are predicted to account for only a small reduction in annual electrical energy yield, of some 2 to 4%.

1. Introduction

Stacked multi-junction solar cells enable thermalisation and transmission losses to be reduced over the constituent single-junction cells, leading to an overall increase in photovoltaic conversion efficiency. This has been utilised in the development of multi-cell combinations of thin-film (amorphous and microcrystalline) silicon, yielding laboratory efficiencies of over 14% [1-4]. Further, increased terminal voltages (> 1.5 volts) are sufficient to drive photochemical water-splitting reactions directly, an application that has attracted considerable recent interest [5-7].

Careful matching of photogenerated currents is needed to optimise efficiency for a given spectrum [1,2], which in the laboratory is normally AM1.5G. However substantial variations in the solar spectrum occur naturally; multi-junction cells seldom operate outdoors in a current-matched condition, impacting on annual energy return. We present a model [8,9] that enables these effects to be evaluated. Realistic outdoor spectra are generated using a linear weighting function, and characterised in terms of average photon energy (*APE*) [10,11]. The short-circuit currents generated by sub-cells as a function of *APE* are calculated by numerical integration, and used to scale the J-V characteristics of reference single cells from which the overall J-V characteristic is obtained. Statistical distributions of spectral irradiance vs. *APE* are used to estimate long-term outdoor performance at a given location.

2. Experimental

2.1. Solar cell deposition

All cells were deposited in p-i-n superstrate configuration by plasma-enhanced chemical vapour deposition onto TCO coated glass. The top cell absorber layer of the a-Si:H/μc-Si:H tandem cell was deposited at 180 °C with a silane concentration (SC) of 10%. For the a-Si:H/a-Si:H tandem cell and the a-Si:H/a-Si:H/μc-Si:H triple cell, the top a-Si:H absorber layer was deposited at 130°C with a SC of 4%, and the following a-Si:H layer at 180°C with a SC of 10%. Representative measured solar cell parameters are provided in table 1. Additional details are given elsewhere [2, 7].

Single junction 'reference' solar cells corresponding to top, middle and bottom were deposited under the same conditions. Layers representing the top, and (top + middle) sub-cells, used to filter the solar spectrum when measuring *J-V* characteristics of single cells representing the middle and bottom sub-cells, were deposited on glass slides and placed in the optical path.

Table 1. Representative measured cell parameters. V_{OC} is open-circuit voltage, J_{SC} short-circuit current, *FF* fill-factor and η conversion efficiency.

Cell configuration	a-Si/μc-Si	a-Si/a-Si	a-Si/a-Si/μc-Si
V_{OC} (V)	1.40	1.87	2.28
J_{SC} (mA/cm^2)	11.8	6.96	8.44
FF (%)	67.4	76.3	68.0
η (%)	11.1	9.9	13.1

2.2. Solar cell characterisation
Current–voltage (*J–V*) measurements under standard test conditions (100 mW/cm^2, 25°C) were made using a double source (Class A) AM1.5 sun simulator External quantum efficiency (EQE) measurements were conducted over the wavelength range 300 nm to 1100 nm using chopped light from a monochromator. Individual sub-cell EQEs of tandem and triple cells were determined separately, by using bias light sources to saturate those sub-cells not under measurement.

2.3 Modelling procedure
The modelling procedure has been described in detail elsewhere [8,9]. It consists of three main steps: (i) Solar spectra with *APE*s between 1.75 and 2.05 eV are generated by applying a linear spectral weighting function pivoted at 630 nm to the AM1.5 spectrum. The resulting spectra are similar in shape to those corresponding to a given *APE* [12,13]. (ii) The J_{QE} values for each sub-cell are calculated by numerical integration of the product of the EQE and a given spectrum, between 350 and 1050 nm. (iii) The relevant J_{QE} values are then used to shift each *J-V* curve along the current axis such that $J_{SC} = J_{QE}$. It is then straightforward to calculate points on the multi-junction *J-V* curve by adding the voltages for each sub-cell at the same current value.

3. Results and discussion

3.1. Sub-cell currents
The EQE data, from which J_{QE} values are computed for given model spectra, are presented in figure 1. The wavelength regimes over which each sub-cell generates photocurrent can be clearly identified. Figure 2 shows sub-cell currents vs. *APE* obtained when the solar spectrum is modelled for the three configurations studied here. The performance of multi-junction cells under given spectral conditions is constrained by the smallest of the currents generated by the component sub-cells. The a-Si:H/μc-Si:H tandem cell (figure 2(a)) is top-cell limited below *APE* = 1.90 eV, and bottom- cell limited above this. A peak in short-circuit current, of approximately 11.3 mA/cm^2 is predicted at around 1.90 eV. The a-Si:H/a-Si:H tandem cell (figure 2(b)) exhibits rather different behaviour. While the sub-cell currents are equal (around 7 mA/cm^2) at *APE* = 1.82 eV, the short-circuit current is predicted to increase monotonically with increasing *APE* over the range investigated. This is a consequence of differing

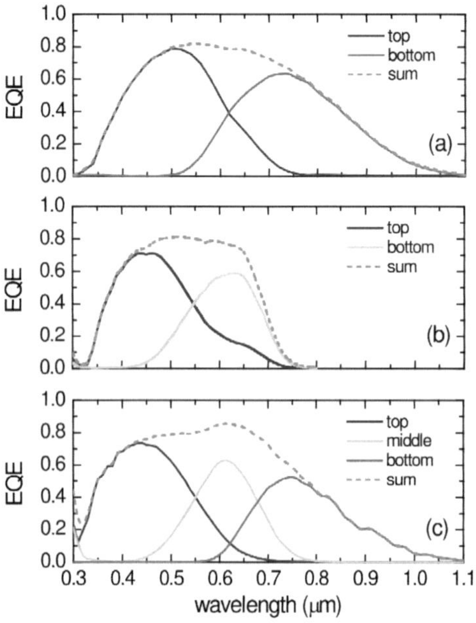

Figure 1. EQE measurements on: (a) a-Si:H/μc-Si:H tandem cell; (b) a-Si:H/a-Si:H tandem cell; (c) a-Si:H/a-Si:H/μc-Si:H triple cell.

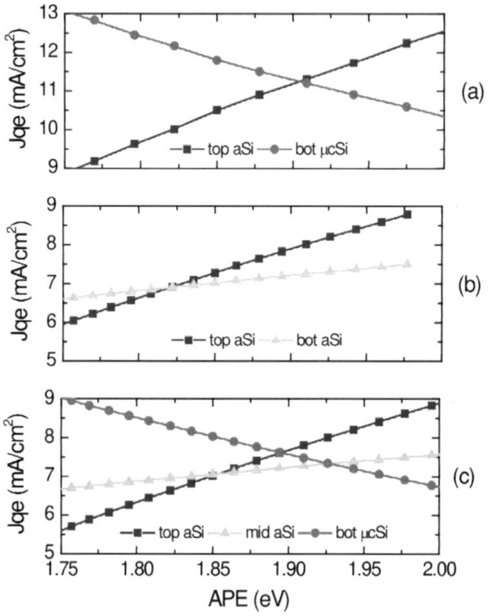

Figure 2. Modelled sub-cell currents for multi-junction cells: (a) a-Si:H/μc-Si:H; (b) a-Si:H/ a-Si:H; (c) a-Si:H/a-Si:H/μc-Si:H.

band-gaps of the bottom cells in these two cases, with the former requiring a thicker top-cell to generate sufficient current to achieve optimal matching

In the a-Si:H/μc- Si:H tandem cell, lower-energy photons generate a substantial fraction of the μc-Si:H bottom-cell current, as the higher-energy photons are absorbed by the thick (300 nm) top cell. This 'shadowing' causes the bottom-cell current to fall with increasing *APE*. In the a-Si:H/a-Si:H tandem cell however, the top cell is thinner (90 nm) and sufficient higher-energy photons are transmitted to the bottom-cell to increase the photogenerated current with increasing *APE*. As can be seen from figure 1(b), the a-Si:H bottom cell does not absorb significantly at wavelengths above 750 nm, so lower-energy photons have little bearing on performance of the a-Si:H/a-Si:H tandem. As *APE* increases beyond the range normally encountered outdoors, the a-Si:H bottom-cell current would ultimately begin to fall as the top-cell absorbs an increasing proportion of the total photon flux.

It can be seen from figure 2(c) that the a-Si:H/a-Si:H/μc-Si:H triple cell passes from top-cell limited to middle-cell limited (but still increasing short-circuit current) to bottom-cell limited (decreasing short-circuit current) as the *APE* is gradually increased. The triple-cell thus shares some of the properties of both tandem cells, but detailed behaviour will depend on layer thicknesses.

3.2 Fill-factor and efficiency

The variations in short-circuit current described in section 3.1 give a good indication of how the PV conversion efficiency of the tandem and triple cells will vary as a function of *APE*. However, while the open-circuit voltage does not vary greatly with *APE* [14], fill-factor may be quite strongly affected. Figure 3(a) reveals that *FF* increases significantly when the tandem cell becomes mismatched, particularly when bottom-cell limited, in agreement with [15]. This moderates the down-turn in efficiency when the short-circuit current decreases either side of the matching point. It should also be noted that the current-matched condition does not correspond exactly to the maximum-power condition. We have previously shown [9] that maximum power is predicted to occur when the tandem cell is mis-matched by 0.8 mA/cm^2, in favour of the top-cell. The a-Si:H/a-Si:H tandem cell (figure 3(b)) shows an increased *FF* when bottom-cell limited but a slight decrease when top-cell limited. Overall, efficiency increases quite strongly with increasing *APE*. The a-Si:H/a-Si:H/μc-Si:H triple-cell (figure 3(c)) simulation curves are similar to the a-Si:H/μc-Si:H tandem-cell, but *FF* variations are comparatively minor, leading to a narrower peak in efficiency. Figures 3(a) and 3(c) indicate that the power-matching for these cells occur at a bluer *APE* (1.93 eV) than the AM1.5G spectrum (1.88 eV).

3.3 Annual-average conversion efficiency

Figure 4 presents the PV conversion efficiencies of the three multi-junction cells as a function of *APE*. In order to determine whether these variations affect longer-term energy yield, a comparison between the efficiency curves and the annual-average spectral irradiance distribution must be made. This reveals variations in the position of the peak of the distribution depending on geographical location. A generic curve based on literature data has been added to figure 4 for comparison. For optimum annual energy yield, the peak in spectral irradiance distribution should be aligned with the peak in conversion efficiency. Figure 4 indicates that for the a-Si:H/μc-Si:H and a-Si:H/a-Si:H/μc-Si:H cells, the majority of the annual spectral irradiance is converted to electricity at a normalised efficiency of >98% of the peak value. More detailed calculations suggest that a 2-4% loss in annual energy due to spectral mis-match is a realistic estimate [9]. For the a-Si:H/a-Si:H cell there is no peak in efficiency, and provided the efficiency increases roughly linearly the gain at higher *APE* will tend to cancel the loss at lower *APE*. Thus spectral mis-match is predicted to have smaller impact in this case.

Figure 4 may be used to estimate the consequences of deploying cells or modules optimised for use at a specific value of *APE*, at a location where the most probable *APE* is not aligned with module peak efficiency. When the different spectral bandwidths are taken into account, distributions measured at a range of sites, including NREL (Colorado USA) [11,12], Ispra (Italy) [12], Loughborough (UK) [11]

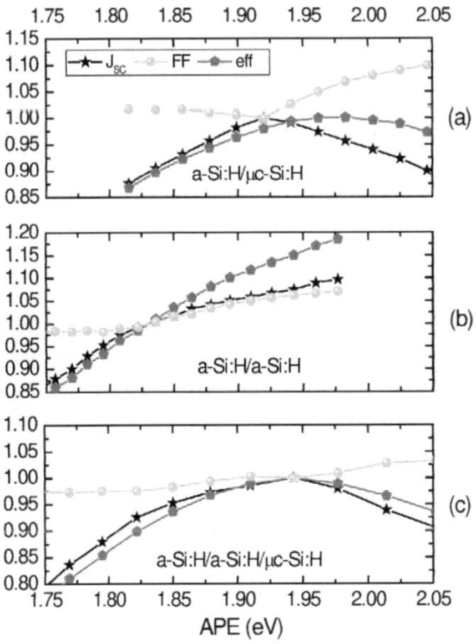

Figure 3. Modelled short-circuit current, fill-factor and efficiency for tandem and triple cells. All scales are normalised to 1.0 at the current-matched point.

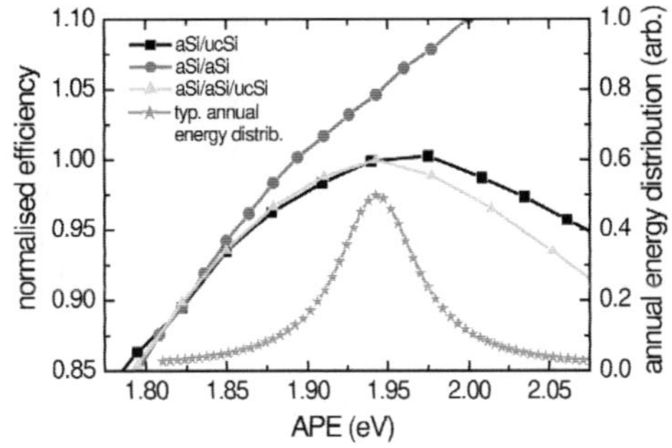

Figure 4. Modelled efficiencies for tandem and triple cells, normalised to overlap low-energy section of curves. Annual average irradiance distribution enables comparison.

and Kusatsu (Japan) [16] span a peak *APE* range of some 0.05 eV, similar to the widths of the distributions. For a tandem or triple cell matched for maximum efficiency under AM1.5G spectrum, the resulting variation in annual electrical energy production across these sites is of a similar magnitude to that due to the widths of the distributions.

Gottschalg et al [17] have differentiated band-gap effects and current mis-match effects for single and multiple cells, as 'primary' and 'secondary', respectively. Our model results indicate that the influence of these factors on long-term outdoor electrical output from multi-junction cells is complex. Spectral effects in multi-junction cells are generally minor compared with other site-related effects, such as temperature variations and degradation/annealing cycles [18], though some studies [19] suggest a greater significance.

3.4 Comparison of model predictions with laboratory and outdoor data

A number of the predictions made here, relating to variation of fill factor and the distinction between current-matching and power-matching, have been demonstrated in the lab for the case of a-Si:H/μc-Si:H tandem cells [14, 15, 20]. This supports the view that the approximations inherent in our model are not an over-simplification. Outdoor data are inherently quite noisy due to limitations in controlling or cancelling the influence of variables other than the set being studied, plus the non-unique nature of *APE* as a measure of spectral quality, which can make it challenging to identify trends. However, a peak in efficiency vs. *APE*, of a similar magnitude and profile to that predicted by our model, has been identified in outdoor test data from a-Si:H/μc-Si:H modules [16]. The tandem and triple-junction 'amorphous silicon' modules studied by Jardine et al [10] and Betts et al [11] behave quite similarly to our a-Si:H/μc-Si:H and a-Si:H/a-Si:H/μc-Si:H cells. Both show a peak in normalised current vs. *APE*, with that of the triple cell being more distinct, when compared with an amorphous silicon single junction. Krishnan et al [21] have performed modelling using actual spectra recorded in the Netherlands, for a-Si:H/a-SiGe:H/μc-Si:H triple cells. They have examined two cells, one well-matched to AM1.5G, and one poorly-matched. For the well-matched cell, the photogenerated current data agree quite well with our figure 3(c), although there are differences in detail, particularly regarding middle-cell current. This can be anticipated since the bandgap is smaller for the a-SiGe:H alloy than for our a-Si:H layer. Overall, results in the literature are in keeping with model predictions.

4. Conclusions

A semi-empirical model has been used to study the influence of solar spectral variations on current-matching in tandem and triple-junction thin film silicon solar cells, in terms of average photon energy. In a-Si:H/μc-Si:H tandem cells, current mis-match reduces photovoltaic conversion efficiency due to top-cell limitation at low *APE* and bottom-cell limitation at high *APE*. This is mitigated by an increase in fill-factor either side of the current-matched condition, resulting in a broad peak in efficiency vs. *APE* whose maximum occurs under slightly bottom-cell limited operation. As the majority of solar terrestrial spectral irradiance falls within a comparatively narrow *APE* range, within ±0.04 eV of the peak cell response, spectral variation is predicted to be a minor concern, accounting for a reduction in annual electrical energy yield of some 2 to 4%.

For a-Si:H/a-Si:H tandem cells, photo-generated currents in both sub-cells increase with increasing *APE* over the range of interest. The short-circuit current thus increases monotonically, with a reduction in gradient rather than a peak being observed at the current-matched point. *FF* is not strongly influenced, with a slight increase occurring under bottom-cell limitation. A peak in efficiency with *APE* is thus not anticipated for this combination. The model predicts that a-Si:H/a-Si:H/μc-Si:H triple cells will exhibit spectral behaviour similar to a-Si:H/μc-Si:H tandem cells, but with smaller variations in fill-factor. The modelled efficiency vs. *APE* peak is slightly narrower, leading to a greater (but still small) reduction in annual electrical energy yield than for the tandem cell.

Acknowledgments
The authors are grateful to K. Wilken for preparation of certain a-Si:H/a-Si:H cells used in this work, and to F. Finger for discussions.

References

[1] Yan B, Yue G, Sivec L, Yang J, Guha S and Jiang C 2011 *Appl. Phys. Lett.* **99** 113512
[2] Smirnov V, Lambertz A, Tillmanns S and Finger F 2014 *Can. J. Phys.* **92** 932
[3] Schüttauf J-W, Bugnon G, Stuckelberger M, Hänni S, Boccard M, Despeisse M, Haug F-J, Meillaud F and Ballif C 2014 *IEEE J. Photovoltaics* **4** 757
[4] Matsui T, Maejima K, Bidiville A, Sai H, Koida T, Suezaki T, Matsumoto M, Saito K, Yoshida I and Kondo M 2015 *Jpn. J. Appl. Phys.* **54** 08KB10
[5] Reece S Y, Hamel J A, Sung K, Jarvi T D, Esswein A J, Pijpers J J H and Nocera D G 2011 *Science* **334** 645
[6] Urbain F, Smirnov V, Becker J-P, Rau U, Ziegler J, Kaiser B, Jaegermann W and Finger F 2015 *Sol. Energy Mater. Sol. Cells* **140** 275
[7] Urbain F, Smirnov V, Becker J-P, Lambertz A, Yang F, Ziegler J, Kaiser B, Jaegermann W, Rau U and Finger F 2016 *Energy Environ. Sci.* **9** 145
[8] Reynolds S and Smirnov V 2012 *J. Phys.: Conf. Ser.* **398** 012006
[9] Reynolds S and Smirnov V 2015 *Energy Procedia* **84** 251
[10] Jardine C N, Betts T R, Gottschalg R, Infield D G and Lane K 2002 *Proc. 'PV in Europe – From PV Technology to Energy Solutions'* (Rome, Italy)
[11] Betts T R, Jardine C N, Gottschalg R, Infield D G and Lane K 2003 *Proc. 3rd World Conference on Photovoltaic Energy Conversion* (Osaka, Japan) p 1756
[12] Norton M, Gracia Amillo A M and Galleano R 2015 *Solar Energy* **120** 337
[13] Minemoto T, Nakada Y, Takahashi H and Takakura H 2009 *Solar Energy* **83** 1294
[14] Blank B, Ulbrich C, Merdzhanova Tz., Zahren C, Pieters B E, Gerber A and Rau U 2015 *Sol. Energy Mat. Sol. Cells* **143** 1
[15] Ulbrich C, Zahren C, Gerber A, Blank B, Merdzhanova Tz., Gordijn A and Rau U 2013 *Int. J. Photoenergy* **2013** 314097
[16] Minemoto T, Toda M, Nagai S, Gotoh M, Nakajima A, Yamamoto K, Takatura H and Hamakawa Y 2007 *Sol. Energy Mat. Sol. Cells* **91** 120
[17] Gottschalg R, Betts T R, Infield D G and Kearney M J 2005 *Sol. Energy Mat. Sol. Cells* **85** 415
[18] Gottschalg R, Betts T R, Eeles A, Williams S R and Zhu J 2013 *Sol. Energy Mat. Sol. Cells* **119** 169
[19] Virtuani A and Fanni L 2014 *Prog. Photovolt: Res. Appl.* **22** 208
[20] Bonnet-Eymard M, Boccard M, Bugnon G, Sculati-Meillaud F, Despeisse M and Ballif C 2013 *Sol. Energy Mat. Sol. Cells* **117** 120
[21] Krishnan P, Schüttauf J W A, van der Werf C H M, Houshyani Hassanzadeh B, van Sark W G J H M and Schropp R E I 2009 *Sol. Energy Mat. Sol. Cells* **93** 691

Spin multiplole moments as collective quantum phenomena

M Georgiev* **and H Chamati**

Institute of Solid State Physics, Bulgarian Academy of Sciences,
Tsarigradsko Chaussee 72, 1784 Sofia, Bulgaria

E-mail: mgeorgiev@issp.bas.bg

Abstract. In this paper we investigate the occurrence of multipole moments of a collection of spins in a quantum magnetic system. To get insights into the behaviour of such systems we start by considering small spin clusters composed of dimers, trimers, as well as mixed spin systems with four spins. We focus our attention on the quantum multipole states, that gave rise to the widely discussed spin nematic and spin liquid phases. The magnetic properties of these systems are computed and the main differences between bosonic and fermionic multipolar states are discussed.

1. Introduction

For the past two decades quantum magnetic systems have attracted the attention of many condensed matter physicists and still are one of the main topics of interest (see e.g. [1–3] and references therein). The occurrence of exotic non-magnetic phases involving higher degree of symmetry, such as nematic phases, have been widely discussed, theoretically as well as experimentally [4–6]. On the experimental side the layered compound $NiGa_2S_4$, the compounds $LaNb_2O_7$ and FeSe have been pointed as materials exhibiting such ordering. Spin-half frustrated ferromagnets on a square lattice [7, 8] and a triangular lattice [9] of ^{3}He thin films exhibit a bond-type spin nematic order. Other enticing possibilities are the different nematic orders of spin-one system on a square lattice [10, 11] and a trianglar lattice [12, 13].

In this report we restrict our attention on the determination of multipole moments in quantum spin clusters with finite number of magnetic moments. We show, by means of quantum mechanical treatment, that at any level beyond the singlet, only the quadrupole state survive. For higher symmetries we have always local dipole components, which makes them observable by experimental means.

We consider spin-$\frac{1}{2}$, spin-1 and mixed dimers, spin-$\frac{1}{2}$ trimer and as a final example a mixed four spin cluster. We report the properties of such systems to get fundamental understanding of the quantum nature of this non-magnetic states before tackling more complex systems with direct application in condensed matter physics. We anticipate that for such systems we have non-zero multipole moments, even when the total spin quantum number is zero or half. Showing that unlike the single spin space, the Hilbert spaces of a spin clusters with the same spin quantum number possess additional symmetries.

The rest of the report is organized as follows: In Section 2 we present the basic notations and general definition to be used later. In Sections 3 and 4 we investigate the properties of clusters with identical of mixed spins. Section 5 presents a summary of our results.

2. General Statements

Below we denote the basis in a 2D vector space by $\{|c^+\rangle, |c^-\rangle\}$, the spherical basis in a 3D space by $\{|d^+\rangle, |d^0\rangle, |d^-\rangle\}$ and the Cartesian basis – $\{|d^x\rangle, |d^y\rangle, |d^z\rangle\}$, with $|d^x\rangle = \frac{i}{\sqrt{2}}\big(|d^+\rangle - |d^-\rangle\big)$, $|d^y\rangle =$

$\frac{1}{\sqrt{2}}\big(|d^+\rangle + |d^-\rangle\big)$, $|d^z\rangle = -\mathrm{i}|d^0\rangle$. The matrices $\{\lambda^i\}_{i=1}^{8}$ and the $\{\gamma^i\}_{i=1}^{15}$ are the SU(3) and SU(4) generators [14], respectively. We denote the spin-half operator by $\hat{\sigma}$, the spin-one operator by $\hat{S}$ and by $\mathbb{1}_n$ the unitary operator acting in n dimensional Hilbert space. Also, i denotes the imaginary unity, s the spin quantum number and m the magnetic quantum number. We set $\hbar = 1$ and the magnetic easy axis is chosen to coincide with the z axis. For brevity we define the set $\mathbb{K} = \{x, y, z\}$.

For a preselected basis in Hilbert space of dimension $2s + 1$ an arbitrary spin can be uniquely determined via the following relations

$$[\hat{s}^\alpha, \hat{s}^\beta] = \mathrm{i}\varepsilon_{\alpha\beta\nu}\hat{s}^\nu \qquad \hat{s}^2|s,m\rangle = s(s+1)|s,m\rangle, \qquad \hat{s}^z|s,m\rangle = m|s,m\rangle, \tag{1}$$

Beside the dipolar ordering described by dipole moments, quantum systems possess additional symmetries associated with different physical properties, such as quadrupole and octupole moments to name a few [15].

The multipole operators are uniquely determined in terms of the spin of the quantum system. Here we concentrate on bosonic systems.

For integer spins s, at any level beyond the singlet state, one can find a set of time reversal invariant non-magnetic states $|s, q^\alpha\rangle$ with eigenvalues

$$\hat{s}^2|s,q^\alpha\rangle = s(s+1)|s,q^\alpha\rangle, \qquad \hat{s}^\alpha|s,q^\alpha\rangle = 0|s,q^\alpha\rangle. \tag{2}$$

This eigenvectors are entanglement states and therefore describe multipole moments as a collective quantum phenomena, which is an example demonstrating the physical richness of boson systems. Notice for arbitrary spin one works with SU(2s+1) groups. A clear example is the spin-one, where the extension of SU(2) to SU(3) naturally leads to the following representations $\hat{S} = (\lambda^7, -\lambda^5, \lambda^2)$ and $\hat{Q} = (-\lambda^3, \lambda^8, -\lambda^1, -\lambda^6, -\lambda^4)$. Note that in this case all time reversal invariant non-magnetic states are quadrupole states. For a single spin with $s > 1$, or spin clusters this is no longer true and as we shall see one non-magnetic state may refer to a different multipole moments. More details about spin-one quadrupole states and the related nematic order can be found in references [4, 13].

3. Dimers and trimers

3.1. Spin-half dimer

We discuss the most simple quantum system composed of a pair of half spins, with total spin $\hat{s} = \hat{\sigma}_1 + \hat{\sigma}_2$. It is interesting that a pair of half spin may couple to form a dipole, as well as a quadrupole and even an octupole. We have one magnetic parameter $m = \pm 1, 0$ and two non-magnetic parameters, the level

Table 1. The eigenvalues of triplet, multipole and singlet states.

$\hat{s}^2$	s	m	$\hat{s}^x$	$\hat{s}^y$	$\hat{s}^z$	Eigenstates	Label	$\hat{\sigma}_1 \cdot \hat{\sigma}_2$
2	1	±1	–	–	–	$\|c^\pm c^\pm\rangle$	$\|T^\pm\rangle$	$\frac{1}{4}$
2	1	0	–	–	0	$\frac{1}{\sqrt{2}}\big(\|c^+c^-\rangle + \|c^-c^+\rangle\big)$	$\|T^0\rangle$	$\frac{1}{4}$
2	1	–	0	–	–	$\frac{\mathrm{i}}{\sqrt{2}}\big(\|T^+\rangle - \|T^-\rangle\big)$	$\|q^x\rangle$	$\frac{1}{4}$
2	1	–	–	0	–	$\frac{1}{\sqrt{2}}\big(\|T^+\rangle + \|T^-\rangle\big)$	$\|q^y\rangle$	$\frac{1}{4}$
2	1	0	–	–	0	$-\mathrm{i}\|T^0\rangle$	$\|q^z\rangle$	$\frac{1}{4}$
0	0	0	0	0	0	$\frac{1}{\sqrt{2}}\big(\|c^+c^-\rangle - \|c^-c^+\big)$	$\|S\rangle$	$-\frac{3}{4}$

parameter $s = 1, 0$ and the eigenvalue of the operator $(\hat{\sigma}_1 \cdot \hat{\sigma}_2)$. We have two magnetic states, which are components of the symmetric triplet group $\{|T^\pm\rangle, |T^0\rangle\}$ and an antisymmetric singlet state $|S\rangle$. In table 1 one can find the various non-magnetic states of the system, where we adopt the notation $|q^\alpha\rangle$

instead of $|s,q^\alpha\rangle$. An arbitrary state function is then written as $|c\rangle = c^x|q^x\rangle + c^y|q^y\rangle + c^z|q^z\rangle + c^s|S\rangle$, with unit vector $\mathbf{c} \in \mathbb{C}^4$, $\mathbf{c} = (c^x, c^y, c^z, c^s)$. Clearly, any state with $\mathbf{c} \in \mathbb{R}^4$ is non-magnetic. Setting $\mathbf{c} = \big(-\frac{1}{\sqrt{2}}(\sin\phi + \mathrm{i}\cos\theta\cos\phi), \frac{1}{\sqrt{2}}(\cos\phi - \mathrm{i}\cos\theta\sin\phi), \frac{\mathrm{i}}{\sqrt{2}}\sin\theta\big)$, we obtain the coherent states of the system, where the triplet level represent the spin-one spin

$$|c\rangle = \tfrac{1}{2}e^{-\mathrm{i}\phi}(1+\cos\theta)|T^+\rangle + \tfrac{1}{\sqrt{2}}\sin\theta|T^0\rangle + \tfrac{1}{2}e^{\mathrm{i}\phi}(1-\cos\theta)|T^-\rangle. \tag{3}$$

An illustration of possible collective spin states, describing an octopole, a quadrupole and a dipole moments, are shown in figure 1, where the director is represented by red line. The quadrupole and octupole components are written as $\hat{Q}^0 = \frac{1}{\sqrt{3}}(|q^x\rangle\langle q^x| + |q^y\rangle\langle q^y| - 2|q^z\rangle\langle q^z|)$ and $\hat{O}^0 = \frac{1}{\sqrt{6}}(|q^x\rangle\langle q^x| + |q^y\rangle\langle q^y| + |q^z\rangle\langle q^z| - 3|S\rangle\langle S|)$. Within the SU(4) group and the basis $\{|q^x\rangle, |q^y\rangle, |q^z\rangle, |S\rangle\}$ for the total spin operator we have $\hat{s} = (\gamma^7, -\gamma^5, \gamma^2)$ and for the quadrupolar operator $\hat{Q} = (-\gamma^3, \gamma^8, -\gamma^1, -\gamma^6, -\gamma^4)$. The SU(2) algebra and s are preserved and in addition we get $\hat{s}^\alpha|S\rangle = 0|S\rangle$, $\gamma^7|q^x\rangle = 0|q^x\rangle$, $\gamma^5|q^y\rangle = 0|q^y\rangle$, $\gamma^2|q^z\rangle = 0|q^z\rangle$. The remaining seven generators can be used to construct the octupolar operator, where $\hat{O}^0 = \gamma^{15}$ [14]. Therefore, unlike single spin-one this dimer is a good example for the existence of additional symmetries and it clearly reveals the quantum collective nature of the multipole moments.

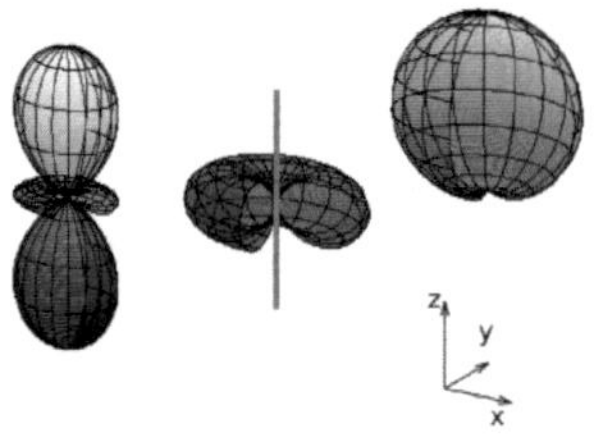

Figure 1. Representation of octupole moment on the left $|\langle\hat{O}^0\rangle| = \frac{1}{\sqrt{6}}(\sin^2(\theta) - 3\cos^2(\theta))$, quadrupole moment on the middle $|\langle\hat{Q}^0\rangle| = \frac{2}{\sqrt{3}}\sin^2(\theta)$ and on the right a fully polarized state(dipole moment) toward the z direction with $|\langle\hat{s}\rangle| = \frac{1}{4}(1+\cos\theta)^2$. The director is illustrated as a red line.

3.2. Spin-one dimer

Naturally the spin-1 spin can form a quadrupole by itself. The quantum numbers for this system are $s = 2, 1, 0$ and $m = \pm 2, \pm 1, 0$. This system has a symmetry higher than octupole, such as hexadecapole and so on. When a pair of spin-1 are coupled to each other we have, [4, p. 337],

$$\hat{Q}_1\cdot\hat{Q}_2 = 2\hat{P}_{12} - \hat{S}_1\cdot\hat{S}_2 - \tfrac{2}{3}\mathbb{1}_9, \qquad \hat{Q}_1\cdot\hat{Q}_2 = 2(\hat{S}_1\cdot\hat{S}_2)^2 + \hat{S}_1\cdot\hat{S}_2 - \tfrac{8}{3}\mathbb{1}_9. \tag{4}$$

Together with the biquadratic form the permutation operator $\hat{P}_{12}$ defines a fundamental constrain between dipole-dipole and quadrupole-quadrupole correlations. Estimating the quantum properties of this pair we work with the magnetic parameter m, the level parameter s and the eigenvalues of the bilinear forms $\hat{S}_1\cdot\hat{S}_2$ and $\hat{Q}_1\cdot\hat{Q}_2$. The value $s = 2$ now defines the quintet level with some of the possible eigenstates presented in table 2.

We have six eigenstates associated to pure multipole moments, three symmetric and three anti-symmetric, see table 3. The other two superpositions stand for the symmetric multipole states $|2,q_1\rangle = \frac{1}{\sqrt{2}}\big(|d^xd^y\rangle + |d^yd^x\rangle\big)$ and $|2,q_2\rangle = \frac{1}{\sqrt{2}}\big(|d^yd^y\rangle - |d^xd^x\rangle\big)$, which preserves the spin quantum number, but are not eigenvectors of $\hat{s}^\alpha$ operators. We remark also that all eigenstates in table 3 are entanglement states and therefore the discussions of the nematic order is not feasible. Despite the property of spin-1 in developing a pure quadrupole, here by the quantum nature of the dimer, we cannot conclude whether a pure quadrupole in the spin-one subspace exists or not. In fact such an estimation can be done by using the variational approach. For instance, the state $|d^+d^z\rangle$ corresponds to spin $\langle\hat{S}_1\rangle$ pointing along the z axis

Table 2. The eigenvalues for quintet, triplet and singlet states are presented.

$\hat{s}^2$	s	m	Eigenstates	Label	$\hat{S}_1\cdot\hat{S}_2$	$\hat{Q}_1\cdot\hat{Q}_2$
6	2	± 2	$\lvert d^\pm d^\pm\rangle$	$\lvert\mathscr{Q}^{2\pm}\rangle$	1	$\frac{1}{3}$
6	2	± 1	$\frac{1}{\sqrt{2}}\big(\lvert d^\pm d^0\rangle + \lvert d^0 d^\pm\rangle\big)$	$\lvert\mathscr{Q}^{\pm}\rangle$	1	$\frac{1}{3}$
6	2	0	$\frac{1}{\sqrt{6}}\big(\lvert d^+ d^-\rangle + \lvert d^- d^+\rangle + 2\lvert d^0 d^0\rangle\big)$	$\lvert\mathscr{Q}^{0}\rangle$	1	$\frac{1}{3}$
2	1	± 1	$\frac{1}{\sqrt{2}}\big(\lvert d^\pm d^0\rangle - \lvert d^0 d^\pm\rangle\big)$	$\lvert T^{\pm}\rangle$	-1	$-\frac{5}{3}$
2	1	0	$\frac{1}{\sqrt{2}}\big(\lvert d^+ d^-\rangle - \lvert d^- d^+\rangle\big)$	$\lvert T^{0}\rangle$	-1	$-\frac{5}{3}$
0	0	0	$\frac{1}{\sqrt{3}}\big(\lvert d^+ d^-\rangle + \lvert d^- d^+\rangle - \lvert d^0 d^0\rangle\big)$	$\lvert S\rangle$	-2	$\frac{10}{3}$

Table 3. The eigenvalues for all non-magnetic eigenstates. The three antisymmetric states are presented in third row.

$\hat{s}^2$	s	$\hat{s}^x$	$\hat{s}^y$	$\hat{s}^z$	Eigenstates	Label	$\hat{S}_1\cdot\hat{S}_2$	$\hat{Q}_1\cdot\hat{Q}_2$
6	2	0	–	–	$\frac{1}{\sqrt{2}}\big(\lvert d^y d^z\rangle + \lvert d^z d^y\rangle\big)$	$\lvert 2,q^x\rangle$	1	$\frac{1}{3}$
6	2	–	0	–	$\frac{1}{\sqrt{2}}\big(\lvert d^x d^z\rangle + \lvert d^z d^x\rangle\big)$	$\lvert 2,q^y\rangle$	1	$\frac{1}{3}$
6	2	–	–	0	$\frac{1}{\sqrt{6}}\big(\lvert d^x d^x\rangle + \lvert d^y d^y\rangle - 2\lvert d^z d^z\rangle\big)$	$\lvert 2,q^z\rangle$	1	$\frac{1}{3}$
2	1	0	–	–	$\frac{1}{\sqrt{2}}\big(\lvert d^y d^z\rangle - \lvert d^z d^y\rangle\big)$	$\lvert 1,q^x\rangle$	-1	$-\frac{5}{3}$
2	1	–	0	–	$\frac{1}{\sqrt{2}}\big(\lvert d^x d^z\rangle - \lvert d^z d^x\rangle\big)$	$\lvert 1,q^y\rangle$	-1	$-\frac{5}{3}$
2	1	–	–	0	$\frac{1}{\sqrt{2}}\big(\lvert d^x d^y\rangle - \lvert d^y d^x\rangle\big)$	$\lvert 1,q^z\rangle$	-1	$-\frac{5}{3}$
0	0	0	0	0	$\frac{1}{\sqrt{2}}\big(\lvert d^x d^x\rangle + \lvert d^y d^y\rangle + \lvert d^z d^z\rangle\big)$	$\lvert S\rangle$	-2	$\frac{10}{3}$

and quadrupole with a director parallel to z axis, where $\langle\hat{S}_1\cdot\hat{S}_2\rangle = 0$ and $\langle\hat{Q}_1\cdot\hat{Q}_2\rangle = -\frac{2}{3}$. Now changing only the orientation of the director, so that we have the state $\lvert d^+ d^y\rangle$ with a director $\mathbf{d}_2^y$, it changes only the expectation value of the quadrupole interaction, or $\langle\hat{Q}_1\cdot\hat{Q}_2\rangle = \frac{1}{3}$. Furthermore, the state $\lvert d^z d^x\rangle$ is a quadrupole state with orthogonal directors $\mathbf{d}_1^z$ and $\mathbf{d}_2^x$ and correlators $\langle\hat{S}_1\cdot\hat{S}_2\rangle = 0$ and $\langle\hat{Q}_1\cdot\hat{Q}_2\rangle = -\frac{2}{3}$, while the quadrupole state $\lvert d^x d^x\rangle$ describe two quadrupoles with parallel directors oriented along the x axis with correlators $\langle\hat{S}_1\cdot\hat{S}_2\rangle = 0$ and $\langle\hat{Q}_1\cdot\hat{Q}_2\rangle = \frac{4}{3}$. On figure 2 we have introduce the geometrical interpretation of arbitrary chosen states for the sake of clarity. Although these states do not revel the real quantum nature of the system, they might help in understanding the exchange mechanism.

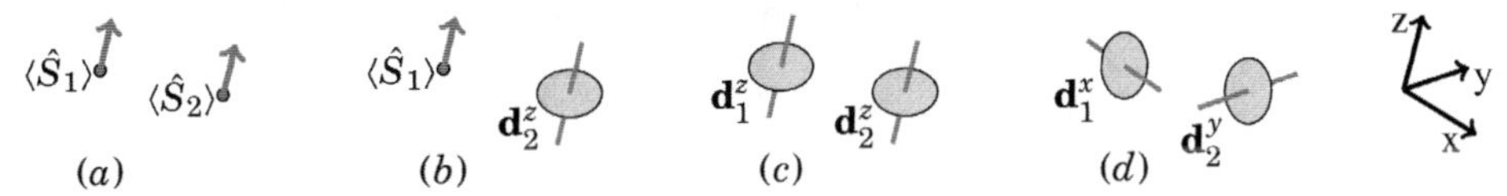

Figure 2. Geometrical representation of (a) ferromagnetic order $\lvert\mathscr{Q}^{2+}\rangle$; (b) mixed state $\lvert d^+ d^z\rangle$; (c) ferroquadrupole order, with state function $\lvert d^z d^z\rangle$; (d) anti-ferroquadrupole order, the corresponding state is $\lvert d^x d^y\rangle$.

3.3. Mixed dimer

Fermions always possess a magnetic moment and therefore we can find non-magnetic states only by considering the average of the total spin. There are two types of non-magnetic states, those that preserve the spin quantum number and those that do not. Using the eigenvectors of table 4 we can obtain six non-magnetic states, with the corresponding quantum number s. A good example is the superposition $\frac{1}{\sqrt{2}}(|Q^{3+}\rangle + |Q^{3-}\rangle)$ with $\langle\hat{s}\rangle = 0$ and $\langle\hat{\sigma}_1 \cdot \hat{S}_2\rangle = \frac{3}{2}$. It is fascinating that even for $s = \frac{1}{2}$ the system exhibits multipole moments. The operator $\hat{Q} \neq 0$ and $\frac{1}{\sqrt{2}}(|D^+\rangle \pm |D^-\rangle)$ are the corresponding non-magnetic states. Moreover, the non-magnetic state $\frac{1}{\sqrt{2}}(|c^+d^x\rangle + |c^-d^y\rangle)$ for which $\langle\hat{s}\rangle = 0$ and $\langle\hat{\sigma}_1 \cdot \hat{S}_2\rangle = 0$ does not preserve the spin quantum number and therefore we cannot deduce that we have collective multipole. This state actually signals the formation of a quadrupole in boson space. Such case can be considered as a spin-half, but notice that we do not have a two level system. Another example is the function $|c^+d^y\rangle$ with $\langle\hat{s}\rangle = \frac{1}{2}$ and $\langle\hat{\sigma}_1 \cdot \hat{S}_2\rangle = 0$. The similar is the case (b) illustrated in figure 2. However, there exists four non-magnetic states that preserve s and should not be omitted $\frac{1}{\sqrt{2}}\big(|c^-d^z\rangle - \mathrm{i}|c^+d^y\rangle\big)$, $\frac{1}{\sqrt{2}}\big(|c^-d^y\rangle + \mathrm{i}|c^+d^z\rangle\big)$, $\frac{1}{\sqrt{2}}\big(|c^+d^x\rangle + |c^-d^z\rangle\big)$ and $\frac{1}{\sqrt{2}}\big(|c^+d^z\rangle - |c^-d^x\rangle\big)$. Note that this states are not time reversal invariant. Finally we end up with twelve non-magnetic states that preserve s.

Table 4. Eigenvalues of quartet (middle row) and doublet states are presented.

$\hat{s}^2$	s	m	Eigenstates	Label	$\hat{\sigma}_1 \cdot \hat{S}_2$
$\frac{15}{4}$	$\frac{3}{2}$	$\pm\frac{3}{2}$	$\|c^\pm d^\pm\rangle$	$\|Q^{3\pm}\rangle$	$\frac{1}{2}$
$\frac{15}{4}$	$\frac{3}{2}$	$\pm\frac{1}{2}$	$\frac{1}{\sqrt{3}}\big(\|c^\mp d^\pm\rangle + \sqrt{2}\|c^\pm d^0\rangle\big)$	$\|Q^\pm\rangle$	$\frac{1}{2}$
$\frac{3}{4}$	$\frac{1}{2}$	$\pm\frac{1}{2}$	$\frac{1}{\sqrt{3}}\big(\sqrt{2}\|c^\mp d^+\rangle - \|c^\pm d^0\rangle\big)$	$\|D^+\rangle$	-1

3.4. Spin-half trimer

This system is analogous to the one discussed in the previous Section, where the total spin quantum number takes the values $s = \frac{1}{2}, \frac{3}{2}$.

In this case the doublet level consists of twelve eigenstates. We assume six of them as a general, which can be used to construct the remaining ones, see table 5. This is not unexpected, we have three couples and therefore three pairs of doublet states which are local singlet states. Clearly, for all α we have

Table 5. Quartet and doublet states of a three spin-half cluster together with the eigenvalues of all bilinear forms are presented.

$\hat{s}^2$	s	m	Eigenstates	Label	$\hat{\sigma}_1 \cdot \hat{\sigma}_2$	$\hat{\sigma}_1 \cdot \hat{\sigma}_3$	$\hat{\sigma}_2 \cdot \hat{\sigma}_3$
$\frac{15}{4}$	$\frac{3}{2}$	$\pm\frac{3}{2}$	$\|c^\pm c^\pm c^\pm\rangle$	$\|Q^{3\pm}\rangle$	$\frac{1}{4}$	$\frac{1}{4}$	$\frac{1}{4}$
$\frac{15}{4}$	$\frac{3}{2}$	$\pm\frac{1}{2}$	$\frac{1}{\sqrt{3}}\big(\|c^- c^\pm c^+\rangle + \|c^+ c^\pm c^-\rangle + \|c^\pm c^\mp c^\pm\rangle\big)$	$\|Q^\pm\rangle$	$\frac{1}{4}$	$\frac{1}{4}$	$\frac{1}{4}$
$\frac{3}{4}$	$\frac{1}{2}$	$\pm\frac{1}{2}$	$\frac{1}{\sqrt{2}}\big(\|c^+ c^- c^\pm\rangle - \|c^- c^+ c^\pm\rangle\big)$	$\|D^\pm_{12}\rangle$	$-\frac{3}{4}$	–	–
$\frac{3}{4}$	$\frac{1}{2}$	$\pm\frac{1}{2}$	$\frac{1}{\sqrt{2}}\big(\|c^+ c^\pm c^-\rangle - \|c^- c^\pm c^+\rangle\big)$	$\|D^\pm_{13}\rangle$	–	$-\frac{3}{4}$	–
$\frac{3}{4}$	$\frac{1}{2}$	$\pm\frac{1}{2}$	$\frac{1}{\sqrt{2}}\big(\|c^\pm c^+ c^-\rangle - \|c^\pm c^- c^+\rangle\big)$	$\|D^\pm_{23}\rangle$	–	–	$-\frac{3}{4}$

$(\hat{\sigma}^\alpha_i + \hat{\sigma}^\alpha_j)|D^\pm_{ij}\rangle = 0$ and thus the eigenvalues $(\hat{\sigma}_1 \cdot \hat{\sigma}_3 + \hat{\sigma}_2 \cdot \hat{\sigma}_3)|D^\pm_{12}\rangle = 0|D^\pm_{12}\rangle$, $(\hat{\sigma}_1 \cdot \hat{\sigma}_2 + \hat{\sigma}_2 \cdot \hat{\sigma}_3)|D^\pm_{13}\rangle = 0|D^\pm_{13}\rangle$, as well as $(\hat{\sigma}_1 \cdot \hat{\sigma}_2 + \hat{\sigma}_1 \cdot \hat{\sigma}_3)|D^\pm_{23}\rangle = 0|D^\pm_{23}\rangle$. Furthermore $\frac{1}{\sqrt{2}}(|D^+_{12}\rangle + |D^+_{13}\rangle) = \frac{1}{\sqrt{6}}(|c^+c^+c^-\rangle + |c^+c^-c^+\rangle - 2|c^-c^+c^+\rangle)$, is an eigenstate of the operators $(\hat{\sigma}_2 \cdot \hat{\sigma}_3)$ and $(\hat{\sigma}_1 \cdot \hat{\sigma}_2 + \hat{\sigma}_1 \cdot \hat{\sigma}_3)$ with eigenvalue $\frac{1}{4}$ and -1, respectively. The total magnetic quantum number is $m = \frac{1}{2}$ and the correlators take the values $\langle\hat{\sigma}_1 \cdot \hat{\sigma}_2\rangle = -\frac{1}{2}$, $\langle\hat{\sigma}_1 \cdot \hat{\sigma}_3\rangle = -\frac{1}{2}$ and $\langle\hat{\sigma}_2 \cdot \hat{\sigma}_3\rangle = \frac{1}{4}$. On the other hand, the eigenstate with the same eigenvalue and $m = -\frac{1}{2}$ is $\frac{1}{\sqrt{2}}\big(|D^-_{12}\rangle + |D^-_{13}\rangle\big) = \frac{1}{\sqrt{6}}\big(|c^-c^-c^+\rangle + |c^-c^+c^-\rangle - 2|c^+c^-c^-\rangle\big)$. Therefore, the system has in total sixteen non-magnetic states that preserve s. Moreover, there are eighteen mixed states. For instance, in order to describe the system in which the spins $\hat{\sigma}_1$ and $\hat{\sigma}_2$ pair into a quadrupole we can use either of the sixth states $|q^\alpha c^\pm\rangle$.

4. Mixed four spin cluster

We consider a four spin cluster, with alternating spin-$\frac{1}{2}$ and spin-1 spins. The maximum number of eigenstates per level is seven, forming the septet group $\{|\mathscr{S}^{3\pm}\rangle, |\mathscr{S}^{2\pm}\rangle, |\mathscr{S}^{\pm}\rangle, |\mathscr{S}^0\rangle\}$. We also have seven groups of quintet states $\{|\mathscr{Q}_i^{2\pm}\rangle, |\mathscr{Q}_i^{\pm}\rangle, |\mathscr{Q}_i^0\rangle\}_{i=1}^7$, seven groups of triplet states $\{|T_i^\pm\rangle, |T_i^0\rangle\}_{i=1}^7$ and one singlet state. Hence the cluster exhibits sixty four unique eigenstates. Among all the magnetic states we have fifteen non-magnetic states satisfying (2), that describe quantum collective multipoles. The other forty eight eigenvectors can be used to construct the remaining non-magnetic multipole states that preserve s, but only thirty are unique. For example, the eigenvectors $|1, q_1^x\rangle = \frac{i}{\sqrt{2}}(|T_1^+\rangle - |T_1^-\rangle)$, $|1, q_1^y\rangle = \frac{1}{\sqrt{2}}(|T_1^+\rangle + |T_1^-\rangle)$ and $|1, q_1^z\rangle = -\mathrm{i}|T_1^0\rangle$ are non-magnetic states with zero eigenvalues. They also share the eigenvalues of the first triplet state shown in table 6. The representation of the components of the triplet $|T_1\rangle$ group

Table 6. Septet, quintet, triplet and singlet eigenstates and the corresponding eigenvalues for the bilinear forms(fourth and fifth column) together with $\hat{M}$, permutation operators and composition operators $(\hat{S}, \hat{M}) = (\hat{S}_2 \cdot \hat{S}_4 + \hat{M})$, $(\hat{\sigma}, \hat{M}) = (\hat{\sigma}_1 \cdot \hat{\sigma}_3 + \hat{M})$ and $(\hat{\sigma}, \hat{S}, \hat{M}) = (\hat{\sigma}_1 \cdot \hat{\sigma}_3 + \hat{S}_2 \cdot \hat{S}_4 + \hat{M})$.

	$\hat{s}^2$	s	m	$\hat{\sigma}_1 \cdot \hat{\sigma}_3$	$\hat{S}_2 \cdot \hat{S}_4$	$\hat{M}$	$(\hat{S}, \hat{M})$	$(\hat{\sigma}, \hat{M})$	$(\hat{\sigma}, \hat{S}, \hat{M})$	$\hat{P}_{13}$	$\hat{P}_{24}$
$\|\mathscr{S}^0\rangle$	12	3	0	$\frac{1}{4}$	1	2	3	$\frac{9}{4}$	$\frac{13}{4}$	1	1
$\|\mathscr{Q}_1^0\rangle$	6	2	0	$-\frac{3}{4}$	1	0	1	$-\frac{3}{4}$	$\frac{1}{4}$	-1	1
$\|\mathscr{Q}_2^{2+}\rangle$	6	2	2	–	–	–	–	–	$\frac{1}{4}$	–	1
$\|T_1^0\rangle$	2	1	0	$\frac{1}{4}$	−2	0	−2	$\frac{1}{4}$	$-\frac{7}{4}$	1	1
$\|T_2^0\rangle$	2	1	0	$-\frac{3}{4}$	−1	0	−1	$-\frac{3}{4}$	$-\frac{7}{4}$	-1	−1
$\|T_3^+\rangle$	2	1	1	$\frac{1}{4}$	–	–	−2	–	$-\frac{7}{4}$	1	–
$\|S\rangle$	0	0	0	$\frac{1}{4}$	−1	−2	−3	$-\frac{7}{4}$	$-\frac{11}{4}$	1	−1

is $|T_1^+\rangle = (|c^+d^+c^+d^-\rangle + |c^+d^-c^+d^+\rangle - |c^+d^0c^+d^0\rangle)$, $|T_1^-\rangle = (|c^-d^-c^-d^-\rangle + |c^-d^-c^-d^-\rangle - |c^-d^0c^-d^0\rangle)$ and $|T_1^0\rangle = \frac{1}{\sqrt{6}}\left(|c^+d^+c^-d^-\rangle + |c^+d^-c^-d^+\rangle + |c^-d^+c^+d^-\rangle + |c^-d^-c^+d^+\rangle - |c^+d^0c^-d^0\rangle - |c^-d^0c^+d^0\rangle\right)$. Also, we have $(\hat{\sigma}_1^\alpha + \hat{\sigma}_3^\alpha)|1, q_1^\alpha\rangle = 0|1, q_1^\alpha\rangle$ and $(\hat{S}_2^\alpha + \hat{S}_4^\alpha)|1, q_1^\alpha\rangle = 0|1, q_1^\alpha\rangle$. Therefore, we have two local multipoles. For example, the SU(3) spin-one subspace representation $|1, q_1^z\rangle = -\frac{i}{\sqrt{6}}\sum_\alpha\left(|c^+d^\alpha c^-d^\alpha\rangle + |c^-d^\alpha c^+d^\alpha\rangle\right)$, corresponding to a triplet paired spin-half spins forming a quadrupole and two spin-one spins associated to a local singlet state, see table 1 and equation (3). Further, the state $|\mathscr{Q}_1^0\rangle = \frac{1}{\sqrt{12}}\left(|c^+d^+c^-d^-\rangle \quad |c^+d^-c^-d^+\rangle - |c^-d^+c^+d^-\rangle - |c^-d^-c^+d^+\rangle + 2|c^+d^0c^-d^0\rangle - 2|c^-d^0c^+d^0\rangle\right)$, is also a time reversal invariant non-magnetic state. On one hand spin-one spins formed a local quadrupole, see table 2, on the other the two half spins define the local singlet, see table 1. Getting back again to table 2 we see that the non-magnetic state $|T_2^0\rangle = \frac{1}{\sqrt{4}}(|c^+d^+c^-d^-\rangle - |c^+d^-c^-d^+\rangle - |c^-d^+c^+d^-\rangle + |c^-d^-c^+d^+\rangle)$ corresponds to a similar case. The spin-1 spins formed a couple being in a $m = 0$ triplet and the spin-half paired in multipole described by a singlet. We further have $(\hat{\sigma}_1 \cdot \hat{S}_2 + \hat{\sigma}_1 \cdot \hat{S}_4)|T_2^0\rangle = 0|T_2^0\rangle$ and $(\hat{\sigma}_3 \cdot \hat{S}_2 + \hat{\sigma}_3 \cdot \hat{S}_4)|T_2^0\rangle = 0|T_2^0\rangle$. The singlet state of the cluster $|S\rangle = \frac{1}{\sqrt{12}}\left(|c^+d^+c^-d^-\rangle - |c^+d^-c^-d^+\rangle + |c^-d^+c^+d^-\rangle - |c^-d^-c^+d^+\rangle - \sqrt{2}|c^-d^+c^-d^0\rangle + \sqrt{2}|c^-d^0c^-d^+\rangle + \sqrt{2}|c^+d^-c^+d^0\rangle - \sqrt{2}|c^+d^0c^+d^-\rangle\right)$, refers to a two local triplet states, see table 6. Besides, for identical spins $i = 1,3$ and $j = 2,4$, either of the bilinear forms $\hat{\sigma}_i \cdot \hat{S}_j|S\rangle \neq 0|S\rangle$, while $(\hat{\sigma}_1 \cdot \hat{S}_2 + \hat{\sigma}_1 \cdot \hat{S}_4)|S\rangle = -1|S\rangle$, $(\hat{\sigma}_3 \cdot \hat{S}_2 + \hat{\sigma}_3 \cdot \hat{S}_4)|S\rangle = -1|S\rangle$ and $(\hat{S}_2^z + \hat{S}_4^z)|S\rangle \neq 0|S\rangle$. Therefore, this is the only level on which all the spins are coupled, forming a multipole. It is also interesting to mention that some of the states which are not eigenstates of the operators $\hat{\sigma}_1 \cdot \hat{\sigma}_3$ and $\hat{S}_2 \cdot \hat{S}_4$ can be eigenstates of the permutation operators associated to these pairs. This particular property can be seen by presenting the vector $|\mathscr{Q}_2^{2+}\rangle = \frac{1}{\sqrt{13}}\big(2|c^+d^+c^-d^+\rangle - 4|c^-d^+c^+d^+\rangle + \frac{1}{\sqrt{2}}|c^+d^+c^+d^0\rangle + \frac{1}{\sqrt{2}}|c^+d^0c^+d^+\rangle\big)$, which is an eigenstate of $\hat{P}_{24}$, but not of $\hat{P}_{13}$. So concerning the relation (4), in some states we are not able to determine the correlation between quadrupoles. This is the case with the triplet $|T_3^+\rangle = \frac{1}{\sqrt{4}}\big(|c^-d^-c^-d^+\rangle - |c^+d^+c^+d^-\rangle + |c^+d^0c^+d^0\rangle - \frac{1}{\sqrt{2}}|c^+d^+c^-d^0\rangle - \frac{1}{\sqrt{2}}|c^-d^0c^+d^+\rangle\big)$. These entanglement states are a sign of frustration. For a chosen convenient microscopic model they may appear as a ground states describing a frustrated system, where neither of the known orders take place.

5. Summary

We have studied the quantum properties of some basic spin clusters revealing the fundamental essence of the spin coupling. An example of this collective phenomena is shown in figure 1 illustrating three different multipole moments. It is clear that by the treatment of quantum mechanic, in a spin system with half integer s only the dipole moments survived. In the boson systems, on the other hand, the existence of multipoles as a quantum occurrence naturally emerge.

It is really fascinating, but not unexpected, that we have non-zero multiple moments even when $s = 0, \frac{1}{2}$, showing that unlike the single spin space, a spin clusters with the same spin quantum number possess additional symmetries. The spin-half dimer is a bright example of this collective phenomena. A coupled fluctuations are shown in figure 1 illustrating three different multipole moments. The discussed different range of pairing (entangled spins) are a simple example of a valence bond. As we have seen, the formation of octupole includes partially developed local dipole moments. Therefore, this property may play an important role for the rearranging of the nearest valence bonds in spin liquids and for the experimental observation of multipolar order.

Acknowledgements

This work was supported by the Bulgarian National Science Fund (DNTS/Germany 01/2).

References

[1] Shimizu Y, Miyagawa K, Kanoda K, Maesato M and Saito G 2003 *Phys. Rev. Lett.* **91** 107001
[2] Pratt F L, Baker P J, Blundell S J, Lancaster T, Ohira-Kawamura S, Baines C, Shimizu Y, Kanoda K, Watanabe I and Saito G 2011 *Nature* **471** 612–616
[3] Banerjee A, Bridges C A, Yan J Q, Aczel A A, Li L, Stone M B, Granroth G E, Lumsden M D, Yiu Y, Knolle J, Bhattacharjee S, Kovrizhin D L, Moessner R, Tennant D A, Mandrus D G and Nagler S E 2016 *Nat. Mater.* **15** 733
[4] Penc K and Läuchli A M 2011 *Introduction to Frustrated Magnetism* vol 164 ed Lacroix C, Mendels P and Mila F (Berlin: Springer) p 331
[5] Läuchli A, Mila F and Penc K 2006 *Phys. Rev. Lett.* **97** 087205
[6] Wang Z, Hu W J and Nevidomskyy A H 2016 *Phys. Rev. Lett.* **116** 247203
[7] Shindou R and Momoi T 2009 *Phys. Rev. B* **80** 064410
[8] Shannon N, Momoi T and Sindzingre P 2006 *Phys. Rev. Lett.* **96** 027213
[9] Momoi T, Sindzingre P and Kubo K 2012 *Phys. Rev. Lett.* **108** 057206
[10] Papanicolaou N 1986 *Phys. Lett. A* **116** 89
[11] Tóth T A, Läuchli A M, Mila F and Penc K 2012 *Phys. Rev. B* **85** 140403
[12] Kaul R K 2012 *Phys. Rev. B* **86** 104411
[13] Smerald A and Shannon N 2013 *Phys. Rev. B* **88** 184430
[14] Pfeifer W 2003 *The Lie Algebras su(N): An Introduction* (Basel: Birkhäuser)
[15] Greiner W and Müller B 1994 *Quantum Mechanics* (Berlin, Heidelberg: Springer)

Soliton dynamics in two coupled ferromagnetic chains

S K Varbev, M T Primatarowa and R S Kamburova

Georgi Nadjakov Institute of Solid State Physics, Bulgarian Academy of Sciences, 1784 Sofia, Bulgaria

E-mail: stanislavvarbev@issp.bas.bg

Abstract. The soliton dynamics in a system of two ferromagnetic chains coupled through the interaction between opposite spins is studied. The on-site anisotropy in the chains is also taken into account. The conditions for the existence of different soliton solutions (bright or dark) in the chains are obtained. The system is reduced to a coupled set of discrete equations with complicated coupling interactions which are linear and nonlinear. We investigate the propagation of a soliton excitation launched in one of the chains with a given velocity. The condition for a perfect switch of the soliton is analysed. Numerical simulations have been performed for a variety of the magnetic parameters and the soliton characteristics.

1. Introduction

The study of solitary waves in solids has been subject of considerable attention for decades. A large amount of theoretical and experimental research has been dedicated to one-dimensional magnetic systems [1]. Models corresponding to real quasi-one-dimensional magnets are broadly investigated as first they were considered as somewhat simpler than the really interesting three-dimensional systems but then turned out to be interesting in their own right. There was the availability of real magnetic compounds which because of extremely small coupling between neighboring chains, could be considered as reasonable realizations of magnetically one-dimensional materials. Soliton solutions for classical sine-Gordon chains [2] and Heisenberg chains with various anisotropies [3-6] were obtained and analyzed. In recent years there is a renewed interest to the topic due to their application. Spin-wave dark solitons were predicted and experimentally generated [7,8]. Solitons in magnetic thin films [9] and in ferromagnets with biquadratic exchange [10] were investigated.

Bright and dark solitons as exact solutions of the nonlinear Schrödinger (NLS) equation have been studied intensively [3,4,11,12]. Another interesting topic of research with practical importance is the propagation of solitons in coupled parallel chains [13-20].

In the present paper we study the dynamics of bright solitons in two anisotropic ferromagnetic chains coupled through a complicated interaction.

2. Hamiltonian of the system

We consider two ferromagnetic Heisenberg chains of N spins with magnitude S described in the nearest-neighbors approximation by the following Hamiltonian:

$$\hat{H} = -J\sum_{n=1}^{N}(\hat{\vec{S}}_n\cdot\hat{\vec{S}}_{n+1}+\hat{\vec{\sigma}}_n\cdot\hat{\vec{\sigma}}_{n+1}) - A\sum_{n=1}^{N}[(\hat{S}_n^z)^2+(\hat{\sigma}_n^z)^2] - \mu H_0\sum_{n=1}^{N}(\hat{S}_n^z+\hat{\sigma}_n^z) + d\sum_{n=1}^{N}\hat{\vec{S}}_n\cdot\hat{\vec{\sigma}}_n, \quad (1)$$

where $J > 0$ is the exchange integral and A is the on-site anisotropy constant which can be positive (easy axis) or negative (easy plane). H_0 is the external magnetic field applied along the z-axis, so that in the ground state of the system all spins are aligned in the z-direction, μ is the magnetic moment per spin, d characterizes the coupling interaction between the two chains.

We use for the scalar products in (1) the rule

$$\hat{\vec{a}} \cdot \hat{\vec{b}} = \frac{\hat{a}^+\hat{b}^- + \hat{a}^-\hat{b}^+}{2} + \hat{a}^z\hat{b}^z, \tag{2}$$

$$[\hat{S}_i^{\pm}, \hat{S}_j^z] = \mp\hat{S}_i^{\pm}\delta_{ij}, \qquad [\hat{S}_i^+, \hat{S}_j^-] = 2\hat{S}_i^z\delta_{ij}, \qquad [\hat{\sigma}_i^{\pm}, \hat{\sigma}_j^z] = \mp\hat{\sigma}_i^{\pm}\delta_{ij}, \qquad [\hat{\sigma}_i^+, \hat{\sigma}_j^-] = 2\hat{\sigma}_i^z\delta_{ij}. \tag{3}$$

All the other commutators are zero.

The equations of motion for $\hat{S}_i^{\pm}$ and $\hat{\sigma}_i^{\pm}$

$$i\hbar\dot{\hat{a}} = [\hat{a}, \hat{H}] \tag{4}$$

yield in our case:

$$\pm i\hbar\dot{\hat{S}}_i^{\pm} = \mu H_0\hat{S}_i^{\pm} - J[\hat{S}_i^z(\hat{S}_{i-1}^{\pm} + \hat{S}_{i+1}^{\pm}) - \hat{S}_i^{\pm}(\hat{S}_{i-1}^z + \hat{S}_{i+1}^z)] + A(\hat{S}_i^z\hat{S}_i^{\pm} + \hat{S}_i^{\pm}\hat{S}_i^z) + d(\hat{S}_i^z\hat{\sigma}_i^{\pm} - \hat{S}_i^{\pm}\hat{\sigma}_i^z),$$
$$\pm i\hbar\dot{\hat{\sigma}}_i^{\pm} = \mu H_0\hat{\sigma}_i^{\pm} - J[\hat{\sigma}_i^z(\hat{\sigma}_{i-1}^{\pm} + \hat{\sigma}_{i+1}^{\pm}) - \hat{\sigma}_i^{\pm}(\hat{\sigma}_{i-1}^z + \hat{\sigma}_{i+1}^z)] + A(\hat{\sigma}_i^z\hat{\sigma}_i^{\pm} + \hat{\sigma}_i^{\pm}\hat{\sigma}_i^z) - d(\hat{S}_i^z\hat{\sigma}_i^{\pm} - \hat{S}_i^{\pm}\hat{\sigma}_i^z). \tag{5}$$

In the quasiclassical approximation, where the components of the spin operators are complex amplitudes, $\alpha_n = \hat{S}_n^+/S$, $\alpha_n^* = \hat{S}_n^-/S$, $\hat{S}_n^z/S = \sqrt{1-|\alpha_n|^2}$, $\beta_n = \hat{\sigma}_n^+/S$, $\beta_n^* = \hat{\sigma}_n^-/S$ and $\hat{\sigma}_n^z/S = \sqrt{1-|\beta_n|^2}$, we have

$$\begin{aligned}
i\hbar\frac{\partial\alpha_n}{\partial t} &= \mu H_0\alpha_n - JS\left[(\alpha_{n+1} + \alpha_{n-1})\sqrt{1-|\alpha_n|^2} - \alpha_n\left(\sqrt{1-|\alpha_{n+1}|^2} + \sqrt{1-|\alpha_{n-1}|^2}\right)\right] \\
&+ 2AS\alpha_n\sqrt{1-|\alpha_n|^2} + dS\left(\beta_n\sqrt{1-|\alpha_n|^2} - \alpha_n\sqrt{1-|\beta_n|^2}\right), \\
i\hbar\frac{\partial\beta_n}{\partial t} &= \mu H_0\beta_n - JS\left[(\beta_{n+1} + \beta_{n-1})\sqrt{1-|\beta_n|^2} - \beta_n\left(\sqrt{1-|\beta_{n+1}|^2} + \sqrt{1-|\beta_{n-1}|^2}\right)\right] \\
&+ 2AS\beta_n\sqrt{1-|\beta_n|^2} + dS\left(\alpha_n\sqrt{1-|\beta_n|^2} - \beta_n\sqrt{1-|\alpha_n|^2}\right).
\end{aligned} \tag{6}$$

The set of differential equations (6) describes our system.

3. Soliton solutions

We shall look for solutions in the form of amplitude-modulated waves

$$\alpha_n(t) = \varphi_n(t)e^{i(kn-\omega t)}, \quad \beta_n(t) = \psi_n(t)e^{i(kn-\omega t)}, \tag{7}$$

where k and ω are the wave number and the frequency of the carrier waves (the lattice constant equals unity) and the envelopes $\varphi_n(t)$, $\psi_n(t)$ are slowly varying functions of the position and time. In the continuum limit, when the soliton width is much larger than the lattice spacing ($L \gg 1$), we use $\varphi \sim O(1)$, $\frac{\partial\varphi}{\partial x} \sim O(2)$ and $\frac{\partial^2\varphi}{\partial x^2} \sim O(3)$ and remain only $O(3)$ terms. Then, equations (6) transform into the following coupled modified NLS equations for the envelopes:

$$\begin{aligned}
i\left(\frac{\hbar\partial\varphi}{S\,\partial t} + 2J\sin k\frac{\partial\varphi}{\partial x}\right) &= (\varepsilon - \hbar\omega S^1)\varphi - J\cos k\frac{\partial^2\varphi}{\partial x^2} + g|\varphi|^2\varphi + \frac{d}{2}(2\psi - |\varphi|^2\psi + |\psi|^2\varphi), \\
i\left(\frac{\hbar\partial\psi}{S\,\partial t} + 2J\sin k\frac{\partial\psi}{\partial x}\right) &= (\varepsilon - \hbar\omega S^1)\psi - J\cos k\frac{\partial^2\psi}{\partial x^2} + g|\psi|^2\psi + \frac{d}{2}(2\varphi - |\psi|^2\varphi + |\varphi|^2\psi),
\end{aligned} \tag{8}$$

where

$$\varepsilon = \mu H_0 S^{-1} - 2g - d\,, \quad g = J(\cos k - 1 - A/J)\,. \tag{9}$$

For $d = 0$ the uncoupled equations (8) possess soliton solutions of different types depending on the sign of the expression $gJ\cos k$. For negative values bright solitons exist while for positive values dark solitons are possible. It is interesting to point out that for the isotropic case ($A = 0$) equations (8) have bright-soliton solutions for $0 < k < \pi/2$ and dark-soliton solutions for $\pi/2 < k \leq \pi$. For the anisotropic case ($A \neq 0$) the situation is more complicated (figure 1).

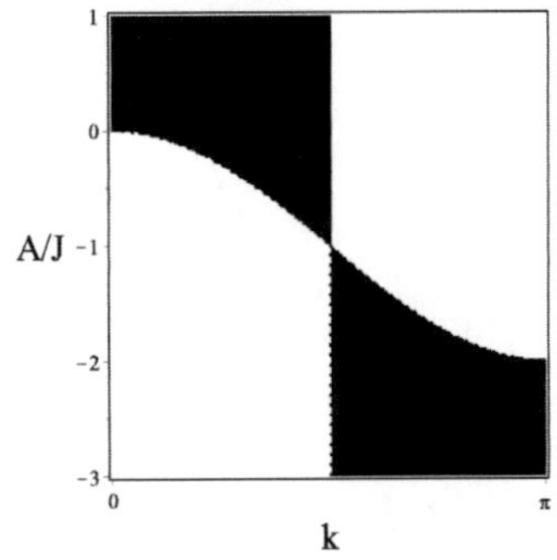

Figure 1. Regions in the Brillouin zone for the existence of bright (shaded region) and dark (light region) solitons depending on A/J.

In what follows, we shall consider bright-soliton solutions which appear for

$$\cos k(\cos k - 1 - A/J) < 0 \tag{10}$$

and $|\varphi(x)|^2 \to 0$, $|\psi(x)|^2 \to 0$ at $x \to \pm\infty$. Note that the condition (10) depends not only on the anisotropy constants but also on the wave number k. Insert

$$\varphi(x,t) = \varphi_0 \text{sech}\frac{x - vt}{L}\,, \quad \psi(x,t) = \psi_0 \text{sech}\frac{x - vt}{L} \tag{11}$$

in (8) one gets

$$\varphi_0^2 = \psi_0^2 = -\frac{2J\cos k}{gL^2}\,, \quad \bar{h}\omega = \varepsilon S - \frac{JS\cos k}{L^2}\,, \quad v = \frac{2JS}{\bar{h}}\sin k\,, \tag{12}$$

where L and v are soliton's width and velocity.

4. Numerical results

We shall investigate the evolution of a soliton which at the initial time $t = 0$ is launched in one of the chains at the position n_0

$$\alpha_n(0) = \frac{1}{L}\sqrt{\frac{2\cos k}{A/J + 1 - \cos k}}\,\text{sech}\frac{n - n_0}{L}e^{ikn}\,, \quad \beta_n(0) = 0 \tag{13}$$

solving numerically the system (6). The simulations are carried out for large enough chains compared to the soliton width L and periodic boundary conditions.

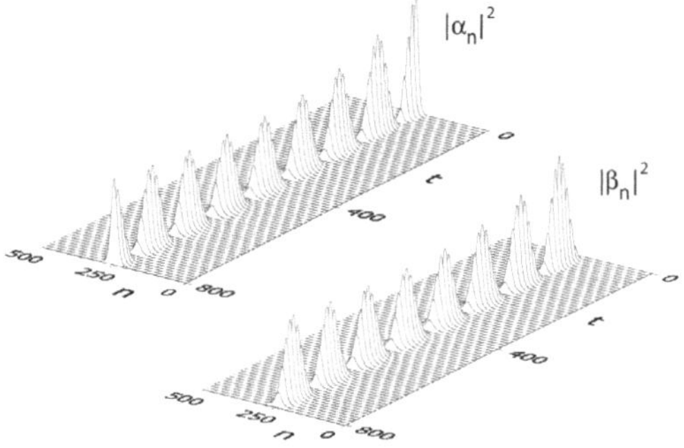

Figure 2. Perfect soliton switching for $J = S = 1$, $L = 10$, $A = 1$ and $|d| = 0.0314$. The time is in units of $\hbar/JS$.

In the linear case the excitation would be transferred from one chain to the other and back with a period $t_0 = \pi/|d|$, where d is the total linear coupling. In our complicated case a soliton can be transferred with the same period (perfect soliton switching) when the condition

$$\left|\frac{J\cos k}{2L^2 d}\right| \ll 1 \tag{14}$$

is fulfilled (figure 2). There is a critical value d_c below which the soliton will be not transferred. It depends on the wave number k i.e. the soliton velocity and when the velocity increases $(k \to \pi/2)$ we have $d_c \to 0$ (figure 3).

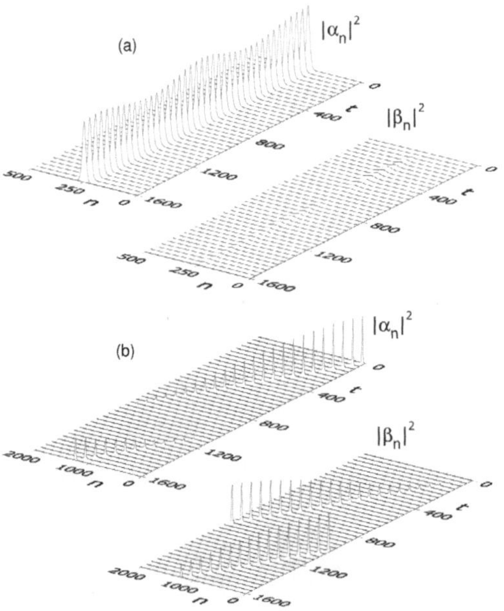

Figure 3. Soliton propagation in the two chains for $A = 1$, $d = 0.00157$ and different values of the wave number k. $k = 0$ (a) and $k = 1.5$ (b).

Let us point that our study is when the soliton amplitude is small ($\varphi_0^2 \ll 1$). First we consider only linear terms in the last bracket in (8). Further we have performed calculations to estimate the influence of the variety of nonlinear interactions in the system on the process of soliton switching. There are two types of nonlinear interchain interactions: Type I connected with the first term and type II connected with the second term proportional to d in the equations (6) or (8). Figures 4 and 5 illustrate the results for both positive and negative values of d. The elimination of the interchain nonlinear terms in equations (6) leads to perfect soliton switching with the period $t_0 = 100$ [figures 4(a) and 5(a)]. The inclusion of the nonlinear terms of type I leads again to perfect soliton switching but with a period $t_0 = 107$ which differs slightly from the period of the linear case [figures 4(b) and 5(b)]. The nonlinear interaction of type I has the role of an effective reduction of the coupling constant d and hence to an increase of the period. The inclusion of the nonlinear terms of type II leads also to perfect soliton switching but with a different soliton form [figures 4(c) and 5(c)]. Due to this type of nonlinear interactions the system (8) has the form of a Manakov system of coupled nonlinear equations which for $g = d/2$ is integrable and has soliton solutions. The parameters of the new solitons φ_1 and L_1

$$\varphi_1^2 = \varphi_0^2 \frac{A - d/2}{A}, \quad L_1 = L \frac{A}{A - d/2}. \tag{15}$$

can be determined using the conservation law for the number of particles of the initial excitation. This explains qualitatively our the numerical results. On figures 4(d) and 5(d) is shown the role of the interplay of both nonlinear interactions between the two chains.

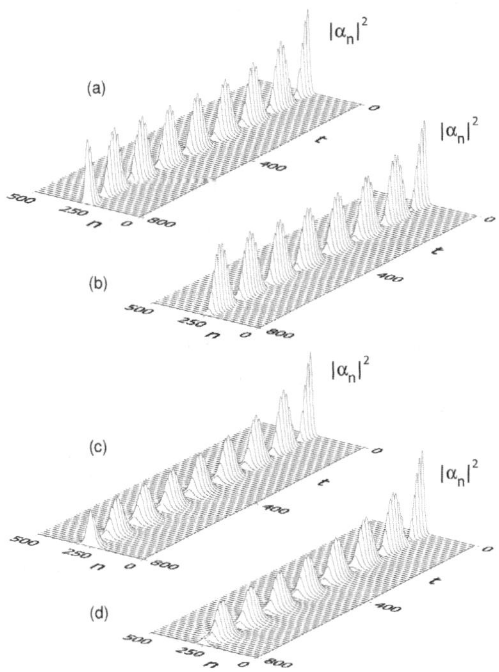

Figure 4. Soliton switching for $A = 0.025$ and $d = 0.0314$ with only linear coupling (a), linear coupling and nonlinear coupling I (b), linear coupling and nonlinear coupling II (c), linear coupling, nonlinear coupling I and II (d).

We can calculate the change of total magnetic moment in the system given by the expression

$$\Delta M = \mu S \int_{-\infty}^{\infty} \left(1 - \hat{S}^z(x,T)\right) dx\,, \tag{16}$$

where μ is the magnetic moment. Using (11) our calculations are performed up to the lowest approximation in ϕ_0^2

$$\Delta M = \mu S L \varphi_0^2\,. \tag{17}$$

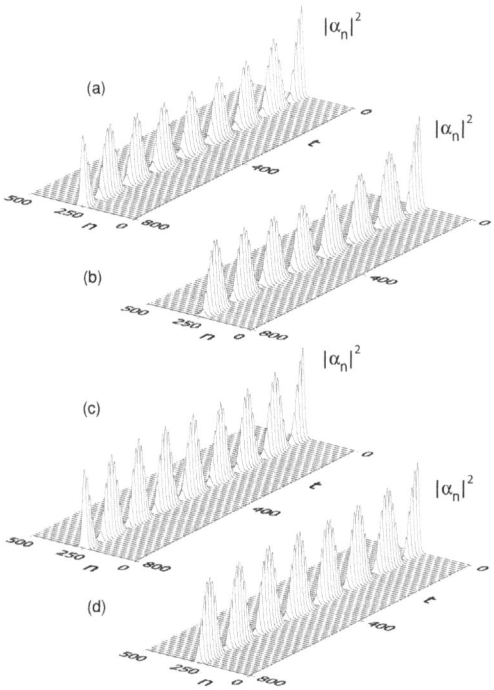

Figure 5. Soliton switching for $A = 0.025$ and $d = -0.0314$ with only linear coupling (a), linear coupling and nonlinear coupling I (b), linear coupling and nonlinear coupling II (c) and linear coupling, nonlinear coupling I and II (d).

5. Conclusion

We have investigated the propagation of a bright soliton in a system of two anisotropic ferromagnetic chains coupled with linear and nonlinear interactions. The condition for a perfect soliton switching is obtained. We have observed that in contrast to the linear interchain coupling for the nonlinear coupling not only its value but also its sign plays a crucial role for the soliton dynamics.

Acknowledgments

This work is supported by the Program for career development of young scientists, BAS (Grant agreement no: DFNP-157).

References

[1] Mikeska H-J and Steiner M 1991 *Adv. Phys.* **40** 191
[2] Tjon J and Wright J 1977 *Phys. Rev.* B **15** 3470
[3] Pushkarov D I and Pushkarov Kh I 1977 *Phys. Lett.* **61A** 339
[4] Huang G, Shi Z-P, Dai X and Tao R 1990 *J. Phys.: Condens. Matter* B **2** 8355
[5] Wallis R F, Mills D L and Boardman A D 1995 *Phys. Rev.* B **52** R3828
[6] Rakhmanova S and Mills D L 1998 *Phys. Rev.* B **58** 11458
[7] Slavin A N, Kivshar Yu S, Ostrovskaya E A and Benner H 1999 *Phys. Rev. Lett.* **82** 2583
[8] Bischof B, Slavin A N, Benner H and Kivshar Yu S 2005 *Phys. Rev.* B **71** 104424
[9] Zaspel C E, Mantha J H, Rapoport Yu G and Grimalsky V V 2001 *Phys. Rev.* B **64** 064416
[10] Ivanov B A, Galkin A Yu, Khymyn R S and Merkulov A Yu 2008 *Phys. Rev.* B **77** 064402
[11] Kivshar Yu S and Luther-Davies B 1998 *Phys. Reports* **298** 81
[12] Primatarowa M T and Kamburova R S 2010 *J. Phys.: Conf. Series* **253** 012022
[13] Jensen S M 1982 *IEEE J. Quantum Electron.* **18** 1580
[14] Trillo S, Wabnitz S and Stegeman G 1989 *IEEE J. Quantum Electron.* **25** 1907
[15] Paré C and Florjańczyk M 1990 *Phys. Rev.* A **41** 6287
[16] Kivshar Y S 1993 *Optics Lett.* **18** 7
[17] Chu P L, Peng G D and Malomed B A 1993 *Optics Lett.* **18** 328
[18] Primatarowa M T and Kamburova R S 2000 *Phys. Status Solidi B* **217** 769
[19] Kamburova R S and Primatarowa M T 2010 *AIP Conf. Proc.* **1203** 261
[20] Primatarowa M T and Kamburova R S 2012 *J. Phys.: Conf. Series* **398** 012041

IOP Conf. Series: Journal of Physics: Conf. Series **794** (2017) 012028 doi:10.1088/1742-6596/794/1/012028

Effective method for calculating modes of multilayer waveguide

I K Ivanov[1] and P D Dimitrov[1]

[1]Faculty of Physics, University of Plovdiv, 24 Tzar Assen Str., 4000 Plovdiv, Bulgaria

E-mail: ik_ivanov@yahoo.com

Abstract. A fast and efficient numerical method for finding the modes of a multilayered waveguide is proposed. Using the complex vector of the Riemann-Silberstein has resulted in a reduction of the order of the differential equations describing the passage of light through the different layers, as well as a double reduction of the searched variables. The approximation of the differential equations is made by the method of Galyorkin by a suitable choice of the base functions. The calculation of the effective indices and their corresponding wave configurations is realized by the inverse-shifting power method with Rayleigh's quantity. The method was successfully applied waveguide systems, generating values close in the effective indices.

1. Introduction

Waveguides are essential building blocks of diverse photonic integrated circuits such as polarizers, switches, filters, rotators, and modulators. To accomplish the desired characteristics, these functional devices are mainly designed by altering the material characteristics by using electro-optic, magneto-optic, and thermo-optic modulations. The determination of eigenmodes is a fundamental problem for waveguide optics. Several techniques are commonly used for the computation of the electromagnetic modes of waveguides, including finite element methods, mode-matching techniques, method of lines, and finite difference methods. Most significant are finite-element frequency-domain (FEFD)-based eigenvalue mode solvers [2, 3], the finite-element beam propagation methods (FE-BPMs), the finite-difference beam propagation method (FD-BPM), and the finite-difference frequency-domain (FDFD)-based eigenvalue mode solvers [6,7].

Many methods completely neglect the anisotropy of the constituent materials. The methods which take into consideration the material anisotropy require the diagonal permittivity tensor expressed in the coordinate system of the waveguide. For waveguide with layers of liquid crystals (LC) with arbitrary molecular orientations, the nine elements of the permittivity tensor are all non-zero.

In this paper we present an algorithm, based of Galjorkin's method of pseudoorthogonal functions for solving full-vector modes of optical waveguides with arbitrary permittivity tensor, i.e., with general three dimensional (3D) anisotropy. This can be observed in [6,7] as we incorporate perfectly matched layers (PML) as the absorbing boundary conditions at the outer boundaries of the computational domain. For the purpose of optimal numerical realization we have proposed the numerical scheme with only one complex array similar to the complex potential of Riemann- Silberstein [1], instead of four field components (two for transverse electric field components E_x, E_y and two for transverse magnetic field components H_x and H_y - total four arrays). The eigenvectors and eigenvalues of the waveguide were calculated using a dynamic shifted inverse power method with the Rayleigh quantity.

2. The model

Multi-layer waveguides are a system of layers with different refractive indices designed to limit the spread of the electromagnetic wave in a single (core) or in a small number of layers. If a material consists of an anisotropic layer (for example, represents a crystal, polymer, etc.) we have a 3D anisotropy. For modeling the properties and distribution of the electromagnetic (EM) waves we use the system of partial differential equations of Maxwell, under appropriate boundary conditions associated with the properties of media of layers through which they pass and reflect. We assume that the properties of the media are modeled by the non-diagonal and non-symmetrical complex permittivity tensor $[\varepsilon]$ and by the diagonal tensor of magnetic permeability $[\mu]$.

For the analysis of the waveguide, the computational domain is normally used (figure 1), where the waveguide cross-section in the transverse (x, y) plane is truncated and surrounded by PML (perfect matched layers), in which the attenuation of the wave does not depend on its direction, but by artificially introducing appropriate electrical and magnetic losses. The incorporation of PML regions allows the analysis of leaky modes.

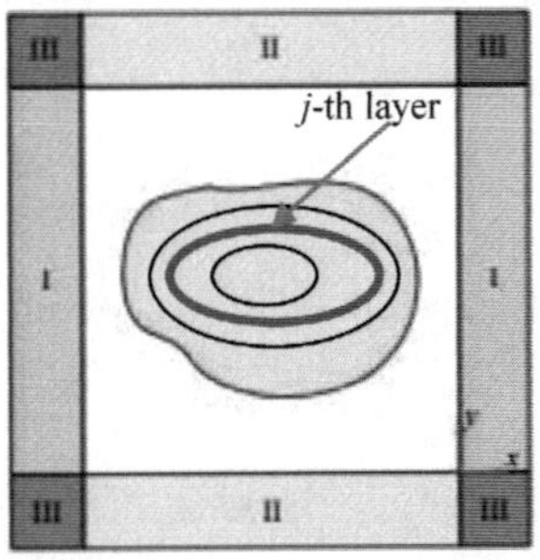

Figure 1. The cross-section of an arbitrary waveguide with the PML, placed at the edges of the computing domain.

A periodic dependence on a wave of time is assumed, i.e. the presence of the factor $\exp(i\omega t)$ where ω is the angular frequency and t the time. The direction of propagation of an electromagnetic wave is parallel to the axis $0z$, i.e. the expression $\exp(-i\beta z)$ describes the propagation, where β is the constant of propagation. It is also assumed, that the tensors of dielectric permittivity $[\varepsilon]_j$ and magnetic permeability $[\mu]_j$ in any "j-th" layer are independent of time t.

For the boundary conditions of the considered anisotropic waveguide in the PML regions, the permittivity and permeability tensors are taken to be:

$$[\varepsilon_{PML}] = \varepsilon_0 \begin{Bmatrix} \frac{s_y s_z}{s_x}\varepsilon_{xx} & s_z\varepsilon_{xy} & s_y\varepsilon_{xz} \\ s_z\varepsilon_{yx} & \frac{s_x s_z}{s_y}\varepsilon_{yy} & s_x\varepsilon_{yz} \\ s_y\varepsilon_{zx} & s_x\varepsilon_{zy} & \frac{s_x s_y}{s_z}\varepsilon_{zz} \end{Bmatrix}, \; [\mu_{PML}] = \mu_0 \begin{Bmatrix} \frac{s_y s_z}{s_x}\mu_{xx} & 0 & 0 \\ 0 & \frac{s_x s_z}{s_y}\mu_{yy} & 0 \\ 0 & 0 & \frac{s_x s_y}{s_z}\mu_{zz} \end{Bmatrix}. \quad (1)$$

Using ideas from papers [6, 7] we may determine the complex PML parameter s, which controls the field attenuation as follows:

$$s = 1 - i\alpha = 1 - i\frac{\sigma_e}{\omega\varepsilon_0 n^2} = 1 - i\frac{\sigma_m}{\omega\mu_0}.. \quad (2)$$

Here σ_e and σ_m are the electric and magnetic conductivities of the PML, ε_0 and μ_0 are permittivity and permeability of free space, and n is the refractive index of the adjacent computing domain. This relation means that the wave impedance of a PML medium exactly equals that of the adjacent medium in the computing window regardless of the angle of propagation.

2.1. Application of the modified vector of Riemann- Silberstein for reducing and simplifying the system of partial differential equations for calculating modes of an anisotropic waveguide.

Our new original proposal is to replace the six unknown calculated field components $\vec{E} = (E_x, E_y, E_z)$ and $\vec{H} = (H_x, H_y, H_z)$ with only three complex field components $\vec{F} = (F_x, F_y, F_z)$ by analogy [1] with Riemann- Silberstein vector. For this purpose, we present the permittivity and permeability tensors as:

$$\varepsilon_0[\varepsilon] = \varepsilon_0 \begin{Bmatrix} \varepsilon_{xx} & \varepsilon_{xy} & \varepsilon_{xz} \\ \varepsilon_{yx} & \varepsilon_{yy} & \varepsilon_{yz} \\ \varepsilon_{zx} & \varepsilon_{zy} & \varepsilon_{zz} \end{Bmatrix} = \varepsilon_0 \begin{Bmatrix} n_{xx} & n_{xy} & n_{xz} \\ n_{yx} & n_{yy} & n_{yz} \\ n_{zx} & n_{zy} & n_{zz} \end{Bmatrix}^2 = \varepsilon_0[n]^2.. \quad (3)$$

$$\mu_0[\mu] = \mu_0 \begin{Bmatrix} \mu_{xx} & 0 & 0 \\ 0 & \mu_{yy} & 0 \\ 0 & 0 & \mu_{zz} \end{Bmatrix} = \mu_0 \begin{Bmatrix} m_{xx} & 0 & 0 \\ 0 & m_{yy} & 0 \\ 0 & 0 & m_{zz} \end{Bmatrix}^2 = \mu_0[m]^2.. \quad (4)$$

Let vector $\vec{F}$ by analogy with Riemann-Silberstein vector be $\vec{F} = \sqrt{\frac{\varepsilon_0}{2}}[n]\vec{E} + i\sqrt{\frac{\mu_0}{2}}[m]\vec{H}$, $i = \sqrt{-1}$.

Then for each of the layers the according to the assumptions $\operatorname{div}(\vec{F}) = 0$ and $\operatorname{rot}(\vec{F}) = -\frac{\omega}{c}[n][m]\vec{F} = -\frac{\omega}{c}[\alpha]\vec{F}$, where $c = \frac{1}{\sqrt{\varepsilon_0\mu_0}}$ is the velocity of light in free space. Consequently, for each layer it is necessary to solve the differential system:

$$\begin{pmatrix} 0 & i\beta & \frac{\partial}{\partial y} \\ -i\beta & 0 & -\frac{\partial}{\partial x} \\ -\frac{\partial}{\partial y} & \frac{\partial}{\partial x} & 0 \end{pmatrix} \begin{pmatrix} F_x \\ F_y \\ F_z \end{pmatrix} = -\frac{\omega}{c} \begin{pmatrix} n_{xx}m_{xx} & n_{xy}m_{yy} & n_{xz}m_{zz} \\ n_{yx}m_{xx} & n_{yy}m_{yy} & n_{yz}m_{zz} \\ n_{zx}m_{xx} & n_{zy}m_{yy} & n_{zz}m_{zz} \end{pmatrix} \begin{pmatrix} F_x \\ F_y \\ F_z \end{pmatrix}. \quad (5)$$

To form the eigenmode equations we need to eliminate $\vec{F}_z$ from the third of equations (5), i.e.,

$$F_z = \left(\frac{c}{\omega n_{zz}m_{zz}}\frac{\partial}{\partial y} - \frac{n_{zx}m_{xx}}{n_{zz}m_{zz}}\right)F_x - \left(\frac{c}{\omega n_{zz}m_{zz}}\frac{\partial}{\partial x} - \frac{n_{zy}m_{yy}}{n_{zz}m_{zz}}\right)F_y.. \quad (6)$$

And after some obvious algebraic simplification we have:

$$\begin{pmatrix} A_{11} & A_{12} \\ A_{21} & A_{22} \end{pmatrix}\begin{pmatrix} F_x \\ F_y \end{pmatrix} = i\beta\begin{pmatrix} F_x \\ F_y \end{pmatrix} \tag{7}$$

$$A_{11} = -\frac{c}{\omega n_{zz} m_{zz}}\frac{\partial^2}{\partial x \partial y} + \frac{n_{zx} m_{xx}}{n_{zz} m_{zz}}\frac{\partial}{\partial x} + \frac{n_{yz}}{n_{zz}}\frac{\partial}{\partial y} + \frac{\omega}{c}\left(n_{yx} m_{xx} - \frac{n_{zx} n_{yz} m_{xx}}{n_{zz}}\right) \tag{8}$$

$$A_{12} = \frac{c}{\omega n_{zz} m_{zz}}\frac{\partial^2}{\partial x^2} + \left(\frac{n_{zy} m_{yy}}{n_{zz} m_{zz}} - \frac{n_{yz}}{n_{zz}}\right)\frac{\partial}{\partial x} + \frac{\omega}{c}\left(n_{yy} m_{yy} - \frac{n_{zy} n_{yz} m_{yy}}{n_{zz}}\right) \tag{9}$$

$$A_{21} = -\frac{c}{\omega n_{zz} m_{zz}}\frac{\partial^2}{\partial y^2} + \left(\frac{n_{zx} m_{xx}}{n_{zz} m_{zz}} - \frac{n_{xz}}{n_{zz}}\right)\frac{\partial}{\partial y} + \frac{\omega}{c}\left(\frac{n_{zx} n_{xz} m_{xx}}{n_{zz}} - n_{xx} m_{xx}\right) \tag{10}$$

$$A_{22} = \frac{c}{\omega n_{zz} m_{zz}}\frac{\partial^2}{\partial y \partial x} + \frac{n_{xz}}{n_{zz}}\frac{\partial}{\partial x} + \frac{n_{zy} m_{yy}}{n_{zz} m_{zz}}\frac{\partial}{\partial y} + \frac{\omega}{c}\left(\frac{n_{xz} n_{zy} m_{yy}}{n_{zz}} - n_{xy} m_{yy}\right) \tag{11}$$

When the waveguide media can be represented by diagonal permittivity tensor $[\varepsilon]$ and with an identity permeability tensor $[\mu]$, (i.e. when $[n]$ is related to the refractive index), which is fulfilled for most optical media, then a significant simplification is achieved:

$$A_{11} = -\frac{c}{\omega n_{zz}}\frac{\partial^2}{\partial x \partial y},\ A_{12} = \frac{c}{\omega n_{zz}}\frac{\partial^2}{\partial x^2} + \frac{\omega n_{yy}}{c},\ A_{21} = -\frac{c}{\omega n_{zz}}\frac{\partial^2}{\partial y^2} - \frac{\omega n_{xx}}{c},\ A_{22} = \frac{c}{\omega n_{zz}}\frac{\partial^2}{\partial y \partial x}. \tag{12}$$

After successfully solving the eigenmode system (7) for each eigenvalue β and eigenvector $\vec{F}$ we can easily find configurations of corresponding mode fields $\vec{E}$, $\vec{H}$ and the density of their electromagnetic energy W by using the expressions:

$$\vec{E} = \frac{1}{\sqrt{2\varepsilon_0}}[n]^{-1}\left(\vec{F} + \vec{F}^*\right),\ \vec{H} = \frac{-i}{\sqrt{2\mu_0}}[m]^{-1}\left(\vec{F} - \vec{F}^*\right),\ W = \left|\vec{F}\right| - \vec{F}.\vec{F}^*, \tag{13}$$

where $\vec{F}^*$ is the complex conjugate of $\vec{F}$.

2.2. Application of Galjorkin's method with a new basis of pseudo-orthogonal functions for solving the differential eigen system.

We look for the solution similar to the method of collocation using Galjorkin's method assuming that solutions $F_x(x,y)$, $F_y(x,y)$ are presented by a linear combination of types :

$$F_x(x,y) = \sum_{i=1}^{N_1-1}\sum_{j=1}^{N_2-1} F_{xij}\varphi_{ij}(x,y) + F_x\left(x_0, x_{N_1}, y\right), \tag{14}$$

$$F_y(x,y) = \sum_{i=1}^{N_1-1}\sum_{j=1}^{N_2-1} F_{yij}\varphi_{ij}(x,y) + F_y\left(x, y_0, y_{N_2}\right), \tag{15}$$

where F_{xij} and F_{yij} are the exact values of the solution (F_x, F_y) at the point of the irregular rectangular grid with coordinates (x_i, y_j). Here N_1 and N_2 are the corresponding number of irregular grid divisions of variables x and y, i.e. $x = \left\{x_0, x_1, \cdots, x_i, \cdots, x_{N_1}\right\}$ and $y = \left\{y_0, y_1, \cdots, y_j, \cdots, y_{N_2}\right\}$. Functions $F_x\left(x_0, x_{N_1}, y\right) = F_x\left(x_0, y\right) + F_x\left(x_{N_1}, y\right), F_y\left(x, y_0, y_{N_2}\right) = F_y\left(x, y_0\right) + F_y\left(x, y_{y_{N_2}}\right)$ are chosen to satisfy the boundary conditions.

Unlike commonly used in finite element methods linear functions (having derivatives different from zero of only the first order) or spline functions (having derivatives different from zero to a maximum of second or third order) the proposed functions have derivatives of any order.

Pseudo-orthogonal functions $\varphi_{ij}(x, y)$ as we presented in [2] are equal to 1 at grid point (x_i, y_j) or equal to 0 at any other point on the grid. They have non-zero derivatives of any row and are defined as:

$$\varphi_{ij}(x, y) = \begin{cases} 0, & x \notin [x_{i-1}, x_{i+1}],\ y \notin [y_{j-1}, y_{j+1}] \\ \sin\left(\frac{\pi}{2}\cdot\frac{x - x_{i-1}}{x_i - x_{i-1}}\right)\sin\left(\frac{\pi}{2}\cdot\frac{y - y_{j-1}}{y_j - y_{j-1}}\right), & x \in [x_{i-1}, x_i],\ y \in [y_{j-1}, y_j] \\ \sin\left(\frac{\pi}{2}\cdot\frac{x_{i+1} - x}{x_{i+1} - x_i}\right)\sin\left(\frac{\pi}{2}\cdot\frac{y - y_{j-1}}{y_j - y_{j-1}}\right), & x \in [x_i, x_{i+1}],\ y \in [y_{j-1}, y_j] \\ \sin\left(\frac{\pi}{2}\cdot\frac{x - x_{i-1}}{x_i - x_{i-1}}\right)\sin\left(\frac{\pi}{2}\cdot\frac{y_{j+1} - y}{y_{j+1} - y_j}\right), & x \in [x_{i-1}, x_i],\ y \in [y_j, y_{j+1}] \\ \sin\left(\frac{\pi}{2}\cdot\frac{x - x_{i-1}}{x_i - x_{i-1}}\right)\sin\left(\frac{\pi}{2}\cdot\frac{y_{j+1} - y}{y_{j+1} - y_j}\right), & x \in [x_i, x_{i+1}],\ y \in [y_j, y_{j+1}] \end{cases} . \quad (16)$$

Therefore, we receive linear algebraic system for determining eigenvalues β and eigenvectors U.

$$\sum_{i=1}^{N_1-1}\sum_{j=1}^{N_2-1} U_{ij} \int_{x_0}^{x_{N_1}}\int_{y_0}^{y_{N_2}} \hat{A}\varphi_{ij}(x, y)\varphi_{km}(x, y)dxdy = i\beta \sum_{i=1}^{N_1-1}\sum_{j=1}^{N_2-1} U_{ij} \int_{x_0}^{x_{N_1}}\int_{y_0}^{y_{N_2}} \varphi_{ij}(x, y)\varphi_{km}(x, y)dxdy, \quad (17)$$

where $\hat{A} = \begin{pmatrix} A_{11} & A_{12} \\ A_{21} & A_{22} \end{pmatrix}$, $U = \begin{pmatrix} F_x \\ F_y \end{pmatrix}$ for each $k = 1, 2, \cdots, N_1 - 1$ and $m = 1, 2, \cdots, N_2 - 1$.

3. Numerical examples

As a first example we solve a 19 layer $Ga_xAl_{1-x}As$-AlAs waveguide structure for which the nine TE mode effective indices are nearly degenerated [4]. The index profile consists of six thin layers of $Ga_{0.8}Al_{0.2}As$ separated by AlAs layers and the three layers of $Ga_{0.6}Al_{0.4}As$ that are also separated by AlAs layers.

Table 1. Calculated TEn mode normalized propagation constants by our method, compared with the results presented in [4].

Mode	Our results at 4096 divisions	Results in[4]
TE0	3.0130388 + 0.000000009i	3.013039
TE1	3.0130253 + 0.000000003i	3.013025
TE2	3.0130091 + 0.000000011i	3.013008
TE3	3.0129832 + 0.000000037i	3.012984
TE4	3.0129566 + 0.000000145i	3.012957
TE5	3.0129342 + 0.000000101i	3.012933
TE6	3.0129119 + 0.000000048i	3.012915
TE7	3.0128931 + 0.000000993i	3.012891
TE8	3.0127597 + 0.000003457i	3.012753

As a second example, we applied our method to the analysis of 3D anisotropy of a nematic LC channel optical waveguide, calculated through the pseudospectral method in [5].

Table 2. Calculated effective indices at various values of angle φ for the first seven modes, compared with results in [5].

Mode	Results for twist angle φ					Results for twist angle φ in [5]				
	0^0	30^0	45^0	60^0	90^0	0^0	30^0	45^0	60^0	90^0
1	1.674	1.674	1.675	1.675	1.674	1.674	1.674	1.674	1.674	1.673
2	1.630	1.629	1.628	1.629	1.630	1.629	1.627	1.627	1.629	1.631
3	1.620	1.623	1.619	1.617	1.614	1.620	1.622	1.620	1.617	1.613
4	1.575	1.576	1.576	1.572	1.570	1.574	1.574	1.574	1.572	1.570
5	1.553	1.544	1.546	1.554	1.561	1.553	1.543	1.546	1.553	1.560
6	1.533	1.538	1.533	1.525	1.513	1.534	1.540	1.534	1.524	1.514
7	1.501	1.502	1.501	1.502	1.501	1.502	1.502	1.502	1.502	1.502

4. Conclusions

This study has developed an efficient modesolver for solving anisotropic multilayer waveguides with full 3D anisotropy.

Using the method of Galjorkin with our new proposed system of pseudo-orthogonal functions leads to a sparse eigen algebraic system with band structure. The advantage of these functions is that they are sufficiently smooth and have derivatives of any order.

We generalized the idea of Silberstein when permittivity ε and permeability μ of the layer's media are tensors $[\varepsilon]$ and $[\mu]$. This is achieved by presenting them as the product of two identical tensors [n] and [m]. So for each layer, it is sufficient enough to solve only a single equation $\vec{F} = -\frac{\omega}{c}[n][m]\vec{F}$.

Instead of using the standard eigenvalue matrix equation for the constant of propagation β involving four transverse field components, we involve only two (as in the complex vector of Riemann- Silberstein) which halves the required computer memory and necessary calculation time.

Numerical results have shown that the proposed scheme is more efficient and still accurate for investigating all mode problems compared to the finite element methods or finite-difference methods that are generally used.

Algorithm has been successfully used to solve guided modes on a liquid-crystal optical waveguide with arbitrary molecular director orientation and for solving of 19-layered waveguide structure, in which case the nine TE mode effective indices are nearly degenerated.

Acknowledgments

This study was supported by Grant NI15-FFIT-005/23.04.2015 and Grant SPI15-FFIT-004/24.04.2015 of the University of Plovdiv (Bulgaria).

References

[1] Silberstein A L 1907 *Annalen der Physik* **327** 3 pp 579-86
[2] Ivanov I K and Dimitrov P D 2016 *BPU9, AIP Conf. Proc.* **1722** 220009
[3] Schulz V 2003 *IEEE Trans. Microw. Theory Tech.* **51** 4 pp 1086–95
[4] Offersgaard J F 1995 *J. Opt. Soc. Am. B* **12** 10 pp 1783-91
[5] Huang Chia Chien 2010 *Optics Express* **18** 25 pp 26632-34
[6] Fallahkhair A B, Li K S and Murphy T E 2008 *J. Lightwave Technol* **26** pp1423–31
[7] Ivanov I K 2015 *Bulgarian Chemical Communications* **47** Special Issue B pp 287-98

Structural and optical properties of $LuVO_4$ single crystals

D Dimitrov[1, 2, 4*], P Rafailov[1], V Marinova[2], T Babeva[2], E Goovaerts[3], Y F Chen[4], C S Lee[5], and J Y Juang[4]

[1]Institute of Solid State physics, Bulgarian Academy of Science, 72 Tzarigradsko Chaussee Blvd., Sofia, Bulgaria
[2]Intstitute of Optical Materials and Technologies, Bulgarian Academy of Science, Acad. G. Bonchev Str., 109, Sofia, Bulgaria
[3]Experimental Condensed Matter Physics, Department of Physics, University of Antwerpen, Universiteitsplein, B-2610, Antwerpen, Belgium
[4]Department of Electrophysics, National Chiao Tung University, 1001 Ta Hsueh Road, Hsinchu, Taiwan
[5]Department of Applied Chemistry, National Chiao Tung University, 1001 Ta Hsueh Road, Hsinchu, Taiwan

E-mail: *dzdimitrov@issp.bas.bg; ddimitrov@iomt.bas.bg

Abstract. The synthesis of large single crystals with good optical quality which is a preliminary condition for the practical applications of these materials frequently is complicated. It is found that large $LuVO_4$ single crystals with high optical quality are possible to be prepared using high temperature solution growth method. It is obtained by X-ray crystallographic analysis that the grown crystals possess centrosymmetric tetragonal structure with the point group symmetry D_{4h} and space group I41/amd (zircon-type structure). The unit cell parameters of a = 7.0236 Å, b = 7.0236 Å, c = 6.2293 Å, volume = 307.30(3) Å^3 are measured. The crystals composition as well as vanadium oxidation state were measured in order to confirm that the crystal phase is mainly $LuVO_4$. Optical transmission and Raman Spectroscopy are further performed on $LuVO_4$ single crystal to reveal the optical quality and structure details

1. Introduction

Rare-earth orthovanadates (RVO_4, where R = Y, La, Ce, Pr, Nd, Sm, Eu, Gd, Tb, Dy, Ho, Er, Tm, Yb and Lu) compounds demonstrate diverse applications in various fields. Due to their exceptional optical properties, as wide optical transparency and large birefringence, RVO_4 are potential candidates for optical isolators, circulators, beam displacers, and components for polarizing optics [1].

Up to now, crystal growth process of RVO_4 has been attempted by several methods. Slow cooling from solution [2], the Czochralski process [3], top-seeded solution growth [4], the laser-heated pedestal growth method [5], the floating-zone (FZ) method [6] and the micro-FZ method [7] have been already reported. All those methods offer single crystals with excellent quality, however with relatively small sizes. The reason is that RVO_4 melts congruently and although the melting temperatures are as high as 1400 °C and above, and consequently large crystals were not obtained.

Among RVO_4 single crystals, $LuVO_4$ is one with a particular interest as a laser host material due to the larger absorption cross section near 800 nm and larger emission cross section at 1.064 μm in comparison with other vanadate crystals [8]. These features are very desirable for diode pumped solid-state lasers, as they make it feasible to achieve highly efficient pumping and to realize low threshold laser operation with high optical-to-optical efficiency. The synthesis of large single crystals with good optical quality which is a prerequisite for the practical applications however is quite complicated. This paper reports the crystal growth process of large, high quality $LuVO_4$ crystals and study of their structural and optical properties.

.

2. Experimental details and discussion

2.1. Crystal growth, structural analysis and composition

Single crystals of $LuVO_4$ were grown by high temperature solution growth method. As a first step, polycrystalline $LuVO_4$ was synthesized by the solid state reaction. Stoichiometric amounts of Lu_2O_3 and V_2O_5 with minimum purity of 99.99% were mixed, compacted, and then calcinated in oxygen at 650 °C for 48 h. The reacted product was ground and mixed with V_2O_5 flux in V_2O_5: $LuVO_4$ =12:1 ratio. The mixture was melted and heated to 1100 °C for 48 h in a platinum crucible of 50 mm diameter and 60 mm depth, covered with a platinum lid. Single crystals were obtained by cooling the solution from 1100 to 700 °C at a cooling rate of 1 °C/h. The residual flux was separated from the as-grown crystals by decanting. The obtained crystals remained on the bottom and the walls of the crucible. The $LuVO_4$ single crystals with high optical quality have tetragonal rectangular shape and typical size of 6 x 8 x 0.5 mm.

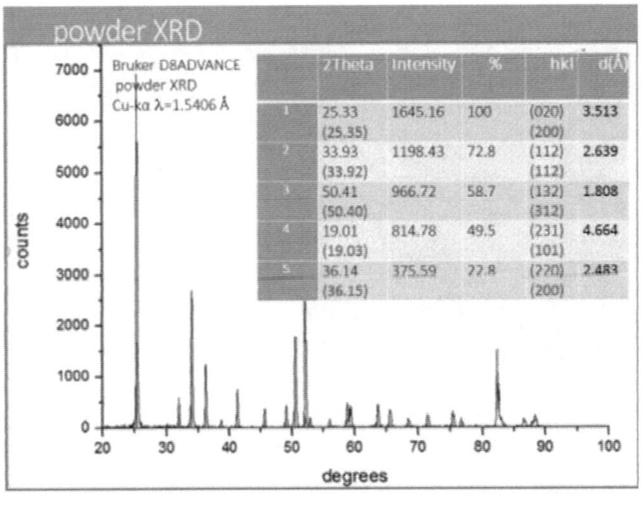

Figure 1: Powder XRD of $LuVO_4$ crystal (inset: main peaks assignment)

X-ray Powder Diffraction, using a D8 Bruker powder diffractometer equipped with a Cu anticathode, an incident slit of 0.51, an anti-scatter slit of 0.51, a detector slit of 1mm, and a graphite monochromator in the diffracted beam was performed on $LuVO_4$ powder prepared from single crystal sample. The powder diffraction pattern and main peaks assignments (inset) are shown in figure1.

A single crystal $LuVO_4$ specimen was used for the X-ray crystallographic analysis The X-ray intensity data were measured at room temperature using Bruker-AXS Smart Apex three-circle diffractometer equipped with a CCD detector which verified that the grown crystals possess centrosymmetric tetragonal structure with the point group symmetry D4h and space group I41/amd (zircon-type structure).

Table 1: Single crystal data

Chemical formula	$LuVO_4$	Unit cell dimensions	a=7.0236(3) Å
Formula weight	289.91		b=7.0236(3) Å
Temperature	296(2) K		c=6.2293(4) Å
Wavelength	0.71073 Å		α=90°,β=90°,γ=90°
Crystal system	tetragonal	Volume	307.30(3) $Å^3$
Space group	I 41/amd	Z	4
F(000)	504	Density(calculated)	6.266 g/cm^3
Absorption coefficient	34.794 mm^{-1}		

In this structure, the vanadium atom is in tetrahedral coordination, while the trivalent R cation is coordinated by eight oxygen atoms. The unit cell parameters of a = 7.0236(3) Å, b = 7.0236 Å, c = 6.2293 Å, volume = 307.30 $Å^3$ are measured. The structure was solved and refined using the Bruker SHELXTL Software Package, using the space group I41/amd, with Z = 4 for the formula unit, $LuVO_4$. A selection of the data obtained is shown in table 1.

The single crystals composition and the valence state of vanadium were further verified with EDAX Energy Dispersive Spectrometry and X-ray Absorption Spectroscopy (XAS) respectively. EDAX data (shown on figure 2) revealed some hafnium impurities in the $LuVO_4$ single crystals prepared by our method.

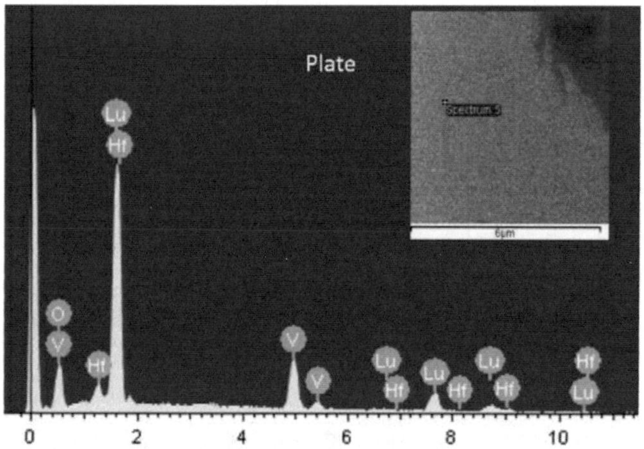

Element	App Conc.	Intensity Corrn.	Weight%	Weight% Sigma	Atomic%
O K	25.18	0.7718	17.45	1.05	60.37
V K	33.33	1.0129	17.60	0.62	19.12
Lu M	104.59	0.9650	57.95	1.30	18.34
Hf M	6.47	0.4941	7.00	1.55	2.17
Totals			100.00		

Figure 2: EDS spectra and composition data of $LuVO_4$ single crystal plate

For $LuVO_4$, the oxidation state of vanadium ion is very close to V^{5+}, since its V-L2,3 X-ray Absorption Spectroscopy (XAS) spectra (figure 3) is very similar to the spectra of reference V_2O_5, V2p and O1s spectra shown on figure 3 top (adapted from [9]) as well as $Ni_3V_2O_8$ [11], $Co_3V_2O_8$ [10], and YVO_4 [11]. Moreover, if there were V^{4+}, an enhancement of the structures below the main peaks will be observed, as observed for example in beta-$Sr_{0.17}V_2O_5$ [12]. However, this effect is absent in our single crystal $LuVO_4$. Therefore, we could conclude that V in $LuVO_4$ is in 5+ state.

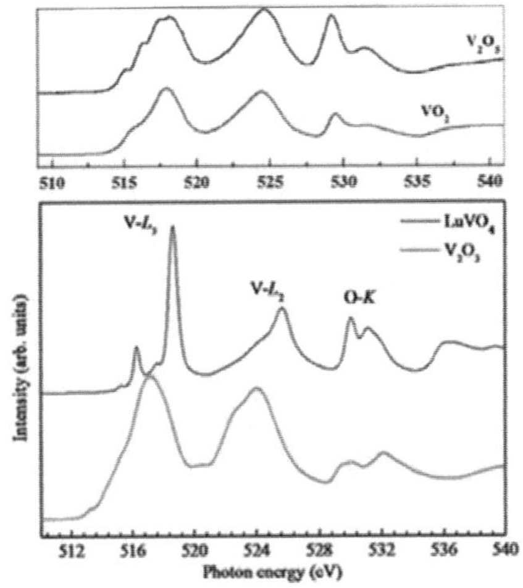

Figure 3: XAS spectra of $LuVO_4$ single crystal

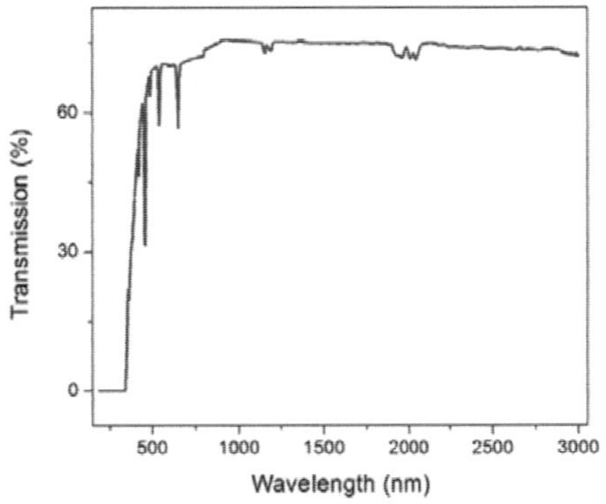

Figure 4: Optical Transmission of $LuVO_4$ single crystal 200-3000 nm

2.2. Optical transmittance and Raman spectroscopy analysis

The $LuVO_4$ single crystals are moreover analysed by Optical spectroscopy and Raman spectroscopy methods.

Optical transmission spectra were measured on crystal polished plate using Cary 5E UV-VIS spectrophotometer. As it seen from figure 4, the obtained crystal shows high transparency in a wide spectral range from 500 to 3000 nm. The absorption bands in the spectral interval 500-750 nm could be due to the hafnium impurities as shown in EDAX/EDS data.

For the Raman spectroscopic measurements single crystals with elongated shapes along the Z axis with naturally grown {100} and {001} surfaces were selected which additionally displayed well shaped {001} surfaces. The Raman spectra were measured in the range of (80 - 1200) cm^{-1} on a HORIBA Jobin Yvon LabRAM HR visible spectrometer equipped with a Peltier-cooled CCD detector. The 1.95 eV line of a He-Ne laser was used for excitation, the absolute accuracy being 0.5 cm^{-1}.

Lutetium vanadate $LuVO_4$ crystallizes in a centrosymmetric tetragonal structure with the point group symmetry D4h and space group I41/amd (zircon-type structure). The structure consists of Lu^{3+} ions and VO_4^{3-} tetrahedra that can be approximately regarded as separate units due to the strong internal bonds within each tetrahedron. Therefore, the lowest-frequency vibrations of $LuVO_4$ comprise translations and librations of the VO_4^{3-} tetrahedrons as rigid units against the Lu^{3+} ions. The higher-frequency phonons, on the other hand, are almost entirely due to internal vibrations of the VO_4^{3-} tetrahedral complexes.

For the polarized Raman measurements, the notations X (100), Y (010) and Z (001) for the main crystal axes as well as X' (110) and Y' (-110) were used. The measured spectra are presented in figure 5 with scattering configurations given in Porto notations. Raman selection rules are clearly discernible despite the considerable depolarisation effects caused by the extremely strong birefringence of the $LuVO_4$ crystal. These effects leading to partial mixing of allowed and forbidden intensity between different Raman modes [Porto] can be minimized by using focusing/collecting optics with small numerical aperture (N.A.) and keeping short the beam path within the simple. These two requirements were successfully met by using a 20× microscope objective with NA = 0.4. The spectra also contain features from impurities or second-phase inclusions that do not obey the Raman selection rules and obviously do not pertain to the vibrational spectrum of $LuVO_4$ crystal. Their frequencies lie in most cases far from $LuVO_4$ signal and with one exception they posed no obstacle to the assignment of the $LuVO_4$ modes.

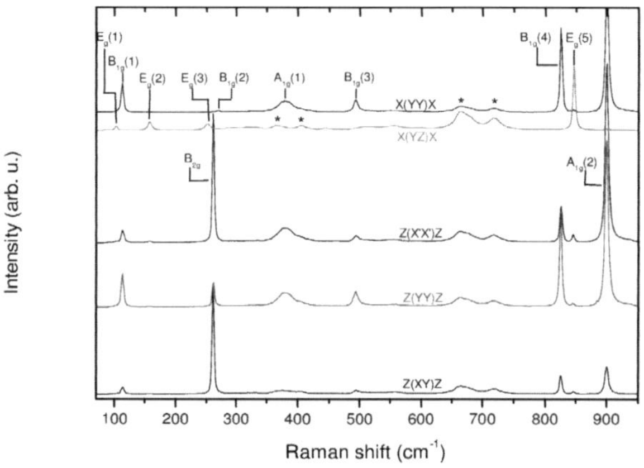

Figure 5: Raman spectra with mode assignment of $LuVO_4$ single crystal in different polarization configurations given in Porto notations. Features stemming from impurities or second-phase inclusions are marked with asterisks

3. Conclusion

Large, optically homogenous single crystals of $LuVO_4$ were prepared by using high temperature solution growth method. The single crystal structure quality details as well as composition and oxidation state of vanadium for $LuVO_4$ crystal phase were confirmed. The samples possess high optical transparency in the spectral interval 500-3000 nm.

Acknowledgements

Financial support by the EU 7th FP REGPOT project INERA (GA No 316309), Ministry of Science and Technology, Taiwan under contracts MOST 103-22221-E-009-079 and MOST 105-2811-M-009-044, Bulgarian Science Fund under the project FNI T-02/26 and bilateral collaboration project supported by the Fund for Scientific Research Flanders (FWO), Belgium and the Bulgarian Academy of Sciences.

References

[1] Oka K, Unoki H, Shibata H and Eisaki H 2006 *J. Crystal Growth* **286** 288
[2] Tanner B K and Smith S H 1975 *J. Crystal Growth* **28** 77
[3] Chow K and McKnight H G 1973 *Mater. Res. Bull.* **8** 1343
[4] Loriers J and Vichr M 1972 *J. Crystal Growth* **13/14** 593
[5] Eedei S and Ainger F W 1993 *J. Crystal Growth* **128** 1025
[6] Muto K, Awazu K 1969 *Japan. J. Appl. Phys.* **8** 1361
[7] Shonai T, Higuchi M, Kodaira K 2000 *Mater. Res. Bull.* **35** 225
[8] Zelmon D E, Northridge J M, Lee J J, Currin K M, and Perlov D 2010 *Appl. Opt.* **49** 4973
[9] Lu Y R, Hsu H H, Chen J L, Chang H W, Chen C L, Chou W C and Dong C L 2016 *Phys. Chem. Chem. Phys.* **18** 5203
[10] Laverock J, Chen B, Preston A R H, Smith K E, Wilson N R, Balakrishnan G, Glans P A and Guo J H 2013 *Phys. Rev. B* **87** 125133
[11] Herrera G, Jiménez-Mier J, Wilks R G, Moewes A, Yang A W and Denlinger J 2013 *J. Mater. Sci.* **48** 6437
[12] Li Q, Kolb B, Román-Pérez G, Soler J M, Yndurain F, Kong L, Langreth D C and Thonhauser T 2011 *Phys. Rev. B* **84** 155103

IOP Conf. Series: Journal of Physics: Conf. Series **794** (2017) 012030 doi:10.1088/1742-6596/794/1/012030

Sub-molecular dynamics in liquids and polymers

Hristo A Solunov

University of Plovdiv „P. Hilendarski", 4000 Plovdiv, Bulgaria.

E-mail: solunov@uni-plovdiv.bg

Abstract. An extension of Adam-Gibbs theory at its molecular level is proposed. New relations, measuring the number of the configurations within the cooperatively rearranging regions, are found. By these relations the transformation of activated-hopping to free-diffusion dynamics can be observed as a measurement of the number of the sub-molecular dynamic units, known in the literature as "beads". Two new methods for measuring the number of the beads in the molecules have been found. One of the methods used the thermodynamics results, and the second one used both the thermodynamics and relaxation ones. The measured number of beads in the twelve liquids is compared with the measuring by other theories. The evolution of the liquid state by measuring the number of the configurations within the cooperatively-rearranging regions is presented. With the suggested equations, the molecular level of Adam-Gibbs theory becomes a completed theory, permitting all molecular quantities, in the theory, to be measured by experimental results.

1. Introduction

1.1 Adam-Gibbs conception for cooperative molecular dynamics

One of the most successful approaches to dealing with the structural relaxation has been the Adam-Gibbs (AG) theory [1]. According to the theory the temperature dependence of the relaxation times $\tau(T)$ is determined by the probabilities of cooperative rearrangements. AG defined "a cooperatively rearranging region as a subsystem of the sample which, upon a sufficient fluctuation in energy, can rearrange into another configuration independently of its environment. The number of molecules or the (monomeric segments in the case of polymers) in such a cooperative region will be denoted by *z*.'' The relation is

$$\tau(T) = A_G \exp\left(\frac{z^*\Delta\mu}{kT}\right), \quad (1)$$

where A_G is a constant, z^* is a critical lower limit size of cooperatively rearranging region (CRR), $\Delta\mu$ is largely the potential energy hindering the cooperative rearrangement per monomer segment, *k* is Boltzmann constant and *T* is temperature. AG expresses the critical size z^* using the configuration entropy of this region

$$s^*_{cz^*} = \frac{\Delta s_\mu}{N_A} z^* = k\ln 2, \quad (2)$$

where ΔS_μ is molar configuration entropy and N_A is Avogadro number. Inserting z^* from equation (2) into equation (1) they obtain

$$\tau = A_G \exp(\frac{C}{T\Delta S_\mu}), \tag{3}$$

where the constant

$$C = N_A \Delta\mu ln2. \tag{4}$$

AG assume the number of the configurations in Boltzmann equation for $s^*_{cz^*}$ to be w^*_z=2 in agreement of their conception for simultaneously surmounting the individual potential barriers by the molecules into the CRR.

It will be mentioned that AG theory has two levels a molecular given by equation (1) and a macroscopic one given by equation (3).

1.2. Wunderlich's introduction of the sub-molecular units in the dynamics of liquids

By investigating a variety of the superrcooled liquids Wunderlich [2] found that, if the molecules (or monomeric segments in polymer) are assumed as constructed by sub-units, named beads, the value of heat capacity differences between liquid and the glass ΔC_p (T_g)=11.3 *J/(K mol of beads)* appeared as „universal'', where T_g is the glass-transition temperature. Wunderlich introduced empirical rules for estimating the number of beads in the molecules and units as -CH_2-, -CH_3, -CH-, or -O- are counted as beads. Wunderlich's beads coincided with Bunn's "chain units," connected with "round rotation bonds'' as one of the factors determining the melting temperature. Accepting the conception of the beads, Bestul and Chang [3] defined the"universal'' rule for the configuration entropy $\Delta S(T_g)$=2.9 *J/ (K mol of beads)*.

In an extended AG theory [4-6] the beads have been accepted as basic dynamics units, and on the basis of "universal'' rules, the number of beads in the CRR at the T_g has been estimated as 4 beads [4]. A method for measuring the number of beads in the molecules by investigating the pressure dependence of relaxation times has been obtained [5].

The theory of the random first-order transition [7] also accepts the beads as basic dynamic units in the liquids, and the entropy of fusion of a Lennar-Jones spherical bead at the melting point has been estimated as 13.8 *J/K mol beads.* In this way the ratio of the mol entropy $\Delta S_\mu(T_m)$ to this value, measured the number of beads in the molecules. Naturally this is an approximation method as the Lennar-Jones potential can be applied to a very close class of liquids.

An approach to measure the number of beads in the molecular liquids has been suggested by Moynihan and Angell [8]. In this bond lattice model of liquids, only one part of the ΔS_μ curve was fitted by the estimated number of beads.

Until recently in an entropy theory [9], the chemical groups, are being used as the basic dynamic units. By reason of the huge mathematical complexity of the theory the calculation has been done for the polymers constructed by –CH-,-CH_2- and –CH_3 groups coinciding with the beads, naturally the results obtained confirm the majority of the results of the extended AG theory.

Although the importance of the identification the basic dynamics units in the molecular dynamics of liquids is great, till now only empirical rules, [2, 3] and approximated [7, 8] or time consuming [5] methods exist.

The aim of this study is to find methods for measuring the number of the basic configuration units in the molecule of liquids, directly from the extended AG theory. On the way to finding a solution of this task it was necessary to estimate the number of the configurations in the CRR. The relations for this number have been found.

2. Theory

2.1. Including the sub-molecular dynamic in the Adam-Gibbs theory

The extended AG theory has some common as well as some different issues, with AG one, discussed in [4-6]. As has been mentioned not the molecules but the beads have been accepted as basic dynamic units, and instead of the equation (2) the configuration entropy of the CRR is given by

$$s_{cz}=\frac{\Delta S_\mu}{nN_A}z, \quad (5)$$

where n is the number of the beads in the molecule, hence n is a new parameter not exist in the original AG theory, measuring the configuration sub-units in the molecule. The number of the occupation states or the configurations in the CRR w_z was not fixed

$$s_{cz}= kln w_z. \quad (6)$$

From equations (5) and (6) we obtain

$$Rln w_z=\frac{\Delta S_\mu}{n}z. \quad (7)$$

2.2. Measuring the size of the CRR in the extended Adam-Gibbs theory

In the extended AG theory the size of the CRR has been found out by using the VTF equation for the temperature dependence of the relaxation times

$$\tau(T) = Aexp(\frac{B}{T-T_0}), \quad (8)$$

where A and B and the T_0 are empirical constants. In the extended AG theory an equation similar to the equation (1) is obtained with replacing z^* with the most-probable size of the CRR in the system $z\ (T)$ and $\Delta\mu$ with the average potential barrier of the beads within this region. Looking in this way at the equation (1) it becomes equivalent to equation (8), both governing the most probable relaxation time and, thus, we obtain

$$z(T) = \frac{T}{T-T_0}, \quad (9)$$

$$\Delta\mu = kB, \quad (10)$$

and A_g=A. It was accepted and proved that equation (9) gives the number of beads in the CRR.

2.3. The number of configurations in the CRR

Coming back to equation (7) we realize that with the experimental values of ΔS_μ, $z(T)$, estimated by equation (9) and the empirical vales of n, the number of configurations in the CRR w_z can be measured. It has been done for different classes of liquids, and the specific constant for every liquid was obtained. It is interesting that the number of configurations per a bead in the CRR $w_z/z(T_g)$, as the average values, were obtained to be equal; for the alcohols it was 1.97, for the oxygen glass it was 1.9 and for the low molecular liquids ≈2 [6]. Having in mind these results, we assume

$$w_z=2z(T_g). \quad (11)$$

The equation (11), obtained, seems to be logical as every rearrangement by means of overcoming a potential barrier separates two equilibrium states, or, two conformations. On the other hand, as the $z(T)$ decreases with the increasing the temperature, while the w_z=constant, this means that part of the configurations are transformed into free-diffusion or non-activated w_{zf} and

$$w_z{=}2z(T_g) = 2z(T) + w_{zf}(T). \tag{12}$$

2.4. Measuring the number of the configuration beads in the molecules
From equations (7) and (11) it follows that

$$n = \frac{\Delta S_\mu (T) z(T)}{Rln2z(T_g)}. \tag{13}$$

This is a general equation for measuring the number of the configuration beads in the molecules. From the relations in [6] it can also be presented as

$$n = \frac{\Delta C_p (T_g) T_g}{T_K\, Rln2z(T_g)}, \tag{14}$$

where T_K is Kauzman temperature excepted to be equal of T_0. Ties equation can be applied to $\Delta C_p (T_g)$ jump between the glass and the liquid for the non-crystallizing liquids, and to both the crystal and the liquid, as well as for the glass and liquid for one and the same substance. It will be mentioned that the configuration entropy, defined by AG, at present, is assumed as exceed entropy and is larger than the net configuration one, due to the vibration entropy difference between the liquid and the crystal [8-10]. Hence the measurement by equation (13) gives the number of beads or the degree of freedom according to the crystal, as is done in the other theories, while by using $\Delta C_p (T_g)$ between the glass and the liquid will eliminate the difference between the vibration entropies and, practically, the results will be then, corresponding to the net configuration entropy.

In the extended AG theory the constant C in equation (3) is given by $C = Unlnw_z$ [6] where $U = N_A \Delta\mu = RB$ or $C = RBnlnw_z$. Hence, n can be measured by

$$n = \frac{C}{BRln2z(T_g)}. \tag{15}$$

3. Application of the theory

3.1. Measuring the number of beads in the molecules of liquids by experimental results
In table I the thermodynamic results for the substances investigated in [8] are plotted. The majority of these liquids are also investigated in [7], and in this way, we can compare the measured values for n on one and the same basis. The measured number of the beads is marked as; n_e estimated by the empirical rules [11], n_t by the thermodynamic method, suggested in this paper, n_A by the excitation model [8], and n_W by the first- order transition theory [7].

The first liquid in table 1 is 3-methylpentane, where in the third place in the molecule one side - CH_3 group is attached to the backbone of it. All -C-C- bonds in the molecule are single, permitting rotation, and this is the reason why n=6 beads, as estimated in [11]. It will be mentioned that in the very similar molecule of 2-methy pentane, where in the same manner one -CH_3 group is attached in the second place in the molecule, n=5 has been assumed in [3]. As can be seen from the table I the beads, nearest to 5-6 numbers in this molecule, are measured by the method, suggested, in this paper. The same concerns for the second liquid 3–brompentane. In alcohols from Ethanol to Glycerol all methods find out an increase in the number of the beads with the increase the number of structural units. And again, the number of beads, measured by method, suggested in this paper, is closest to the

Table 1. Thermodynamic data and the number of the configuration beads

	Substances	T_m K	T_g K	T_K K	$\Delta S_\mu(T_m)$ *J/mol K*	$z(T_g)$	n_e *beads*	n_t *beads*	n_A *beads*	n_W *beads*
1	3-methylpentane CH_3-CH_2-$CHCH_3$-CH_2-CH_3	110.3	77	35.6	48.10	3.6	6	5.9	3	3.5[a]
2	3-brompentane CH_3-CH_2-CHBr-CH_2-CH_3	167.3	108	84.0	50.19	4.5	-	5.5	3	3.6
3	Ethanol CH_3-CH_2-OH	159	97	71.5	31.01	3.8	3	3.3	2	2.3
4	n-propanol CH_3-CH_2-CH_2-OH	148.8	96.2	71.8	36.25	3.9	4	4.1	2	2.6
5	Ethylene glycol HO-CH_2-CH_2-OH	260.5	152.7	119.7	45.42	4.6	-	4.5	3	3.3
6	1.3 Propanediol HO-CH_2-CH_2-CH_2-OH	248.2	145	108.6	52.16	4.0	-	5.4	6	3.7
7	Glycerol HO-CH_2CH-OH-CH_2OH	291	190	136.7	62.8	2.6	6	8.7	7	4.5
8	Toluene C_6H_5-CH_3	172.2	117	99.9	37.25	6.8	4	3.9	1	2.7
9	Ethyl benzene C_6H_5-CH_2-CH_3	188.1	115	89.4	51.47	4.5	5	5.7	3	3.7[a]
10	o-therphenyl $C_{18}H_{14}$	329.4	246	204.1	52.2	5.9	9	6.7	2	3.8[a]
11	Se	494.3	304	210.7	12.46	3.3	-	1.4	1	0.9
12	$ZnCl_2$	591[b]	380	250	17.32	2.9	1	2	-	1.2

[a] The liquids not investigated in [9], measured by the method in the same reference.
[b] The data from [12] and [14].

empirical values and the corresponding structure of the molecules.

In table 1 Toluene is the simplest molecule that comprises a phthalate ring. Privalko [11] estimated the phthalate ring as equivalent to three beads, and by adding -CH_3 group, the molecule of Toluene is estimated to behave, as such of 4 beads. In the following molecule Ethyl benzene one -CH_2- is added and the configurations then increase. The molecule of o-therphenyl is constructed by three phenyl rings and Privalko then assumed every ring to be valid to 3 beads. It should be mentioned that the central ring is attached by one bond to every one side ring, and is not equivalent to them. Generally, the effective beads of the given structural unit would depend on the way it is attached or interacted with the nearest structural units. It is interesting that the number of beads measured for the o-therphenyl by the method, suggested in this study, practically coincides with the number, measured by means of the relaxation volume of a bead n=6.5 [5]. For Se and $ZnCl_2$ a very close number is estimated by all methods. As is can be seen from table 1 in the simplest case the beads coincided with the structural units but in general they are different.

For the 3-brompentane in [10] AG constant C=79.6 *kJ/mol* has been measured and in [13] the VTF constant B=861.2 *K*, and using $z(T_g)$ from table 1 by equation (15) we obtain n=5.1 beads.

From the $\Delta C_p\,(T_g)$=72 *J/mol K* between glass and liquid [14] and the data for 3-brompentane in table 1 by the equation (14) we then measured n=5.1 beads.

3.2. Transformation of the activated-hopping to free-diffusion dynamics

In figure1 the evolution of liquid dynamics measured by the number of the configuration in the 3-brompentane is plotted. It is assumed at T_g dynamics to be 100 % activated (for the most probable CRR) and with increasing the temperature this dynamics is continuously transformed in the free-diffusion one. At the crossover temperature T_c of the mode-coupling-theory (MCT) the free-diffusion

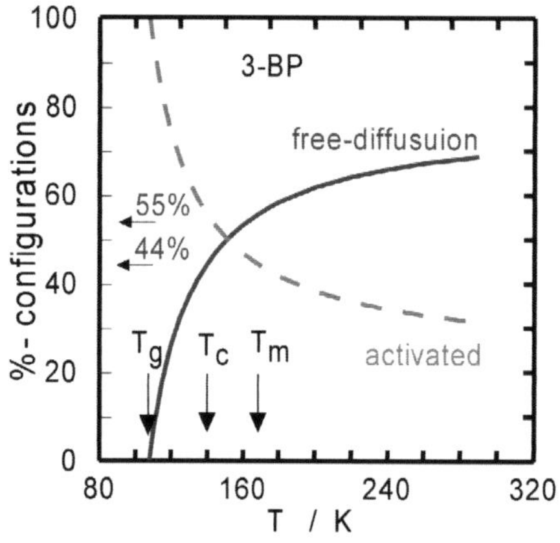

Figure 1. (Color on line). The temperature dependence of the configurations in 3-brompentane(3-BP) CRR. The numbers inside the figure are % s of free-diffusion configurations at T_c and T_m.

dynamics became 48% and at T_m it is equal to 55% from the total dynamics. For the 3-brompentane the T_c =140 *K*. According the MCT at T_c a crossing from the activation-hopping to the free-diffusion dynamics occurs. Hence the result in figure 1 may be accepted to demonstrate such a behavior. It will be mention that above transformation depends on the fragility of the liquids [15].

4. Conclusions

On the basis of extended AG theory a detailed dynamics in liquids emerges. The repeated units, the molecules in liquids or the monomeric segments in polymers, in dynamics break dawn to configuration beads. In the simplest case the beads coincided with the structural units (the chemical groups, the ions and the atoms) but, in general, they are different. During the relaxation time several beads are rearranging forming the cooperatively rearranging regions. The cooperatively rearranging regions can be smaller or larger than the molecules. At the glass transition temperature all rearrangements in the cooperative rearranging regions are activated-hooping. With increasing the temperature, above the glass-transition, the activated-hooping configurations are transformed in free-diffusing ones. At the melting temperature approximately half of the configurations are activated-hooping while the other half are free-diffusion ones.

References

[1] Adam G and Gibbs J 1965 *J. Chem. Phys.* **43** 139
[2] Wunderlich B 1960 *J. Phys. Chem.* **64** 1052
[3] Bestul B A and Chang S S 1964 *J. Chem. Phys.* **40** 373
[4] Solunov Ch 1999 *Eur. Polymer J.* **35** 1543
[5] Solunov H (Ch) 2002 *J. Phys. Condense Matter* **14** 7297
[6] Solunov H (Ch) 2005 *J. Optoelectronics and Advanced Materials* **7** 365
[7] Stevenson J D and Wolynes P G 2005 *J. Phys. Chem. B* **109** 15093
[8] Moynihan T and Angell C A 2000 *J. Non-Cryst. Solids* **274** 131
[9] Dudowich J, Douglas F and Freed F 2014 *J. Chem. Phys.* **141** 234903
[10] Takahara S, Yamamuro O and Matsuto T 1995 *J. Phys. Chem.* **99** 9589
[11] Privalko V P 1980 *J. Phys. Chem.* **84** 3307
[12] Angell C A 1997 *J. Res. Natl. Inst. Stand. Technol.* **102** 171
[13] Richert R and Angell C A 1998 *J. Chem. Phys.* **108** 901
[14] Wang L M Angell A C and Richert R 2006 *J. Chem. Phys.* **125** 074505
[15] Solunov H (Ch) 2006 *J. Non-Cryst. Solids* **352** 4871

IOP Conf. Series: Journal of Physics: Conf. Series **794** (2017) 012031 doi:10.1088/1742-6596/794/1/012031

Interaction of elaiophylin with model bilayer membrane

J Genova and M Dencheva-Zarkova
Institute of Solid State Physics, Bulgarian Academy of Sciences, 72 Tzarigradsko Chaussee, Blvd., 1784 Sofia, Bulgaria

E-mail: julia.genova@issp.bas.bg

Abstract. Elaiophylin is a new macrodiolide antibiotic, which is produced by the Streptomyces strains [1]. It displays biological activities against Gram-positive bacteria and fungi. The mode of action of this antibiotic has been attributed to an alteration of the membrane permeability. When this antibiotic is inserted into the bilayer membranes destabilization of the membrane and formation of ion-penetrable channels is observed. The macrodiolide antibiotic forms stable cation selective ion channels in synthetic lipid bilayer membranes. The aim of this work was to study the interactions of Elaiophylin with model bilayer membranes and to get information on the mechanical properties of lipid bilayers in presence of this antibiotic. Patch-clamp technique [2] were used in the study

1. Introduction

Ultrathin lipid membranes consist of lipid bilayers, self-assembled at the tips of patch pipettes from lipid monolayers. Monolayers are formed by spreading lipid in organic solvent onto the surface of electrolyte solutions [3].

Planar lipid bilayer is one of the unique electrophysiological techniques that is intended to study specific channel properties of the purified complexes in a well-controlled artificial environment [4]. The lipid bilayer enables study of functional activities of ion channels and monitoring ion channels behavior at the single molecule level in the model membranes.

Artificially created lipid vesicles mimic the cells and their membranes. The model bilayer membrane of lipid vesicles presents a simple model for investigation of membrane properties and can be formed from various lipids with controlled composition in controlled environment.

Elaiophylin also known as Azalomycin B, Gopalamicin or Salbomycin (figure 1) is a highly characteristic macrolide antibiotic that was first isolated from a culture of Streptomyces melanosporus [1]. It demonstrates antifungal activity, cytotoxicity against cancer cells and antibacterial activity aganst gram-positive bacteria [1,5,6]. Elaiophylin is thought to exert its antifungal activity by forming an ion-channel assembly [7]. The mode of action of Elaiophylin is attributed to an alteration of the membrane permeability and it is directly related to the effect of the drug on the lipid phase of membranes. The aim of the present study is to investigate the effect elaiophylin on the electric properties of lipid membranes formed from synthetic lipids.

In the present work the effect of Elaiophylin on the properties of lipid bilayers were studied with application of Voltage-clamp method.

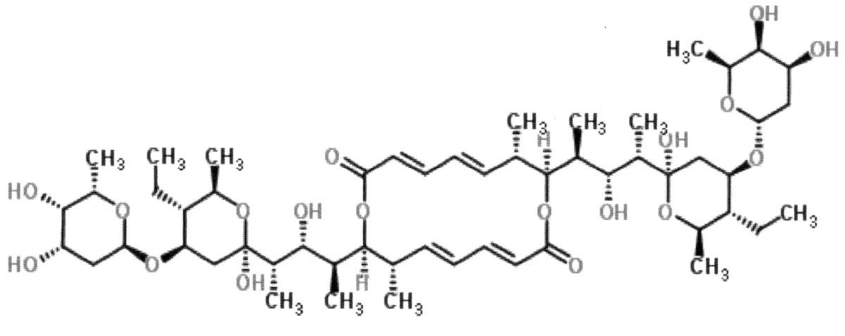

Figure 1. Chemical structure of Elaiophylin molecule.

2. Materials and methods

Soya bean phosphatidylcholine (L-α-Lecithin, Sigma, P5638), HEPES (buffer) and KCl (p.a.) were obtained from Sigma. Ethanol (p.a.) and n-hexane (p.a.) were purchased from Valerus Co. Elaiophylin were from Bioaustralis (BIA-E1028). All the chemicals were used without any further purification.

In the patch-clamp experiments lipid bilayers were self-assembled at the tips of patch pipettes, using the tip-dip patch clamp technique [2], from monolayers formed by spreading L-α-lecithin from soybean in n-hexane (10 mg/ml) onto the surface of electrolyte solution contained in Petri dishes (figure 2).

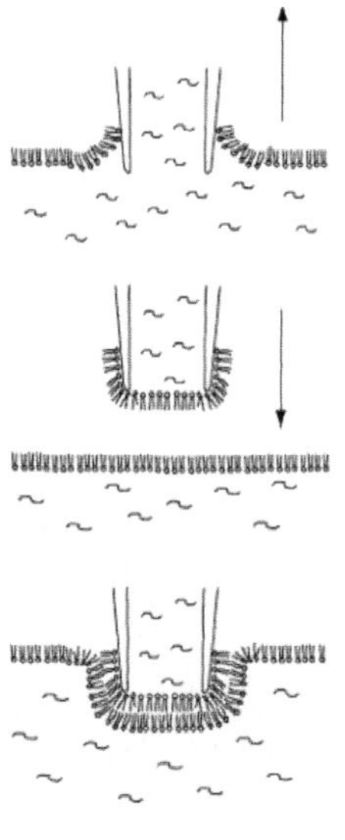

Figure 2. Scheme of lipid bilayer formation at the tip of patch pipette.

Patch pipettes (tip diameter 1-2 μm) and Petri dishes ($10cm^2$ area) were filled with aqueous solutions of KCl (1 M) buffered with HEPES (0.01 M) at pH 7. Only bilayers with seal resistances > 1GΩ were used. Elaiophylin, dissolved in ethanol, was added to the Petri dish to a final concentration of $5x10^{-7}$ g ml^{-1}. Single elaiophylin channel currents were monitored using a patch clamp amplifier Model 2400 (A-M Systems, Inc.). The currents were stored on a PC hard disc with 1 ms time resolution. Lipid bilayers were tested before and after introducing the ions with voltage ramps from -100 to +100 mV. The data were analyzed using current-voltage surfaces' software. All measurements were performed at room temperature (~25°C).

3. Results and discussion

3.1. Patch clamp technique

Elaiophylin forms long-lasting cation selective ion channels in planar lipid bilayer membranes prepared from soybean phosphatidylcholine.

As an explanation of the behaviour of elaiophylin in a lipid membrane, we are discussing, a barrel-like aggregation of several Elaiophylin molecules whereby a channel is formed. As shown in figure 3(a), the planar macrodiolide rings of several Elaiophylin molecules, as their lipophilic parts, could interact with the aliphatic hydrocarbon part of the lipid bilayer. As shown in figure 3(b) an ion-conducting pore spanning half of the bilayer thus, could be formed by self-assimilations of several elaiophylin molecules [8].

a) b)

Figure 3. Models for the elaiophylin channel in the lipid bilayer. (a) Cross-section through the membrane. (b) Suggested "double-pore" structures of the channel.

Membrane current was measured by the voltage-clamp method [2]. The principle of the method is to isolate a patch of membrane electrically from the external solution and to record current flowing into the patch. Single channel recording yields information about unitary conductance and kinetic behaviour of ionic channels. Ion channels form aqueous pores across the lipid bilayer and allow ions of appropriate size and charge to cross the membrane. The channels are "gated" and usually open transiently in response to a specific perturbation in the membrane, such as a change in membrane potential [9].

Figure 4 shows an example of channels recorded in presence of Elaiophylin in lecithin membranes. These recording demonstrate conductance levels, which are typical for elaiophylin channels opening. The membrane current started to increase within about 5 min after addition of the elaiophylin and reached a relatively stable level after approximately 10 min. Currents of single elaiophylin channels were recorded at 0.5 μg/ml of the antibiotic concentration.

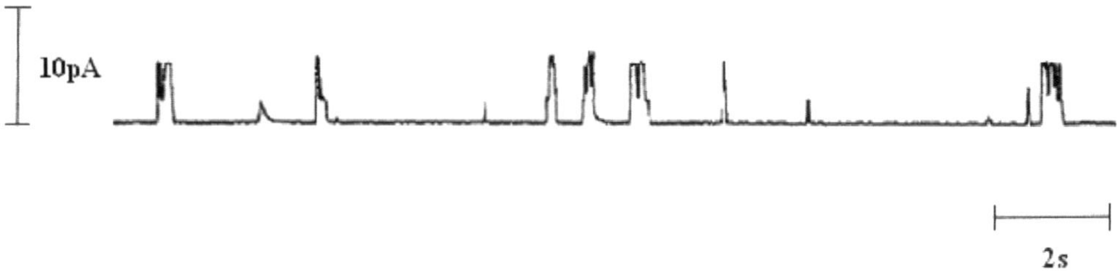

Figure 4. Traces of membrane current in presence of elaiophylin 0.5 μg/ml at the bilayer. Conditions: 1 M KCl, pH 7, membrane voltage—100 mV.

We observed that Elaiophylin has ionophoric activity and reduces the electrical resistance of the lipid membrane. The volt-ampere (I–V) characteristics were linear over a range of ± 100 mV. Multiple conductance states are seen only in the negative voltages' area (figure 5).

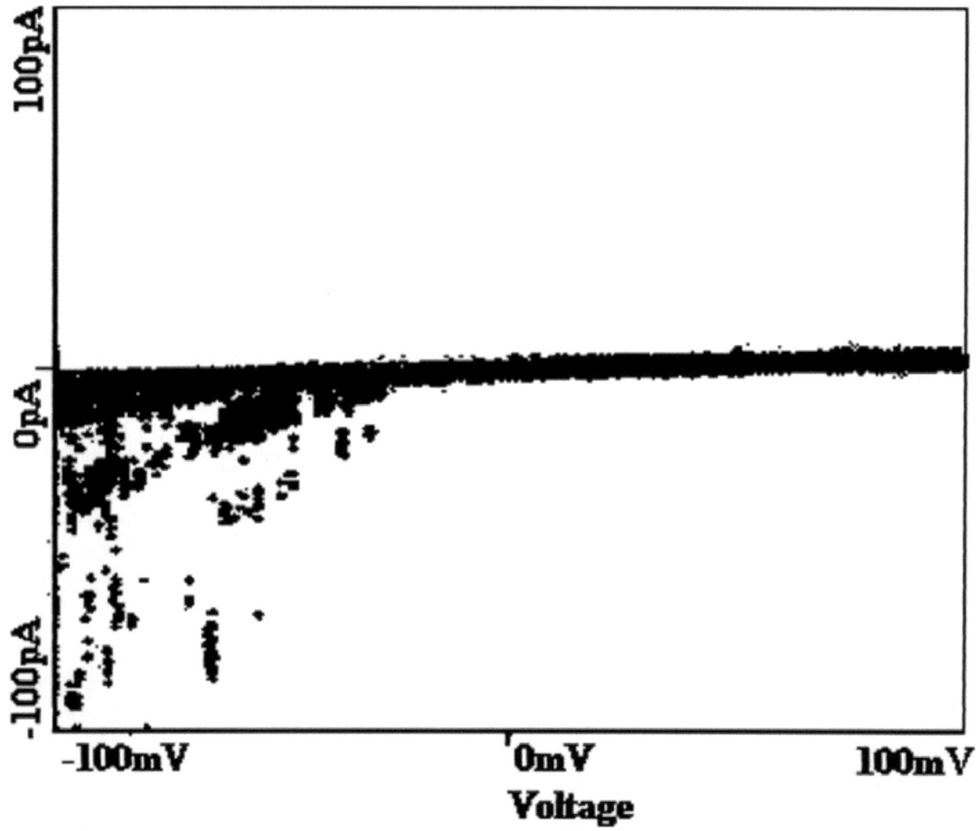

Figure 5. Elaiophylin channels' I-V surfaces in a patch clamped membrane after adding Elaiophylin to the bath (ramp. ±100mV; bath 1M KCl; pH 7.

4. Conclusions

Elaiophylin forms cation selective ion channels in model bilayer membranes with millisecond dwell times. The antibiotic has ionophoric activity and reduces the electrical resistance of the lipid membrane. The obtained volt-ampere characteristics were linear over a range of ± 100 mV.

The experiments in this work provide further evidence in support of the widely held view that Elaiophylin's primary mechanism of killing fungal and cancer cells is a result of formation of ion-permeable channels [5,6].

References

[1] Ritzau M., Heinze S., Fleck W., Dahse H. and Gräfe U. 1998 *J. Nat. Prod.* **61** 1337
[2] Veitinger S., 2014 *Sci. Lab. Leica Microsystems.* **10**
[3] Petrov A.G. 1999 *The Lyotropic State of Matter* Gordon and Breach Publs.L.- NY Chap.5
[4] Zakharian E 2013 *Methods Mol. Biol.* **998** 109
[5] Fang A., Wong G. K. and Demain A. L. 2000 *J. Antibiot. (Tokyo)* **53(2)** 158
[6] Zhao X., Fang Y. and Yang Y. 2015 *Autophagy* **11(10)** 1849
[7] Xu J., Wang X., Ensing B., Li M., Wu L. et al. 2001 *Drug Discov. Today* **6** 1278
[8] Grigoriev P., Schlegel R., Dornberger K. and Gräfe U. 1995 *Biochim. Biophys. Acta* **1237** 1
[9] Alberts B, Johnson A, Lewis J et al. 2002 *Molecular Biology of the Cell* New York: Garland Science From: Ion Channels and the Electrical Properties of Membranes

IOP Conf. Series: Journal of Physics: Conf. Series **794** (2017) 012032 doi:10.1088/1742-6596/794/1/012032

Effect of immobilization conditions on the properties of β-galactosidase immobilized in xanthan/chitosan multilayers

T Yovcheva[1], T Vasileva[2], A Viraneva[1], D Cholev[2], I Bodurov[1], M Marudova[1], V Bivolarski[2] and I Iliev[2]

[1]Department of Experimental Physics, Plovdiv University,
24 Tzar Assen str., 4000 Plovdiv, Bulgaria
[2]Department of Biochemistry & Microbiology, Plovdiv University,
24 Tzar Assen str., 4000 Plovdiv, Bulgaria

E-mail: ilievini@abv.bg

Abstract. The effect of lactose concentration on the activity of the immobilised enzyme β-galactosidase from *Aspergillus niger* has been evaluated, considering future applications for the production of galactooligosaccahrides with prebiotic potential. The following enzyme was immobilized in xanthan and chitosan polyelectrolyte multilayers (PEMs) deposited by dip coating method on polylactic acid positively corona charged pads. The pads were charged in a corona discharge system, consisting of a corona electrode (needle), a grounded plate, and a metal grid placed between them. Positive 5 kV voltage was applied to the corona electrode. 1 kV voltage of the same polarity as that of the corona electrode was applied to the grid. The chitosan layers were crosslinked with sodium tripolyphosphate (Na-TPP). The enzyme showed a temperature optimum at 50 °C and a pH optimum at 5.0. The immobilization was carried out over the different adsorption time and optimum conditions were determined. These results give insights for further optimization of transgalactosydase reactions in order to produce galactooligosaccharides with specific structure and having pronounced better prebiotic properties. For the determination of the surface morphology of the investigated samples an atomic force microscope was used and root mean square roughness was obtained.

1. Introduction

β-Galactosidases (EC 3.2.1.23) also known as lactases are enzymes belonging to glycoside hydrolase families 1, 2, 35, 42 and 59 (GH1, GH2, GH35, GH42 and GH59) [1, 2]. These enzymes catalyze the hydrolysis of terminal non-reducing β-D-galactose residues in β-D-galactoside substrates. In the case of lactose as a substrate the products of hydrolysis – namely glucose and galactose, are transferred to water. When the initial concentration of lactose is more than 15 %, an acceptor of galactose residue can be another carbohydrate present in the reaction mixture – for example lactose, fructose or lactulose, and an transgalactosylation reaction is observed, resulting in production of galactooligosaccharides (GalOSs) with a different degree of polymerization (DP) [3, 4]. A large part of world population has lactose intolerance. According to M. Heyman lactose intolerance is a clinical syndrome of 1 or more of the following: abdominal pain, diarrhea, nausea, flatulence, and/or bloating after the ingestion of lactose or lactose-containing food substances [5]. On the other hand the synthesized by β-galactosidases GalOSs contain a synthesized variety of bonds including β-(1→4), β-(1→6) and β-(1→3), which determines their application as food additives with significant prebiotic potential and other beneficial effects [4, 6].

The amount and structure of the synthesized GalOSs are significantly affected by enzyme source, the acceptor molecules and the reaction conditions (substrate concentration, pH and temperature) [7-9].

Immobilization of the enzyme lactase on polyethylene polymer films enables development of an active packaging film to produce lactose-free milk products. Using layer by layer immobilization chemistry technique increases the amount of enzyme that can be immobilized per unit area of packaging film. This technique makes it possible to increase the functional efficiency of enzyme, enhance the reproducibility of the processes, improve the process control and ensure stable supply of the products in the market [10]. Various applications of immobilized β-galactosidases can be found in the food industry, medicine and research. Other industrial applications of immobilized β-galactosidases include laboratory-scale organic synthesis and analytical and medical applications.

One of the suitable and mostly used natural polymers for enzyme immobilization is chitosan. It is characterized as biocompatible, nontoxic, physiologically inert, and hydrophilic, offer the unique characteristic of a remarkable affinity to proteins and has been widely applied in medicine and biological research. Furthermore, chitosan is positively charged in acidic aqueous solutions and its charge density is high [11].

The effect of the support size on the properties of enzyme immobilization was investigated by using chitosan macroparticles and nanoparticles in [12]. Alginate–chitosan core-shell microcapsules were prepared in order to develop a biocompatible matrix for enzyme immobilization, where the protein is retained either in a liquid or solid core and the shell allows permeability control over substrates and products. In [13] alginate–chitosan core-shell microcapsules were prepared as a novel biocompatible matrix system for β-galactosidase enzyme immobilization where the catalyst is confined to either a liquid or solid core and the transport properties of the substrate and product are dictated by the permeability of the shell. Thus, the biological agent is protected in the inner biocompatible alginate core and the outer chitosan shell dictates the transport properties.

The multilayer immobilized enzyme can be employed in a biocatalyst reactor for production of galacto-oligosaccharides from lactose and the hydrolysis of lactose to glucose and galactose. The method of surface modification by layer-by-layer (LbL) polyelectrolyte multilayers allows very precise control and changes in a wide range of the carrier's physicochemical properties – thickness, charge, hydrophilic–hydrophobic balance [14]. Such multilayer coatings with included functional components (nanoparticles, enzymes, and dyes) may be used in microelectronics, optics, biotechnology, and pharmacy [15]. The LbL technique is based on sequential deposition of oppositely charged polyelectrolytes from their solutions via electrostatic interactions. The assembly is based on spontaneous adsorptions, no stoichiometric control is necessary to maintain surface functionality, and the assembled films have good thermal and mechanical stability [16].

The aim of the present work is to describe the immobilization of the β-galactosidase enzyme in xanthan/modified chitosan/multylayers deposited on polylactic acid corona charged pad. Emphasis was focused on the effect of the number of polyelectrolyte multilayers (PEMs) and the immobilization time on the immobilized enzyme functional properties.

2. Materials and methods

2.1. Enzyme activity

A commercial fungal β-galactosidase (from *Aspergillus niger*) was used in the current kinetic studies. One unit of β-galactosidase activity is defined as the amount of enzyme catalyzing the release of 1 μmol min^{-1} glucose or fructose at 37 °C and pH 5.0. The influence of the substrate concentration on the initial velocity of the enzyme reaction was studied at a range 0.01 M – 1.30 M lactose. β-Galactosidase activity was studied in the presence of lactose at 5% and 10% and mixtures of chitosan (0.1%) and lactose (5% and 10%). The concentrations of the released glucose was determined enzymaticaly [17]. The β-galactosidase activity assays were carried out additionally using ONPG(ortho-Nitrophenyl-β-galactopyranoside) with substrate prepared in acetate buffer solution. One β-galactosidase unit (U) was defined as the amount of enzyme which liberated 1 μmol of ONP(ortho-Nitrophenyl) per min per mg of

IOP Conf. Series: Journal of Physics: Conf. Series **794** (2017) 012032 doi:10.1088/1742-6596/794/1/012032

protein at 37 °C and pH 5.0. Protein concentration was assayed by the method of Bradford [18]. All the analyses were performed at least in triplicate. Programmable scientific calculation "CITIZEN" SRP-45N and SigmaPlot 12.0 (Systat Software, Inc) were used for data analysis.

2.2. Samples preparation
Polylactic acid (PLA) with inherent viscosity (0.55 ÷ 0.75) dL/g, purchased from Lactel Absorbable Polymers (USA), was used for the preparation of the biodegradable pads. The PLA pads were casted from 2% w/v PLA chloroform solution and then dried at 35 °C for 48 hours. After that the pads obtained were stored in an exicator at room temperature (T = 25 °C) and relative humidity (RH = 54%). Immediately prior to the deposition process, the pads were treated in corona discharge in order to achieve positive surface charge excess. The corona discharge system consisted of corona electrode (needle), grounded plate electrode, and a grid placed between them. The pads were placed on the grounded plate electrode and a voltage of 5 kV was applied to the corona electrode. One kV voltage of the same polarity as that of the corona electrode was applied to the grid. The PLA pads were charged for one minute under room conditions (T = (21÷23) °C and RH = (40÷60)%). After charging the electrets, the surface potential was measured by the method of the vibrating electrode with compensation by which the estimated error was better than 5%.

2.3. Polyelectrolyte multilayers deposition
Chitosan (low molecular mass, degree of deacetylation > 75%) and xanthan gum were purchased from Sigma-Aldrich and were used as received. The layer-by-layer (LbL) deposition technique was applied for obtaining of multilayer structures. For obtaining of this structures were used 1% w/v chitosan and 1% w/v xanthan solutions of acetate buffer (pH 4.5 and ionic strength 0.1 M).The deposition was done by the dip-coating process. The first built-up layer was of xanthan because at these conditions the xanthan was negatively charged and the PLA pads were charged in a positive corona. For dip-coating assembling a programmable slide stainer (Poly Stainer IUL, Spain) was used with deposition program: (1) 15 min xanthan solution dipping - adsorption of polyelectrolyte molecules; (2) followed by 5 min washing in the acetate buffer (pH 4.5 and ionic strength 0.1 M); (3) 15 min chitosan solution dipping - adsorption of polyelectrolyte molecules of opposite charge; (4) 20 min dipping in 1.5 % sodium tripolyphosphate (Na-TPP) water solution – the layer was crosslinked; (5) 30 min dipping in 0.1% glutaraldehyde (GA) water solution - the layer was activated; (6) 5 min washing in the same acetate buffer. After that the samples were dipped in 1% solution of β-galactosidase for 1 hour or 2 hours. The procedure was repeated until obtaining the desired numbers of layers (4 or 8) xanthan/chitosan. After obtaining the polyelectrolyte structures, they were dried in hot air and then stored in exicator, at RH = 54%.

2.4. Immobilisation of the Enzyme.
As a support to enzyme immobilization, a chitosan layer was washed in 1 M acetic acid solution with 1.5% w/v Na-TPP crosslinking agent. Thereafter, the chitosan layer was washed with deionized water until neutrality. For activation, chitosan layer was incubated in 0.1% GA solution under constant stirring. The excess GA was washed off and suspended in 0.025 M of acetate buffer at pH 4.5. For immobilisation, the activated chitosan layer was incubated in 1% β-galactosidase solution under constant stirring. The excess enzyme was washed off and suspended in 0.025 M of acetate buffer at pH 5.0.

2.5. Atomic force microscope
Chitosan and xanthan polyelectrolyte multilayers (PEMs) surface morphology roughness was investigated by means of Atomic force microscope (AFM NANOSURF FlexAFM). The samples were scanned with a standard silicon cantilever (type Tap 190Al-G) and the measurements were performed in ambient atmosphere, and dynamic operating mode. The applied force was always minimized, not to deform the samples. The surface topographic images of the PEMs films were taken.

3. Results and discussion

3.1. Effect of type of PEMs on immobilization and enzyme activity.

The effect of the conditions of β-galactosidase immobilization on the enzyme activity was studied. Chitosan was selected as a material for polyelectrolyte multilayer preparation for several reasons – it is cationic biopolymer with intra- and intermolecular hydrogen bonding ability; it has unique approach to modify the surface of chitosan by anionic biopolymer xanthan to improve biocompatibility; and exists the possibility to prepare chitosan membranes with controlled pore size and density. Four different PEMs were obtained by chemical modification of the chitosan layer using NA-TPP and GA. The results are shown in table 1. The enzymatic activity of immobilized β-galactosidase was compared with that of free enzyme using ONPG as a substrate. The storage stability of all immobilized systems for 5 h exposition was studied (table 1). It was found that the most stable PEMs are PLA pad with 4 multilayers dipping in enzyme solution for 1 h. Immobilization efficiency indicated that the significant amount of enzyme was bond to PEMs with 4 layers of modified xanthan/chitosan. In contrast immobilization efficiency is 5 times less in PEMs with 4 multilayers dipping in enzyme solution for 2 h. Obviously, in order to achieve a good immobilization efficiency of β-galactosidase the optimal configuration is PLA pad with 4 xanthan/chitosan multilayers dipping in enzyme solution for 1 h.

Table 1. β-Galactosidase activity in different types of PEMs after incubation in acetate buffer.

No	Type of PEMs	Enzyme activity (U/mg protein)	Enzyme activity (U/mg protein)	Enzyme activity (U/mg protein)	Enzyme activity (U/mg protein)	Enzyme activity (U/mg protein)
		1 h	2 h	3 h	4 h	5 h
1	PLA pad with 4 multilayers dipping in enzyme solution for 1 h	0.022	0.019	0.029	0.021	0.017
2	PLA pad with 8 multilayers dipping in enzyme solution for 1 h	0.056	0.025	0.088	0.067	0.054
3	PLA pad with 4 multilayers dipping in enzyme solution for 2 h	0.119	0.117	0.126	0.118	0.090
4	PLA pad with 8 multilayers dipping in enzyme solution for 2 h	0.084	0.123	0.123	0.083	0.085

3.2. Effect of lactose concentration in 0.1 % chitosan solution on enzyme activity

The operational stability of the studied enzyme was evaluated by the hydrolysis of buffered lactose solutions 5% and 10% w/v, pH 5.0 at 37 °C.

Lactose hydrolysis was performed at 5 and 10 percent concentrations of lactose and with PLA pad with 4 multilayers dipping in enzyme solution for 1 h. As illustrated in table 2, there is not any inhibition effect of different percent of lactose. From the tested concentrations of lactose, the highest value of enzyme hydrolysis was measured in the presence of 5% of this sugar (15.24% of total substrate). When the enzyme reactions were performed in the presence of 0.1%, a chitosan light decrease of hydrolyzing reaction was observed in reaction without any chitosan concentration (15.04% of total substrate). On the other hand lactose hydrolysis performed with PLA pad with 4 multilayers dipping in enzyme solution for 1 h in the presence of 10% lactose resulted in 10.78% of lactose conversion in 3 h reaction time. When the enzyme reactions were performed in the presence of 0.1% chitosan in acetate buffer, a light increase of hydrolyzing reaction was observed in reaction with substrate 10% lactose (11.52% of total substrate).

Table 2. Lactose hydrolysis preformed with immobilized β-galactosidase in PLA pad with 4 multilayers dipping in enzyme solution for 1 h.

Time (h)	Glucose (mg/ml)		Glucose (mg/ml)	
	5.0% Lactose	10.0% Lactose	5.0% Lactose in 0.1% Chitosan	10.0% Lactose in 0.1% Chitosan
0	0.00	0.00	0.00	0.00
0.5	0.38	0.79	0.28	1.40
1.0	1.63	1.77	1.85	2.06
1.5	1.84	2.68	1.93	2.98
2.0	2.31	3.12	2.29	3.59
2.5	2.72	4.02	2.66	4.06
3.0	3.81	5.39	3.67	5.76

3.3. Morphology investigation

The surface topographic images of all the investigated PEMs were taken. The root mean square roughness was calculated. Surface morphology of the PEMs carried out by AFM was judged by their root mean square roughness Sq presented in table 3. In figures 1(a) and 1(b) surface topographic images of PLA pads with 4 multilayers dipping in enzyme solution for 1 h and 2 h are presented.

Table 3. Values of the root mean square roughness for type of PEMs listed in table 1.

No of type of PEMs	1	2	3	4
Root mean square roughness - Sq, nm	2.5	3.3	12.8	12.4

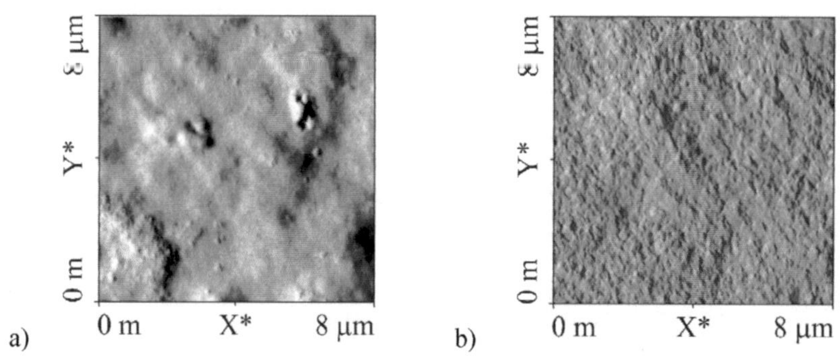

Figure 1. AFM images of PLA pads with 4 multilayers dipping in enzyme solution for: a) 1 h and b) 2 h.

The results obtained (see figure 1 and table 3) clearly show that the roughness of PLA pads dipping in enzyme solution for 1 h is much smaller in comparison with that dipping in enzyme solution for 2 h independently of the layer numbers. The PEMs dipping in enzyme solution for 1 h are the samples with greater β-galactosidase activity (see table 1). The PLA pad with 4 multilayers dipping in enzyme solution for 2 h has the great roughness and the smallest activity. Consequently, smaller roughness is associated with greater activity, which is most likely due to uniformly stable engagement of the enzyme in all layers.

When comparing the activity of the immobilized enzyme with that of the free enzyme, a similar inhibition has been observed due to galactose in the activity of the immobilized enzyme and a decrease in the activation energy of Michaelis–Menten constant of the immobilized enzyme compared to that of the free enzyme at 37 °C. The activity of the enzyme after immobilization was in all cases equal or higher than 50% of the activity of the free enzyme (data not show).

In the present work, we have studied the influence of different concentrations of lactose on the initial velocity of β-galactosidase. Such kind of kinetic information for lactose hydrolysis by β-galactosidases is scarce in the literature, despite of the fact that this disaccharide is a substate by transgalactosylation or used in many studies as a donor/acceptor of galactose/fructose units during the synthesis of GalOSs having specific structures and bioactive properties [19-21].

From studies performed by other authors it is known that the galactosyltransferase reaction for synthesis of GalOSs is favored when the lactose concentration in the reaction mixture is more than 15%. In these conditions the content of free water is low and the transfer of galactose moiety of lactose to suitable acceptor molecules is more likely to occur [7, 22]. It is necessary to perform additional studies concerning the distribution, composition and yield of the obtained transgalactosylation products during the reactions with increasing concentrations of lactose and fructose molecules.

4. Conclusion

The current study is the first time to compare the influence of chitosan on the initial velocities of reaction catalyzed by β-galactosidase. The studied immobilizing enzyme in multilayer biofilms showed substantial differences in its affinity to the lactose concentration. The obtained results will be further applied and extended in experiments dealing with the direction and optimization of the GalOSs production.

References

[1] http://www.cazy.org/Glycoside-Hydrolases.html
[2] Cantarel B L, Coutinho P M, Rancurel C, Bernard T, Lombard V and Henrissat B 2008 *Nucleic Acid Research* **37** 233
[3] Martinez-Villaluenga C, Cardelle-Cobas A, Corzo N, Olano A and Villamiel M 2008 *Food Chem.* **107** 258
[4] Otieno D O 2010 *Compr. Rev. Food Sci. Food Saf.* **9** 471
[5] Heyman M B 2006 *Pediatrics* **118** 1279
[6] Rodriguez-Colinas B, Fernandez-Arrojo L, de Abreu M, Urrutia P, Fernandez-Lobato M, Ballesteros A O and Plou F J 2013 *On the enzyme specificity for the synthesis of prebiotic galactooligosaccharides* in *Advances in Enzyme Biotechnology*, (New Delhi: Springer) p 23
[7] Mahoney R R 1998 *Food Chem.* **63** 147
[8] Kim C S, Ji E S and Oh D K 2004 *Biochem. Bioph. Res. Commun.* **316** 738
[9] Gaur R, Pant H, Jain R and Khare S K 2006 *Food Chem.* **97** 426
[10] Datta S, Christena L R and Rajaram Y R S 2013 *3 Biotech.* **3** 1
[11] Suh J K F and Matthew H W T 2000 *Biomaterials* **21** 2589
[12] Klein M P, Nunes M R, Rodrigues R C, Benvenutti E V, Costa T M H, Hertz P F and Ninow J L 2012 *Biomacromolecules* **13** 2456
[13] Taqieddin E and Amiji M 2004 *Microcapsules Biomaterials* **25** 1937
[14] Decher G and Schlenoff J B 2002 *Multilayer Thin Films* (Weinheim:Wiley VCH)
[15] Picart C, Caruso F and Voegel J-C 2014 *Layer-by-Layer Films for Biomedical Applications* (Weinheim: John Wiley & Sons)
[16] Zhou X C, Wu L Y and Zhou J Z 2004 *Langmuir* **20** 8877
[17] Bivolarski V, Vasileva T, Dzhambazov B, Momchilova A, Chobert J-M, Ivanova I and Iliev I 2013 *Biotechnol. Biotec. Eq.* **27** 3811
[18] Bradford M M 1976 *Anal. Biochem.* **72** 248
[19] Guerrero C, Vera C, Plou F and Illanes A 2011 *J. Mol. Catal. B Enzym.* **72** 206
[20] Padilla B, Ruiz-Matute, Belloch C, Cardelle-Cobas A, Corzo N and Manzanares P 2012 *J. Agric. Food Chem.* **60** 5134
[21] Shen Q, Yang R, Hua X, Ye F, Wang H, Zhao W and Wang K 2012 *Food Chem.* **135** 1547
[22] Monsan P, Paul F, Remaud M and Lopez A 1989 *Food Biotechnol.* **3** 11

Discrimination of sweeteners based on the refractometric analysis

I Bodurov[1], I Vlaeva[2], A Viraneva[1] and T Yovcheva[1]

[1]Department of Experimental Physics, Plovdiv University "Paisii Hilendarski", 24 Tsar Assen str., 4000 Plovdiv, Bulgaria
[2]Department of Mathematics and Physics, University of Food Technologies, 26 Maritsa blvd., 4002 Plovdiv, Bulgaria

E-mail: bodurov@uni-plovdiv.net

Abstract. In the present work, the refractive characteristics of aqueous solutions of several sweeteners are investigated. These data in combination with ones from other sensors should find application for brief determination of sweeteners content in food and dynamic monitoring of food quality. The refractive indices of pure (distilled) water and aqueous solutions of several commonly used natural and artificial sweeteners (glucose, fructose, sucrose, lactose, sorbitol [E420], isomalt [E953], saccharin sodium [E950], cyclamate sodium and glycerol [E422]) with 10 wt.% concentration are accurately measured at 405 nm, 532 nm and 632.8 nm wavelengths. The measurements are carried out using three wavelength laser microrefractometer based on the total internal reflection method. The critical angle is determined by the disappearance of the diffraction orders from a metal grating. The experimental uncertainty is less than ±0.0001. The dispersion dependences of the refractive indices are obtained using the one-term Sellmeier model. Based on the obtained experimental data additional refractive and dispersion characteristics are calculated.

1. Introduction

As for all food additives, the sweeteners are regulated substances, which are a subject to safety evaluation prior to market authorization. Over the centuries various foods such as honey or sugar have been used to sweeten our food. Today we also have a range of new sweeteners, which provide alternatives to sugar. Natural sweeteners provide calories in the form of sugars. Some are found naturally in foods like fruit and milk and others are added to foods during preparation or processing, for example, high fructose corn syrup and others [1-3].

Artificial sweeteners provide zero or very low calories since the molecules are large and are partially indigestible. Most are usually much sweeter than sugar. The Food and Drug Administration (FDA) has approved the use of the only five of them. They are less expensive than most sweeteners and is stable under heating [4].

Therefore, the selective identification and sensitive identification and sensitive determination of different sweeteners are necessary to effectively perform food quality control. Several analytical techniques can feasibly be applied to the analysis of natural sugars and artificial sweeteners. High-performance liquid chromatography (HPLC) [5-9] coupled (or not) to mass spectrometry [10] is the most popular method for determining sweeteners, although this approach is relatively complicated for on-site food quality control applications. Furthermore, enzyme [11, 12] or enzyme-free sensors, such as electrochemical sensors [13], fluorescent sensors [14], spectrometric sensors [15], colorimetric sensors [16] and diffraction resonance sensors [17] are relatively simple, but these methods are usually only used to determine a certain type of sweetener – they may lack sufficient selectivity to be able to identify various sweeteners in a food content.

The aim of the present paper is to make an analysis of the refractive indices of several commercially used natural and artificial sweeteners for its fast and precise detection and discrimination.

2. Materials and experiments

2.1. Materials

Glucose, sorbitol [E420], glycerol [E422], saccharin sodium [E950] purchased from Valerus, Bulgaria and sucrose, fructose product of Fillab, Bulgaria are used. Other sweeteners (cyclamate sodium and isomalt [953] are bought from the local market. All of them have been used without any characterization and modification. The aqueous solutions with concentration of 10 wt.% are made from chemically pure sweetener and a distilled water by stirring. For the sample preparation the used sweetener and the solvent are weighted with accuracy ±0.0001 g by using an analytical balance.

2.2. The refractive index measurement

The refractive index values of the investigated samples are measured by the laser refractometer based on the method of the disappearing diffraction pattern for three different laser wavelengths – 405 nm, 532 nm and 632.8 nm respectively at a room temperature (22 °C). The principle scheme of the refractometric system is illustrated in [18]. The sample is placed between a measuring glass prism and a metal diffraction grating. When the laser light strikes the base of the prism at an angle smaller than the critical one, it penetrates through the sample, reaches the diffraction grating forming a diffraction pattern on a screen. The incident angle gradually changes via rotating stage, since at a critical angle φ_{cr}, total internal reflection is observed. As a result, the diffraction pattern disappears because the incident light cannot reach the diffraction grating. In this way the critical angle φ_{cr} is measured in air at a chosen wavelength. The refractive index was calculated by the following equation:

$$n = N\sin\left[A \pm \arcsin\left(\frac{\sin\varphi_{cr}}{N}\right)\right]. \qquad (1)$$

The sings (+) and (–) correspond to clockwise and counter-clockwise measurement of φ_{cr}, respectively. In our case, the sign is (–). In equation (1) $A = 64.75\,°$ is the reflecting angle of the prism, N is the refractive index of the prism.

The main source of experimental uncertainty is the used rotary stage with $\Delta\varphi = 1$ arcmin resolution. By using equation (1) the experimental uncertainty can be estimated as:

$$\Delta n = N\Delta\varphi_{cr}\cos\varphi_{cr} \qquad (2)$$

In our case, since the obtained values of the critical angle are of the order of 25°, we have $\Delta n = \pm 1\times 10^{-4}$ for the experimental uncertainty.

The optical properties of the materials are usually presented by their dispersion dependences. For the most optical materials, far from the fundamental absorption band, the dispersion dependence of their refractive index can be build using the Sellmeier dispersion equation, if they are non-magnetic [19].

The values of the coefficients s and λ_s - the so called the Sellmeier's coefficients, can be obtained using the following relation [19]:

$$n^2 - 1 = \frac{s\lambda^2}{\lambda^2 - \lambda_s^2}. \qquad (3)$$

In this paper the values of Sellmeier's coefficients are obtained by non-linear fitting the set of experimentally determined values of the refractive index with the relation (3) using Wolfram Mathematica® software [20]. The confidence level is 0.95. By using the coefficients s and λ_s the dispersion dependences of the refractive index are built, by the approximated values in the spectral range 400 nm – 800 nm.

3. Results and discussion

The investigated samples basically can be divided in five groups, depending on their chemical structure and origin as follow:

- monosaccharides (glucose and fructose),
- disaccharides (sucrose and lactose),
- alcohols (sorbitol and isomalt),
- artificial (saccharin sodium and cyclamate sodium),
- glycerol.

As the last sample for testing, we choose glycerol because it is not a typical sweetener, but has a sweet taste due to the large numbers of alcohol groups. It is sometimes added to food products because it retains moisture in them for longer time or to prevent the crystallization of the sucrose.

The refractive index values of the all samples are measured at three different wavelengths – 405 nm, 532 nm and 632.8 nm respectively at a room temperature (22 °C).

The measured values of the refractive indices and the Sellmeier's coefficients at used wavelengths are presented in table 1.

The constructed dispersion dependences are presented on figure 1.

Table 1. Refractive indices at the used laser wavelengths and obtained values of the Sellmeier's coefficients.

	n (405 nm) ±0.0001	**n (532 nm)** ±0.0001	**n (632.8 nm)** ±0.0001	s	λ_s
Glucose	1.3555	1.3471	1.3438	0.786	100.68
Fructose	1.3574	1.3488	1.3454	0.789	102.03
Sucrose	1.3561	1.3490	1.3462	0.795	93.19
Lactose	1.3567	1.3493	1.3464	0.795	94.47
Sorbitol	1.3573	1.3492	1.3460	0.792	99.04
Isomalt	1.3574	1.3493	1.3462	0.793	98.55
Glycerol	1.3530	1.3461	1.3434	0.788	91.98
Saccharin-Na	1.3600	1.3515	1.3482	0.797	100.80
Cyclamate-Na	1.3547	1.3474	1.3446	0.790	94.23

The measured values of the refractive indices are very close to each other at the used wavelengths (table 1). From figure 1 it can be seen that the dispersion curves are very close to each other and some of them intercross.

Since it is not possible to make a clear separation between investigated sweeteners based on the measurement of the refractive index at one wavelength or on the dispersion curves, the Abbe numbers (ν) for all samples are calculated using the simple relationship:

$$\nu = \frac{n_D - 1}{n_F - n_C}, \tag{4}$$

where n_D, n_F and n_C are the refractive indices of the sample at 589.3 nm (D line), 486.1 nm (F line) and 656.3 nm (C line), respectively.

The calculated values of Abbe numbers are presented in table 2.

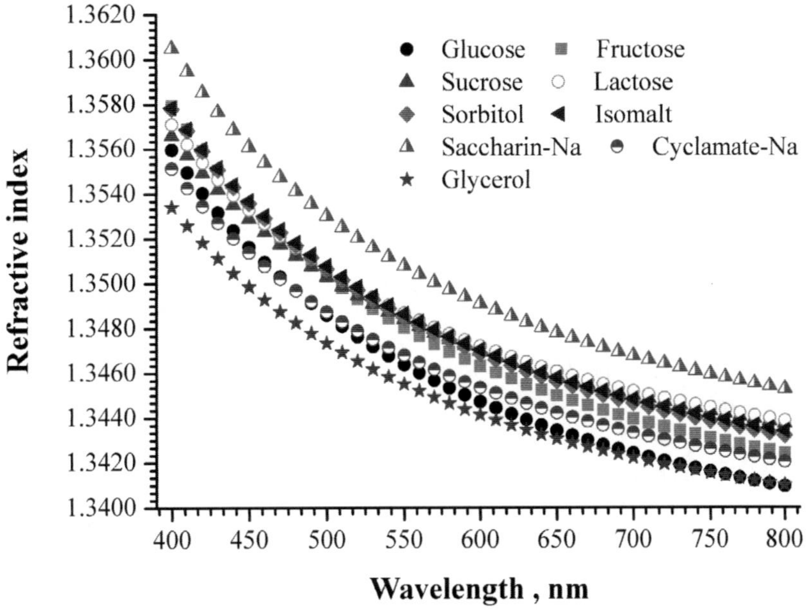

Figure 1. Dispersion dependences of the investigated sweeteners in the visible range.

On figure 2 a 2D mapping of the measured refractive indices at 405 nm versus the Abbe numbers is made. It is convenient to use the refractive index values of the short-wave region (405 nm) due to the strong dispersion. The dependencies at the other wavelengths, where the refractive index is measured, are similar but not so clearly visible.

Table 2. Calculated values of Abbe numbers for the investigated sweeteners.

Sample	v
Glucose	57.0773
Fructose	55.5578
Sucrose	67.0350
Lactose	65.1742
Sorbitol	59.1058
Isomalt	59.1058
Glycerol	68.8057
Saccharin sodium	57.0214
Cyclamate sodium	65.4821

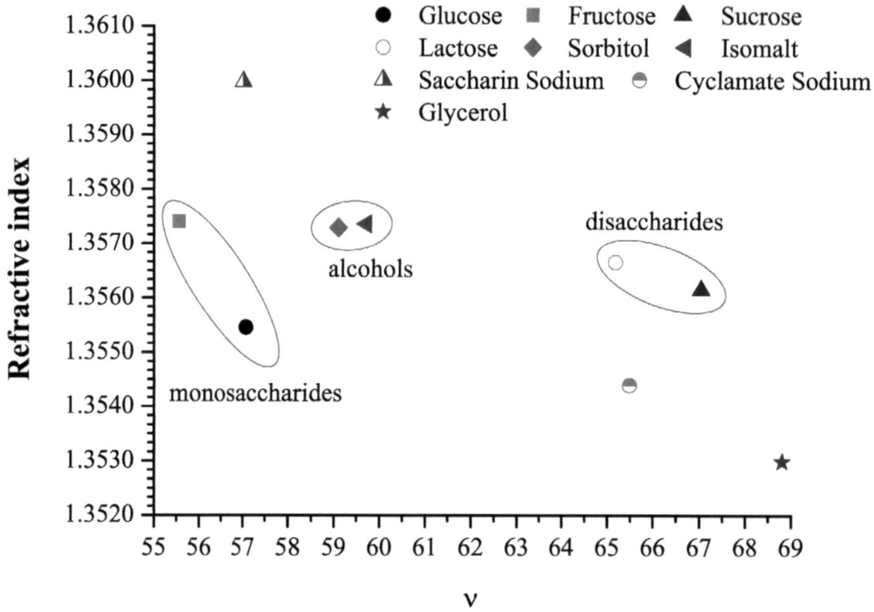

Figure 2. Measured refractive index for investigated samples at 405 nm versus the Abbe numbers.

The investigated sweeteners formed three groups. From figure 2 it can be seen that the investigated monosaccharides have refractive indices between 1.3550 and 1.3575 and Abbe numbers between 55 and 57. The alcohols have refractive index around 1.3575 and Abbe numbers between 59 and 60. The disaccharides have refractive index between 1.3560 and 1.3570 and Abbe numbers in the range 65-67. It was found that the most sensitive parameter for distinguishing natural sweeteners is the Abbe number at the small wavelengths of the visible range. For the artificial sweeteners, there is no differentiation in a separate group due to its distinct chemical composition and structure.

4. Conclusion

This work has made one first attempt to separate several sweeteners – monosaccharides, disaccharides, alcohols and artificial using experimental data from the refractometer measurements that have been carried out in an easy, fast and inexpensive way. The refractive index values are measured by the laser refractometer using the method of the disappearing diffraction pattern for three different wavelengths. The results showed that the refractive index values and dispersion dependences in the visible range are not suitable for discrimination of sweeteners. It was found that discrimination of natural and artificial sweeteners by the Abbe number is not possible. If we know that is a natural sweetener used, this method can be used for its determination.

Acknowledgements

This paper was performed with the financial support of the project NI15-FFIT-005, Department of Scientific Research at Plovdiv University "Paisii Hilendarski".

References

[1] Oehler N 2013 *Sweeteners: Nutritive and Non-Nutritive* (Oregon State University, Extension Service) SP 50-935
[2] Chattopadhyay S, Raychaudhuri U and Chakraborty R 2014 *J. Food Sci. Technol.* **51** 611
[3] Kroger M, Meister K and Kava R 2006 *Compr. Rev. Food Sci. Food Saf.* **5** 35

[4] Smith J and Hong L 2008 *Food Additives Handbook* (Hoboken, New Jersey: John Wiley & Sons Ltd.) p 960
[5] Buchgraber M and Wasik A 2009 *J. AOAC Int.* **92** 208
[6] Wasik A, McCourt J and Buchgraber M 2007 *J. Chromatogr. A* **1157** 187
[7] Chen Q C, Mou S F, Liu K N, Yang Z Y and Ni Z M 1997 *J. Chromatogr. A* **771** 135
[8] Chen Q C and Wang J 2001 *J. Chromatogr. A* **937** 57
[9] Zhu Y, Guo Y Y, Ye M L and James F S 2005 *J. Chromatogr. A* **1085** 143
[10] Yang D J and Chen B 2009 *J. Agric. Food Chem.* **57** 3022
[11] Chou S F, Chen J H, Chou L W, Fan J J and Chen C Y 1995 *J. Food Drug Anal.* **3** 121
[12] Odaci D, Timur S and Telefoncu A 2004 *Food Chem.* **84** 493
[13] Feng Li A B, Yan Fengb, Limun Yangb, Liang Li B, Chenfei Tangb and Bo Tanga 2011 *Biosens. Bioelectron.* **26** 2489
[14] James T D, Shinmori H and Shinkai S 1997 *Chem. Commun.* **1** 71
[15] Capitan-Vallvey L F, Valencia M C and Arana Nicolas E 2004 *Food Addit. Contam.* **21** 32
[16] Musto C J, Lim S H and Suslick K S 2009 *Anal. Chem.* **81** 6526
[17] Zhang X, Dou F and Liu H 2010 *Nanotechnology* **21** 335501
[18] Bodurov I, Yovcheva T and Sainov S 2015 *Opt. Appl.* **45** 199
[19] Tan W C, Koughia K, Singh J and Kasap S O 2006 *Fundamental optical properties of materials* in *Optical Properties of Condensed Matter and Applications* (UK: John Wiley & Sons Ltd.) p 7
[20] Wolfram Research Inc. 2010 Mathematica Version 8.0 (Champaign Illinois)

IOP Conf. Series: Journal of Physics: Conf. Series **794** (2017) 012034 doi:10.1088/1742-6596/794/1/012034

Using differential scanning calorimetry, laser refractometry, electrical conductivity and spectrophotometry for discrimination of different types of Bulgarian honey

I Vlaeva[1], K Nikolova[1], I Bodurov[2], M Marudova[2], D Tsankova[1], S Lekova[3], A Viraneva[2] and T Yovcheva[2]

[1]University of Food Technologies, 26 Maritsa blvd., 4002 Plovdiv, Bulgaria
[2]Plovdiv University "Paisii Hilendarski", 24 Tzar Assen str., 4000 Plovdiv, Bulgaria
[3]University of Chemical Technology and Metallurgy, 8 St. Kliment Ohridski blvd., 1756 Sofia, Bulgaria

E-mail: i_vlaeva@uft-plovdiv.bg

Abstract. The potential of several physical methods for investigation of the botanical origin of honey has been discussed. Samples from the three most prevalent types of honey in Bulgaria (acacia, linden and honeydew) have been used. They have been examined by laser refractometry, UV, VIS and FTIR spectroscopy, electric conductivity measurement and differential scanning calorimetry. The purpose of this study was to reveal the physical characterizations of honeys from different flora produced in Bulgaria and to identify honeys with a high apitherapy potential for future studies.

1. Introduction

Production of a natural honey is a laborious process, which is time consuming and involves a lot of cost. Therefore, honey is often a subject to falsification by adding sugar and other impurities. Furthermore, from a commercial perspective monofloral honey is preferred over polyfloral one, as well as certain types of monofloral honey are preferred over others. Therefore, it is important to have reliable methods for their discrimination. Several methods have been used for the determination of the floral origin of honey and among them the pollen recognition and sensory analysis are the most popular ones. However, the technique of analysis of honey's pollen content is a long process and has some limitations. The other methods are mainly based on the analysis of honey's aroma compounds, sugar profile, flavonoid pattern, non-flavonoid phenolics, organic acids, isotopic relations, and protein and amino acid compositions and marker presence [1, 2]. However, some of these methods are generally too time-consuming, complex, and labour intensive for routine quality control application or require very specialized personnel to interpret the results [3]. The advantages of the technique of spectroscopy (UV, visible, near and middle infrared, fluorescent) with respect to other methods are the non-invasive approach, the relatively easy and quick data acquisition. Recently, both near infrared and middle infrared spectroscopy have been successfully used for the classification of unifloral and multifloral honeys [4-6]. Some authors have used UV and VIS spectrometry for the same purpose [7-10].

The electrical conductivity is a good criterion for the botanical origin of honey and therefore it is very often used in routine honey control [11, 12]. The moisture content of honey or conversely the soluble solids content, is determined by measuring the refractive index of honey [13, 14]. Differential scanning calorimetry (DSC) is a powerful technique for characterization of the thermal behaviour of honeys and for detecting the effect of adulteration on the physicochemical and the structural properties of samples [15, 16]. Therefore, it is of prime importance to find the best parameters able to distinguish pure honeys from syrups and from adulterated honeys.

The aim of this article is a comparative analysis of several physical methods (UV, VIS and FTIR spectroscopy, laser refractometry, electrical conductivity and DSC analysis) which are widely used for examination of honey in view of its classification on the base of their "botanical origin" and will be applied to three of the most common types of Bulgarian honey – acacia, linden and honeydew honey. For the appropriate visualization of the three classes of honey Principal Components Analysis (PCA) is used. The simulations were carried out in MATLAB environment.

2. Materials and methods

2.1. Samples preparation
Three different types of Bulgarian honey (acacia, linden and honeydew) were purchased from the local market and from private producers. Ten different samples from various manufacturers of each type of honey were purchased. Each of the thirty samples were pre-treated thermally, having been placed in a water bath of 50°C for 30 minutes, after which it was cooled down to room temperature.

2.2. Absorption spectroscopy
The three types of spectrum (UV, VIS, FTIR) were acquired as follows. UV spectra were taken with a spectrophotometer Cary 100 ranging from 190 nm to 380 nm at 1 nm sampling space. VIS spectra were taken with a spectrophotometer Helios Omega ranging from 380 nm to 780 nm at 1 nm sampling space and using the software package VISIONlite ColorCalc [9]. FTIR spectra were taken with Mid-FTIR spectrometer Varian 660 ranging from 600 cm^{-1} to 4000 cm^{-1} at 1.929 cm^{-1} sampling space.

2.3. Electrical conductivity (EC)
The EC was measured using a conductometer DiST by HANNA Instruments in a 20 g/100 ml solution of honey (dry matter basis) in deionised water, at temperatures varying in the range of 5°C to 45°C by step of 5°C. According to the harmonised methods of the European Honey Commission EC is measured at 20°C and expressed in mS/cm [17].

2.4. Differential scanning calorimetry (DSC)
The thermal characteristics of the honey samples were investigated at the temperature range of -70°C to 100°C by differential scanning calorimeter DSC 204 F1 Phoenix (NETZSCH, Germany) equipped with intracooler. The DSC was calibrated with indium. In order to avoid condensation of water, argon gas was used to purge the furnace chamber at 20 ml/min. The samples (5-10 mg) was weighed into 40 μl aluminium standard crucible and hermetically sealed with aluminium standard lead. An empty aluminium crucible was used as a reference. Firstly, the samples were cooled from room temperature to – 70°C at cooing rate of – 40°C/min. The heating was done with standard rate of 5°C/min.

2.5. Laser refractometry
The refractive indices, *n* for all the samples were measured by the method of the disappearing diffraction pattern using a laser refractometer [18] at wavelengths of 405 nm, 532 nm and 635 nm. The total experimental uncertainty of the measured refractive indeces was less than 3×10^{-4}.

IOP Conf. Series: Journal of Physics: Conf. Series **794** (2017) 012034 doi:10.1088/1742-6596/794/1/012034

2.6. Principal Components Analysis (PCA)

The aim of the method is to reduce the dimensionality of multivariate data (e.g., wavelengths, temperatures) whilst preserving as much of the relevant information as possible. PCA is a linear transformation that transforms the data (observations of possibly correlated variables) to a new coordinate system such as the new set of variables, the principal components (PCs) are linear functions of the original variables. PCs are uncorrelated, and the greatest variance by any projection of the data comes to lie on the first coordinate, the second greatest variance - on the second coordinate, and so on. This is achieved by computing the covariance matrix for the full data set. Then the eigenvectors and eigenvalues of the covariance matrix are computed and sorted according to the decreasing eigenvalue [19, 20]. All principal components are orthogonal to each other. The full set of principal components is as large as the original set of variables. Usually the sum of the variances of the first few principal components exceeds 80% of the total variance of the original data [21].

3. Results and discussion

UV absorbance spectra of the three types of honey in the wavelengths range from 190 nm to 380 nm are shown in figure 1(a). There are two peaks at about (215-240) nm and (284-340) nm with can be attributed with content of hydroxymethylfurfural [22].

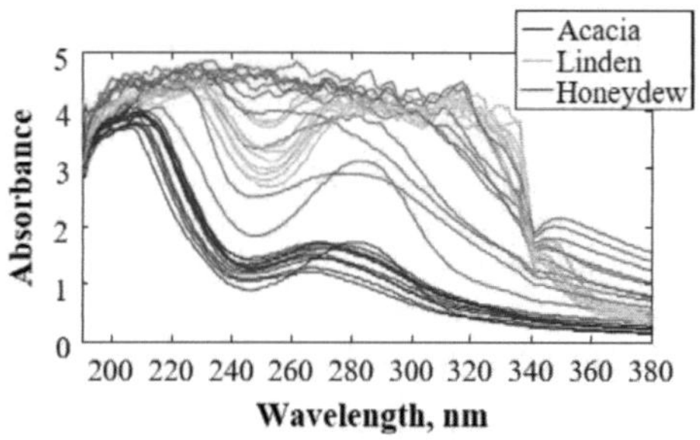

(a)

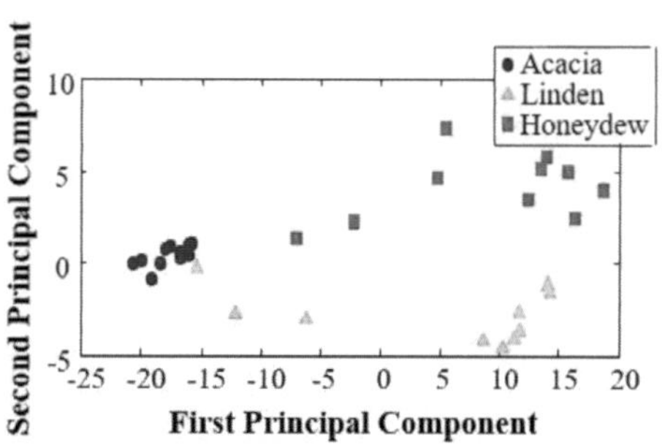

(b)

Figure 1. UV absorbance spectra (a), PCA of the spectral characteristics (b).

The experimental data are subject to pre-treatment. In order to remove some apparent interference received data are limited from above by a predetermined value (AbsorbMax). The resulting absorbance curves are smoothed by the method of creeping averaging, using the equation:

$$a_{i+l/2} = \frac{1}{l+1} \sum_{k=0}^{l} a_{i+k} , \qquad (1)$$

where l is the width of a linear filter accepting even-numbered values.

PCA method of multivariate analysis was applied to distinguish honey. The calculations and visualization were carried out in MATLAB environment. Figure 1(b) shows the scattered plot of the first and the second PCs of UV spectral characteristics. As evident, the three types of honey are well distinguished from PCA of the UV spectra.

VIS absorbance spectra of the three types of honey in the wavelengths range from 380 nm to 780 nm are shown in figure 2(a) and PCA of the VIS characteristics – in figure 2(b). The spectra present wide peak at the band of (400-500) nm that can be connected with the colour of the samples [22]. The honeydew has the absorbance peak at about (450-470) nm [23]. As it can be seen the three types of honey are well distinguished from PCA of the VIS spectra.

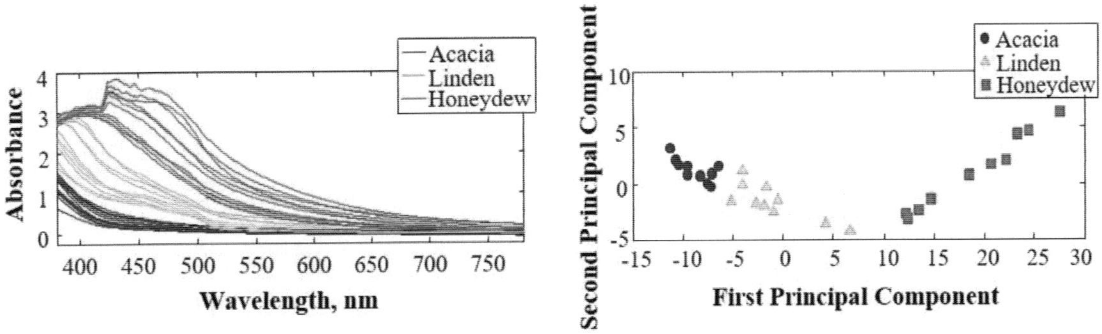

(a) (b)

Figure 2. VIS absorbance spectra (a), PCA of the spectral characteristics (b).

FTIR absorbance spectra of the three brands of honey in the wave numbers range from 4000 cm^{-1} to 600 cm^{-1} are shown in figure 3a and PCA of the FTIR characteristics – in figure 3(b).The mid-infrared bands could be interpreted using previous reported studies [24]. The first peak (3600-3000) cm^{-1} corresponds to O-H stretching of water and carbohydrates, and the second peak at (2950-2800) cm^{-1} – to the C-H stretching bound of carbohydrates (methyl and methylene groups). The scattered plot in figure 3(b) shows the mixing of the three classes of honey. The infrared spectroscopy allows distinguishing honeydew honey from monofloral honey, but cannot differentiate different monofloral honey groups.

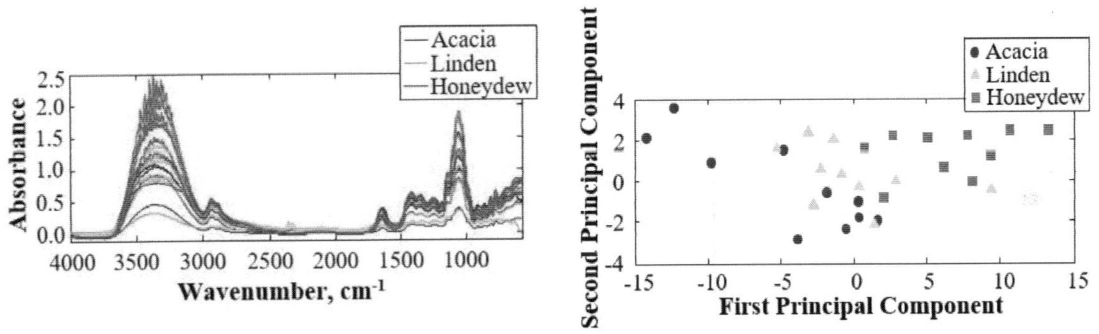

(a) (b)

Figure 3. FTIR absorbance spectra of honey (a), PCA of the spectral characteristics (b).

The dependence of PCA of the EC characteristics is shown in figure 4. EC values (recorded at 20°C) of honey samples tested in our study are 0.248 mS/cm for acacia honey samples, (0.499-0.640) mS/cm - for linden honey and (1.100-1.291) mS/cm - for honeydew honey samples. The EC value depends on the ash and acid content in honey: the higher their content, the higher the resulting conductivity [25]. The EC is a good criterion for the botanical origin of honey. A lower limit has been proposed for blossom than for honeydew honeys [25]. Our investigation on the EC of Bulgarian honey shows a good distinguishing of the three types of honey, based on this metric.

The information on honey, which is derived from DSC, is not sufficient for the reliable determination of the botanical origin of honey, but it can be used for quality estimation. A linear relation between the glass transition temperature and the moisture content was established, which is in a good agreement with the numerous reports about the water activity (a_w) effect on T_g and the plasticizing properties of water [26].

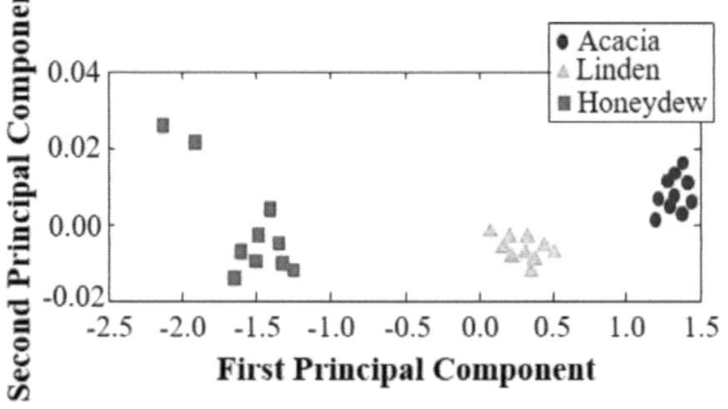

Figure 4. PCA of the EC characteristics of honey.

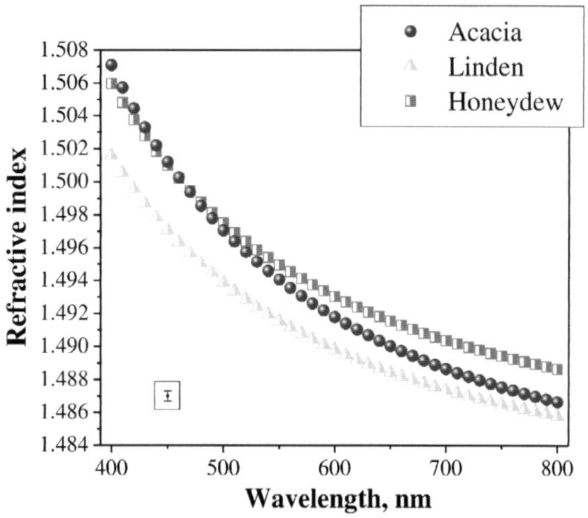

Figure 5. Dispersion dependences for the honey samples.

The refractive indices values for the all samples are accurately measured at 25°C at three different laser wavelengths – 405 nm, 532 nm and 635 nm. Far from the fundamental absorption band, the most appropriate dispersion relation for liquids is the one term Sellmeier relationship:

$$n^2 - 1 = \frac{s\lambda^2}{\lambda^2 - \lambda_s^2}, \tag{2}$$

where n is the refractive index at wavelength λ, s and λ_s are the Sellmeier's coefficients.

By applying equation (2) for three combinations of refractive indices pairs, we determined the Sellmeier's coefficients (s and λ_s) [27]. The measured average values of the n at the used three wavelengths and the obtained values of the Sellmeier coefficients are listed in table 1.

Table 1. Average values of the refractive indices and obtained values of the Sellmeier coefficients for three types Bulgarian honey.

Honey type	n (λ=405 nm)	n (λ=532 nm)	n (λ=635 nm)	s	λ_s
Linden	1.5021	1.4921	1.4892	1.1926	89.07
Acacia	1.5056	1.4953	1.4902	1.1910	100.56
Honeydew	1.5045	1.4961	1.4916	1.1997	92.77

Using the already determined dispersion coefficients, the dispersion dependences are built in the range (400-800) nm and plotted in figure 5. The sign of error bar in the lower left of the figure 5 represents the experimental uncertainty of the refractive indices values. The n is the highest for honeydew and it is the lowest for linden honey.

It can be seen from figure 5 that the most appropriate values for the discrimination between a linden honey and an acacia honey are the refractive indices values in the range (400-550) nm while the n values for acacia and honeydew are almost the same. A good separation between acacia honey and honeydew, but not between acacia honey and linden honey, can be made by measurement of the n in the range (700-800) nm. The honeydew and the linden honey are well separated by their refractive indices in the whole visible range. The results show that the refractive indices gives a good opportunity for distinguishing the botanical origin of honey. But it should be noted that for good separation between different types of honey a measurement of the refractive index at one wavelength is not sufficiently secure. Our studies indicate that the refractive index value can do sufficiently good separation of the investigated three types of honey using precision laser refractometer at one wavelength in the range (550-700) nm.

The obtained experimental results have shown that not all the presented investigations are suitable for discrimination of three different types of Bulgarian honey. Such separation can be done only by PCA of UV and visible spectra and PCA of EC characteristics. It should be noted that the honeys cannot be divided by group according to its botanical origin by using only peaks in the visible and UV spectra. This can be obtained by PCA. Another suitable method for good separation among investigated types of honey is to obtain the dispersion dependencies in the visible range by the laser refractometry.

4. Conclusion

In this work, the several physical methods (UV, VIS and FTIR spectroscopy, laser refractometry, electrical conductivity and DSC analysis) for discrimination of three of the most common types of Bulgarian honey – acacia, linden and honeydew honey are applied. It was shown that the most appropriate methods, which enable differentiation in botanical origin among investigated types of honey, are PCA of UV and visible spectra, PCA of EC characteristics and laser refractometry in visible range. It was established that PCA of FTIR spectra and DSC method could not be used to differentiate botanical origin of honey.

Acknowledgements

The paper presents research and development, supported by the Scientific Fund of Internal Competition of the University of Food Technologies – Plovdiv under the Research Project No.7/16-H.

References

[1] Anklam E 1998 *Food Chem.* **63** 549
[2] Hermosin I, Chicon R and Cabezudo M 2003 *Food Chem.* **83** 263
[3] Bogdanov S and Martin P 2002 *Mitt. Geb. Leb. Hyg.* **93** 232
[4] Davies A, Radovic B, Fearn T and Anklam E 2002 *J. Near Infrared Spectrosc.* **10** 121
[5] Ruoff K, Luginbühl W, Bogdanov S, Estermann B, Ziolko T and Amado R 2003 *Eur. Congres for Authenticity of Food*, Nyon, Switzerland
[6] Lichtenberg-Kraag B 2003 *Apidologie* **34** 479
[7] Li Y and Yang H 2012 *Int. Sch. Res. Network, ISRN Spectrosc.* **2012** 4
[8] Tsankova D and Lekova S 2015 *J. Chem. Technol. Met.* **50** 638
[9] Tsankova D, Lekova S, Nikolova K and Terziiski G 2015 *Annals of Faculty Eng. Hunedoara – Intern. J. Eng.* **XIII** 275
[10] Tsankova D and Lekova S 2015 *Intern. Conf. "Automatics and Informatics'2015"* 107
[11] Kaskoniene V, Venskutonis P and Ceksteryte V 2010 *Food Sci. Technol.* **43** 801
[12] Lašáková D, Nagy J and Kasperová J 2009 *Folia veter.* **53** 31
[13] Ouchemoukh S, Louaileche H and Schweitzer P 2007 *Food Control* **18** 52
[14] Nyau V, Mwanza E P and Moonga H B 2013 *African J. Food, Agricult. Nutrit. Devel.* **13** 7416
[15] Cordella C, Faucon J, Cabrol-Bass D and Sbirrazzuoli N 2003 *J. Thermal Anal. Calor.* **71** 279
[16] Tomaszewska-Gras J, Bakier S, Goderska K and Mansfeld K 2015 *J. Apic. Scie.* **59** 109
[17] Bogdanov S, Martin P and Lullmann C 1997 *Apidologie (extra issue)* 1
[18] Bodurov I, Yovcheva T and Sainov S 2015 *Opt. Appl.* **XLV** 199
[19] Hotelling H 1933 *J. Educational Psychology* **24** 498
[20] Jolliffe I 2002 *Principal Component Analysis* 2nd ed (Springer Series in Statistics, New York: Springer-Verlag New York)
[21] Statistics Toolbox™ User's Guide, R2014a, Copyright 1993–2014 (The MathWorks, Inc.)
[22] Escuredo O, Fernández-González M and Seijo María C 2012 *Agriculture* **2** 25

[23] Sun D-W 2008 *Modern Techniques for Food Authentication* 1st ed (San Diego, California, USA: Academic Press Elsevier) p 201
[24] Domínguez-Martínez I, Meza-Márquez O, Osorio-Revilla G, Proal-Nájera J and Gallardo-Velázquez T 2014 *J. Korean Soc. Appl. Biol. Chem.* **57** 133
[25] Bogdanov S 2002 Harmonized methods of the International Honey Commission, *International Honey Commission* p 1
[26] Costa PA, Moraes ICF, Bittante AMQB, Sobral PJA, Gomide CA and Carrer CC 2013 *Int. J. Food Stud.* **2** 118
[27] Tan WC, Koughia K, Singh J and Kasap SO 2006 Fundamental optical properties of materials, *Optical Properties of Condensed Matter and Applications*, ed Jai Singh (New York: Wiley-Interscience) p 7

IOP Conf. Series: Journal of Physics: Conf. Series **794** (2017) 012035 doi:10.1088/1742-6596/794/1/012035

Classification of different types of beer according to their colour characteristics

Kr T Nikolova[1], R Gabrova[2], D Boyadzhiev[3], E S Pisanova[3], J Ruseva [2] and D Yanakiev[3]*

[1]Medical University, 55 Marin Drinov Blvd., Varna, Bulgaria
[2] Univerity of Food Technologies, 26 Maritsa Blvd., 4000 Plovdiv, Bulgaria
[3] University of Plovdiv, 24 Tzar Assen Str., 4000 Plovdiv, Bulgaria

E-mail: kr.nikolova@abv.bg

Abstract. Twenty-two samples from different beers have been investigated in two colour systems - XYZ and SIELab - and have been characterised according to their colour parameters. The goals of the current study were to conduct correlation and discriminant analysis and to find the inner relation between the studied indices. K-means cluster has been used to compare and group the tested types of beer based on their similarity. To apply the K-Cluster analysis it is required that the number of clusters be determined in advance. The variant K = 4 was worked out. The first cluster unified all bright beers, the second one contained samples with fruits, the third one contained samples with addition of lemon, the fourth unified the samples of dark beers. By applying the discriminant analysis it is possible to help selections in the establishment of the type of beer. The proposed model correctly describes the types of beer on the Bulgarian market and it can be used for determining the affiliation of the beer which is not used in obtained model. One sample has been chosen from each cluster and the digital image has been obtained. It confirms the color parameters in the color system XYZ and SIELab. These facts can be used for elaboration for express estimation of beer by color.

1. Introduction

It is known that beer is the third most often consumed drink worldwide after water and tea and the first wide consumed alcoholic drink. Beer is a complex mixture consisting mainly of water and ethanol with about 0.5% of dissolved solids [1]. The analysis of beer is very important to evaluate its organoleptic characteristics, quality, nutritional aspects and safety. Healthy ingredients are identified [2] including antioxidants [3], mainly polyphenols [4], essential vitamins – particularly B vitamins [5-6] and minerals.

A wide variety of methods have been developed and optimized to characterize and quantify different types of beer. Beers are differentiated mainly according to their visual appearance and their fermentation process. The visual appearance of beer depends on both its color and its turbidity. This is not only because scattering reduces the transparency of beer, but also because the suspended particles themselves can contribute to light absorption [7].

Depending on the yeast used and the fermentation temperature, three main beer categories can be identified: top-fermented, bottom-fermented, and naturally-fermented. The third category beer is produced mainly in Belgium. In our investigation we use bottom-fermented beers, which are also called "Lager", are produced by adding saccharomycesuvarum (or pastorianus) at cooler temperatures, namely between 7 and 12°C.

The objective of this paper was to analyze a database of colorimetric parameters obtained for different types of beer and using K-means cluster to compare and group the tested types of beer on the base of their similarity. The defined objective requires the solution of the following problems:

- Creation of database, including the different types of beer from Bulgarian hypermarkets;
- Determination the color parameters in two different colorimetric systems, SIE Lab and XYZ;
- Establishment of statistically significant differences between the indices of beer groups and development of suitable models describing their characteristics.

2. Materials and methods

2.1. Samples

The database includes 22 types of beer from 3 countries. Five Bulgarian beers are also used for estimation of the obtained model. The samples are purchased from the Bulgarian supermarkets. The information from the investigated samples is presented in table 1.

Table 1. The beer collection.

Country	Code	Class	Alcohol, %	Country	Code	Class	Alcohol, %
Bulgaria	1	Lager	4.5	Bulgaria	13		4.3
Bulgaria	2		4.95	Serbia	14		4.7
Bulgaria	3		4.3	Holland	15	Lager	5
Bulgaria	4		5	Bulgaria	16	Dark ale	6.5
Bulgaria	5		4.8	Bulgaria	17	fresh	2.3
Serbia	6		4.6	Bulgaria	18	fresh	2.3
Bulgaria	7	Lager	5	Bulgaria	19	lemon	1.8
Bulgaria	8		2.2	Serbia	20	fresh	4.5
Holland	9		4.8	Bulgaria	21		4.8
Bulgaria	10	Dark ale	5.2	Bulgaria	22		5
Bulgaria	11	Lager	4.5				
Bulgaria	12		5.2				

2.2. Used methods

Determining the color characteristics of the beers is made by using two independent methods:- spectrophotometer and digital imaging. The spectrophotometer Thermo Scientific Helios Omega UV-VIS is used in the visible region with the cuvete of thickness of 10 mm and capacity of 1.5 ml.

To obtain a primary image of beer samples an optoelectronic system is used, containing a color CCD camera of the company Basler Vision Technologies (type Basler scA1390-17gc), a computer system with an interface to connect to the camera lighting emitting daylight (fluorescent lighting), consisting of 4 lamps with adjustable brightness; luminescent ring from Philips.

Four samples were chosen to obtain digital images – one for each group of light beer, dark beer, beer with lemon and beer with fruits. For this purpose a layer with thickness 10 mm is obtained.

Digital images of samples of beer in bmp format were obtained in a different luminance (from 940 to 1440 Lux at step 100). Images are photographed in the same conditions on standard cards of Color Checker Passport Photo, Company X-rite - card Classic Target, consisting of 24 colors with defined color characteristics and White Balance Target - 18% color gray. The deviation between the color components for each color of the cards, obtained by the image and those, specified in the passport are determined. In order to be able to obtain the best approximation of the color parameters of the beer samples to the real color parameters. A correction of the color characteristics of the images of samples is performed with the received deviations of color values of the reference colors. The correction shall be implemented in the Lab color model, which requires conversion of the RGB coordinates of the image in XYZ, and subsequent conversion into Lab coordinates.

2.3. Statistical analysis

The program "Statistica 10.0" has been used for the treatment of the data. The distribution was found to be Gaussian by the Kolmogorov-Smirnov criterion [8-9]. To establish statistically significant difference between the groups, the Scheffe criterion was applied for multiple comparisons [10]. Discriminatory analysis is used to model the group with *a priori* equal probabilities to fall into the groups [11-12]. The values for the Mahalanobis distances between the groups formed.

The K-means clustering has been applied on the base of all investigated parameters for comparing and grouping the beers [13-14]. This method classifies a given data set through a certain number of clusters fixed *a priori* [15-16]. Such an approach has been used in reference [17]. We determined the number of clusters to be 4. As a measure of similarity between samples squared Euclidean distance is used [18]. Also the cluster centers have been calculated after all objects have been assigned to a given cluster. The MATLAB™ 7.0 software has been used for determining the color characteristics of digital imagine.

3. Results and discussion

The database includes 22 types of beer from 3 different countries. For each of the samples studied, four independent measurements have been performed. The Scheffe criterion shows significant statistical differences in the studied samples. The presence of considerable difference in the color characteristics of beer provides the basis for a subsequent modeling of its origin. To model the beers by type (light, dark, fresh or with lemon) a step by step linear discriminatory analysis was used. A model with grouping parameter "beer type" was obtained and it includes the following parameters by the order of introduction into the model: color coordinate X, Z, x, chroma and hue angle. The classification of the probes according to the type is 100%.

Cluster analysis was applied with the method of K–means clusters. After some steps made with different number of clusters, most suitable was the grouping in 4 clusters. The samples, included in them had similar values of a set of the given parameters, which differentiate them from the samples in the other 3 clusters. Table 2 shows both the percentage distribution of the samples into the 4 clusters and the samples participation in each of the 4 clusters.

Table 2. Distribution of types beer in clusters.

Variants in clusters	Samples, number	Distribution of the samples of beer,%
1,2,3,4,5,6,7,9,11,12,13,14	12	55
8, 18, 20, 21	4	18
15,17, 19, 22	4	18
10, 16	2	9

The samples in the first cluster pose high lightness, greater than 92, and color parameter with a negative value. The percentage of samples is almost the same – around 18% - in the second and in the third clusters, which include 4 samples each. The beers in the second cluster have almost equal color coordinates x, y and b, have the same lightness. All of the samples are beers with fruits with exception of sample 21, which is unpasturised beer and its color parameters are similar to those of fresh beers.

The samples in the third cluster pose lemon, they are turbid with the hue angle around 76°and lightness between 54-60. In this group uncorectly is the state of sample 15, because its lightness (L) is in the interval 68-78 and hue angle is 78°. Dark beers which belong to the fourth cluster have very low value for L-between 40-50 -and the Hue angle is 65°-66°. The beer samples can be divided in groups by using only the hue angle. The first cluster was characterized with by hue angle between 90°- 96°, the second cluster cannot be differentiated from the first according to this parameter. The third one has a hue angle of around 80°, the cluster with dark beer posses the smallest value for hue angle at about 66°. The difference between the clusters is presented in figure 1.

For a better visualization of the results a subsequent canonical analysis was performed. On the basis of the first two canonical variables, the position of the separate samples for the model with included color parameters is presented in figure 2.

Thus the input set of correlating data was transformed into a new set with a smaller number of uncorrelated artificial variables, so-called „factors" or „principal canonical variables", which explain the greatest part of the data variation. The reduction in the number of original variables was achieved by grouping the correlated variables in a common factor and the separation of uncorrelated ones in different factors.

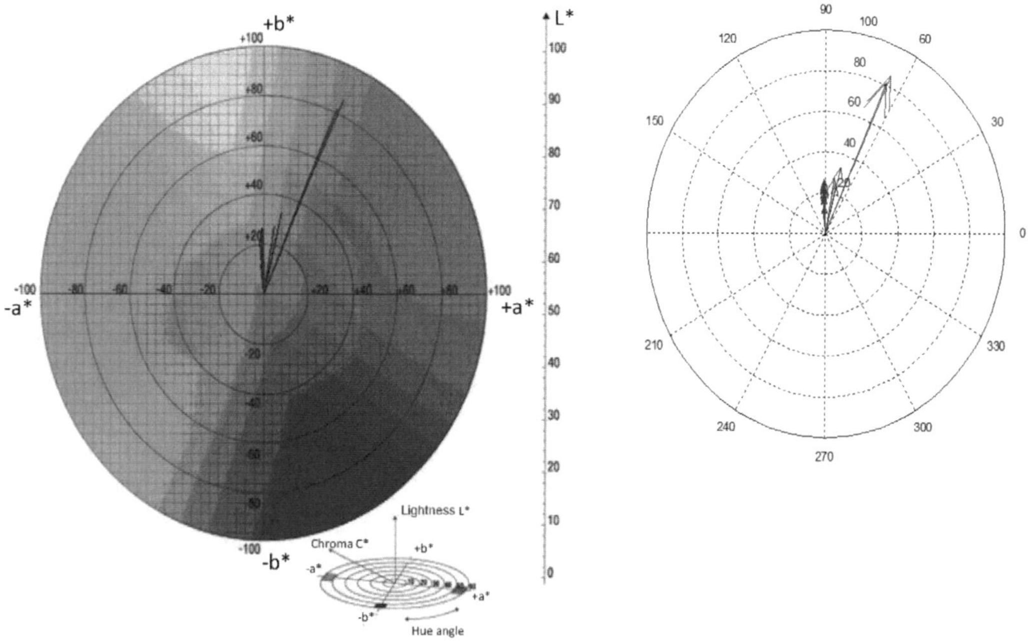

Figure 1. The dependence between color parameters *a* and *b* and hue angle for different clusters.

The figure confirms the stated hypothesis for the presence of significant differences between the separate types of beer. The analysis of the Mahalanobis distances between the four basic groups shows that the lager and light beers and the beers with fruits are close to each other and are relatively far from the dark ale beers. This is clearly seen from the figure 2, shown if we trace the projections of the clouds of the various sorts upon the first canonical variable which plays an important role in discrimination of the groups – the dark beers are projected on the positive, while all of the other samples are projected on the negative direction. For each of the sample groups of beers the Mahalanobis distances were calculated, and canonical analysis was simultaneously performed, because Mahalanobis distances provide an idea of the specific features of the examined group in the original space, whereas canonical representations are in a bidimensional space. They serve for a better visualization of the particular groups, since their canonic variables are linear combinations of the initial color indices. Mahalanobis distances represent individual groups, better than canonical variables, because they give an idea of the dynamics of the change in the distance between the group centroids.

On creating the discriminant functions for the sake of a better illustration of the results obtained, a canonical analysis is carried out consisting in the construction of new (latent) canonical variables. This is done according to the rule: each subsequent variable is chosen in a way as to contain together with the preceding variables a maximum of the existing information. In this way they are arranged so that the first is the "most precious", followed by the second "less precious" etc. The method of canonic correlation analysis is used to determine the coefficients in the linear combinations.

Greater differences were observed between the centroids of groups from beer with lemon and these from dark beer. The four samples are for the control of the adequacy of the created model for the description of the beer types. From the remaining five types of beer with a known type, the samples which are insufficient to form separate groups, two can be classified as light beers and two as the beer with fruits. The results from the classification according to the obtained models are shown in table 4. The investigated samples are classified 100%correctly.

3.1. *Parameters of digital image*

Determining the color coordinates of each beer sample is made using the MATLAB^{TM} 7.0 software. The program includes: identifying the areas of each color map by using gradient algorithm, first image is binarized and filtered; determining the deviation between the color components for each color of the cards, obtained by the image and those, specified in the passport; correction of color characteristics of samples of beer with received deviations.

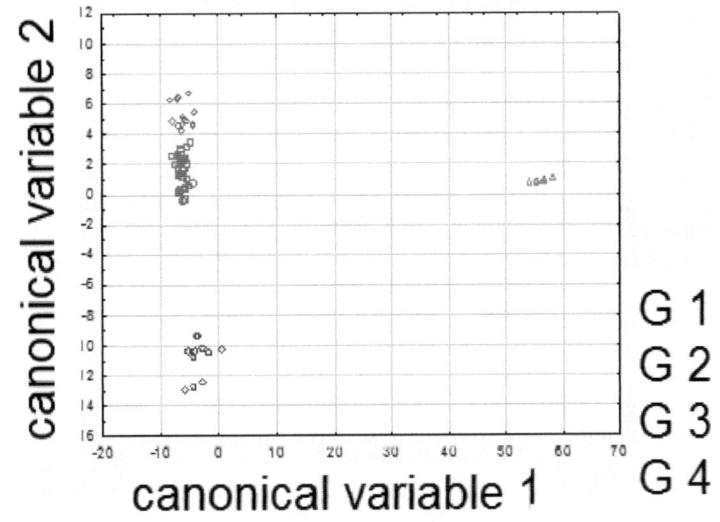

Figure 2. Disposition of the four types of beer in the plane of the first two canonical variables.

Table 3. Mahalanobic distances for different beer groups.

Groups	G 1	G 2	G 3	G 4
G 1	0.000	165.595	263.810	3673.606
G 2	165.595	0.000	34.273	3862.837
G 3	263.810	34.273	0.000	3879.297
G 4	3673.606	3862.837	3879.297	0.000

Table 4. Verification of the model for beer type with independent samples.

Sort	Classified as
Zagorka	Light beer
Kamenitza	Light beer
Kamenitza fresh	Kamenitza fresh
Kamenitza lemon	Kamenitza fresh

3.1. *Parameters of digital image*

The resulting color coordinates of the samples based on the images correspond to those obtained by the spectrophotometer with the exception of experiments in illuminance 1440 Lux, which may be due to the strong glare. The maximum deviation of the Lab coordinates is 4.2%, disregarding the experiment at 1440 Lux. Minimum deviations are obtained in illuminance of 1140 Lux.

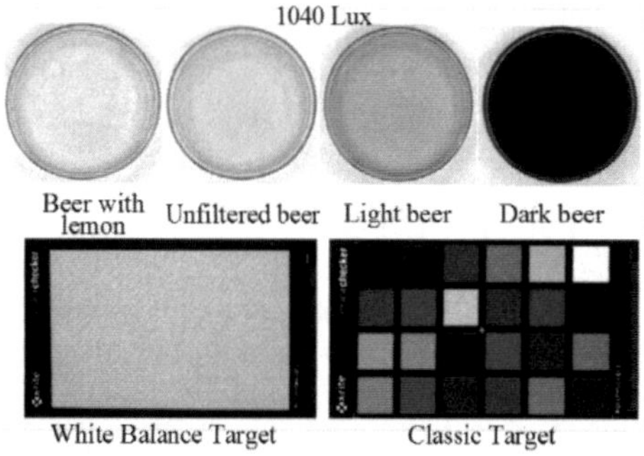

Figure 3. Digital image

4. Conclusion

The analysis of the data base gives the opportunity to characterize different types of beer by using the discriminant analysis. The models and the associated Mahalanobis distances enable the classification of unknown samples. By applying cluster analysis the obtained results give us the opportunity to point out the different types of beer, which have indices around the average for each group and which are similar. They have been featured in accordance with the examined indices. It was pointed out which of them had the biggest influence on their unification.The obtained color characteristics on the base of digital imagine correspond to these obtained by spectrophotometer with minimal deviation for 1140 Lux.

Acknowledgments

This study was supported by Project SP15-FFIT-004 of the University of Plovdiv (Bulgaria).

References

[1] Cortacero-Ramirez S, Castro M H B, Segura-Carretero A, Cruces-Blanco and Fernandez-Gutierrez A 2003 *Trends in Analytical Chemistry* **22** 440
[2] Bamforth C W 2002 *Nutrition Research* **22** 227
[3] Zhao H, Chen W, Lu J and Zhao M 2010 *Food Chem.* **119** 1150
[4] Piazzon A, Forte M and Nardini M 2010 *J. Agric. Food Chem.* **58** 10677
[5] Pietercelie A, Allardin D and Van Nedervelde L 2011 *Cerevisia* **36** 41
[6] Hucker B, Wakeling L and Vriesekoop F 2011 *J. Agric. Food Chem.* **59** 12278
[7] Mignani A, Ciaccheria L, Mencaglia A, Ottevaere H, Baca E and Thienpont H 2012 *Sensors and Actuators B* **23** 141
[8] Geoffrey J 1992 *Discriminant Analysis and Statistical Pattern Recognition* ed Mc. Lachlan (New Jersey, Canada: John Wiley & Sons. Inc) p 132
[9] Smeyers J J 1998 *Handbook of Chemometrics and Qualimetrics Part A. Part B* ed Verbeke (Amsterdam, The Netherlands, Elsevier) p 145
[10] Ivanov K 2001 *Statistical methods in social science* (Sofia, Bulgaria: New Bulgarian University) p 124
[11] Bondar A and Statuha G 1976 *Experiment in chemical technologies* (Moscow, Russia) p 194
[12] Lakin L 1990 *Biometry* (Moscow, Russia) p 105
[13] Vundev D 2003 *Applied Statistic* (Sofia, Bulgaria: SU Kliment Ohridski) p 76
[14] Duran B and Odelle P 1977 *Cluster analysis* (Moscow, Russia: Statistics) p 115
[15] Hartigan J *1985 Journal of Classification* **2** 63

[16] Hartigan J A and Wong M A 1979 *Applied statistics* **28** 100
[17] Milev M, Nikolova Kr, Ivanova Ir and Dobreva M 2015, *AIP Conf. Proc.* **1690** 020021
[18] Ward J H 1963 *J. Amer. Stat. Assoc.* **58** 236

IOP Conf. Series: Journal of Physics: Conf. Series **794** (2017) 012036 doi:10.1088/1742-6596/794/1/012036

Precise femtosecond laser ablation of dental hard tissue: preliminary investigation on adequate laser parameters

Todor Hikov[1], Emilia Pecheva[1], Paul Montgomery[2], Frederic Antoni[2], Audrey Leong-Hoi[2] and Todor Petrov[1]

[1] Institute of Solid State Physics, Bulgarian Academy of Sciences, Sofia, Bulgaria
[2]Laboratoire ICube,University of Strasbourg -CNRS, Strasbourg, France

E-mail: emily@issp.bas.bg

Abstract. This work aims at evaluating the possibility of introducing state-of-the-art commercial femtosecond laser system in restorative dentistry by maintaining well-known benefits of lasers for caries removal, but also in overcoming disadvantages such as thermal damage of irradiated substrate. Femtosecond ablation of dental hard tissue is investigated by changing the irradiation parameters (pulsed laser energy, scanning speed and pulse repetition rate), assessed for enamel and dentin. The femtosecond laser system used in this work may be suitable for cavity preparation in dentin and enamel, due to the expected effective ablation and low temperature increase when using ultra short laser pulses. If adequate laser parameters are selected, this system seems to be promising for promoting a laser-assisted, minimally invasive approach in restorative dentistry.

1. Introduction

Since the first use of lasers in the medical field, the contact-free application of laser light for removal of hard dental tissue has been studied as a means for replacing conventional drilling instruments. For conventional pulsed laser ablation with pulses from 100 ps to μs, a strong thermal shock wave is accompanied by cracking of the remaining bulk material and inefficient, uncontrolled material removal. Other damages to the tooth surfaces include cracking, melting, charring, fissuring or crazing and inefficient material removal [1]. High laser power causes significant excess heat deposition in the tooth with potential for irreversible intra pulp damage, which may cause pain for the patient [2]. However, in recent years the development of high-average-power, high-repetition-rate subpicosecond lasers is resulting in an increasing interest in laser surgical applications due to their precise and highly effective ablation capabilities and none or minimal thermal and shock wave collateral damage [3]. By decreasing the laser pulse duration to the subpicosecond time regime and increasing the laser peak intensity, it is possible to induce a different mechanism of laser-tissue removal. In this "electrostatic ablation" mechanism, the laser light is absorbed in a small volume of material and electrons and then ions are ejected rapidly, reducing the heat that is conducted to the surrounding tissue [4]. The change in ablation mechanism from thermally induced in the conventional long pulse regime to the electrostatic mechanism, results in distinct changes in tooth morphology, substantially reduced collateral damage to the tissue, and a decrease in the energy fluence required for the fast removal of a significant amount of dental tissue.

In this study, we present the preliminary results of minimally invasive ablation of dentin and enamel using a state-of-the-art commercial femtosecond (fs) laser system (FELS). Different laser parameters were tested in order to choose optimal parameters for the ablation in which neither collateral damage, nor overheating of the dental tissue was obtained.

2. Materials and Methods

Extracted human molars were sectioned transversely into 2 mm slices, polished with a SiC #1200 grinding paper with water irrigation and subsequently with a 3 µm diamond suspension. Each tooth slice was positioned on an XYZ computer-controlled table (Aerotech, USA) equipped with a precise optical scanning system with a maximum linear scanning velocity of 2000 mm/s.

A pulsed laser (Satsuma Amplitude, France) operating at 1030 nm wavelength and having pulse duration of 350 fs was used. The maximum average power was 5 W, the laser pulse energy was up to 10uJ and the pulse repetition rate was up to 500 kHz. The FELS was equipped with a precise scanning system with a maximum speed of 2000 mm/sec and used to move the laser beam in the XY direction.

Ablated areas in dentin and enamel were thoroughly imaged with an optical microscope (OM; Alicona G4 InfiniteFocus, Austria, Measure Suite software) and coherence scanning interferometry (CSI, Leitz-Linnik microscope, objective x50, NA=0.85, FSA algorithm [5]). Structural characterization was performed with a micro-Raman spectroscope (LabRamAramis, JobinYvon Horiba, λ=532 nm, spectral range 100-3000 cm^{-1}, resolution 1 cm^{-1}).

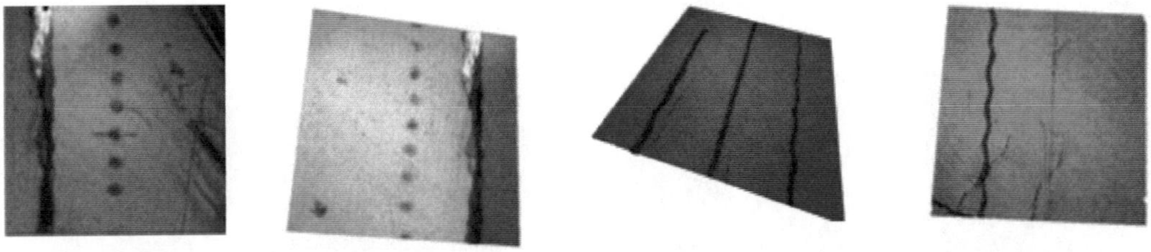

100% energy 90% energy
single spots - 1.20 m/sec, 50 kHz

100-90-80% energy (from left to right); 70-60% energy (from left to right);
3 pulses overlap in the focal spot of 20 µm - 60 cm/sec, 100 kHz

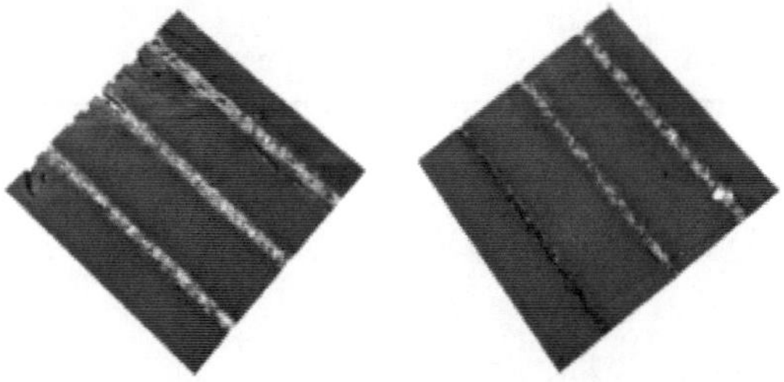

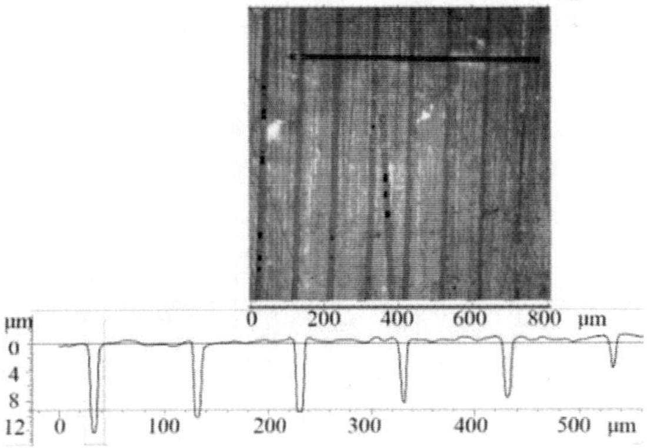

100-90-80% energy (from right to left); 70-60-50% energy (from right to left);
5 pulses overlap in the focal spot of 20 µm - 40 cm/sec, 100 kHz

Enamel ablated with overlapping pulses at 100 kHz, energy decreases from 100% to 30% (from left to right; step of 10). The corresponding line profile is given and shows the depth of the ablated lines

Figure 1. OM images of laser ablated enamel with the corresponding parameters of irradiation given under the images.

3. Results and Discussion

Satisfactory ablation rates, minimal temperature increase and absence of collateral damage were considered as an effective ablation. Thus, the laser pulse energy, pulse repetition rate and scanning speed were varied in order to find optimal parameters for the ablation of dental tissue. The laser beam was focused on the tooth slice using a telecentric lens to produce a focal spot diameter of 20 µm. The maximum laser pulse energy of 10 µJ was taken as 100% and decreased in steps of factors of 10 until no traces were obtained on the tooth surface upon ablating the dentin and enamel with consecutive pulses. The scanning velocity was also varied from one that produced single non-overlapping spots to

100% energy single spots 90% energy 80% energy
single shots - 1.20 m/sec, 50 kHz

100-90-80% (from left to right); 70-60-50% (from left to right); 40-30% (from left to right);
4 pulses overlap in the focal spot of 20 µm- 50 cm/sec, 100 kHz

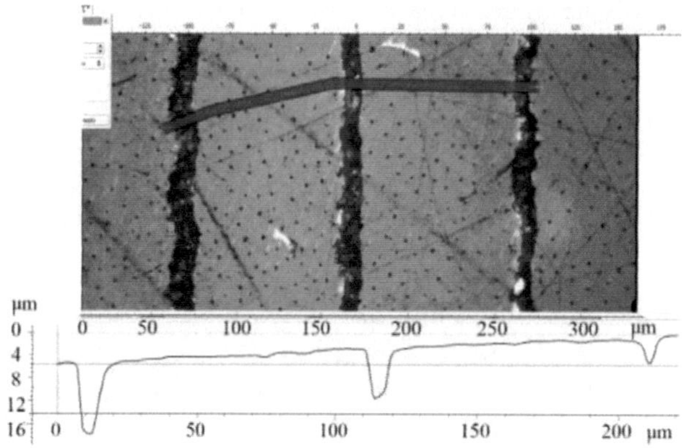

Dentin ablated with overlapping pulses at 100 kHz, energy decreases from 100-90-80% (from left to right), 3 pulses overlap in the focal spot of 20 µm - 60 cm/sec, 100 kHz.
Line profile shows depth

Figure 2. OM images of laser ablated dentin with the corresponding parameters of irradiation given under the images.

speeds that yielded the formation of lines of overlapping spots. Various OM images of ablated enamel and dentin are shown in figures 1 and 2 with the corresponding laser parameters.

Enamel and dentin consist of the mineral carbonated hydroxyapatite ($Ca_{10}(PO_4)_6(OH)_2$; CO_3-HA), water and organic materials (collagen proteins and lipids) [6]. Enamel has 85 vol% CO_3-HA (of which 2–4 % CO_3), 12 vol% water and 3 vol% organics, which makes it much harder than the dentin. On the contrary, dentin has only 47 vol% CO_3-HA and more water (20 vol%) and organic materials (33 vol%) as constituents. After varying the above-specified laser parameters, finally they were set at 100% pulsed energy, 500 mm/sec scanning speed and 100 kHz repetition rate, and these were considered to be the optimal parameters for an efficient and controlled removal of dental tissue with the FELS used in this work. The ablation threshold and ablation rate were measured as a function of the laser fluence.

Figure 3. Ablated craters 1x1 mm^2 in enamel (left) and dentin (right) imaged with OM.

The ablation threshold was 2.0 J/cm^2 for enamel and 1.6 J/cm^2 for dentin. Different ablation thresholds of the two materials come from their different composition. OM images of ablated craters with a size of 1x1 mm^2 in dentin and enamel are shown in figure 3. Laser-tissue interaction depends on the effect of the laser wavelength on the different tissues treated. More energy is needed for the highly mineralized and crystalline enamel and less energy for the dentin, which has a high organic content. After the laser ablation with the optimum parameters, no cracks were present in either the enamel or the dentin and the craters had well-defined precise vertical sides and edges. The craters in both materials were homogeneously machined and collateral damages around them such as discoloration, melting or carbonization as a result of the laser interaction with the dental tissue were not found.

CSI was used for imaging the boundary between the smooth dentin surface and the ablated crater, and to obtain roughness parameters. Figure 4 shows the 3D image of the boundary area. The R_a roughness of the mirror polished dentin surface was 164 nm while the ablation introduced micrometer roughness, and a R_a of 2.31 μm. The increased roughness of the dental tissue may be advantageous since the larger surface area yields improved adhesion strength of dental fillings after the dentist has prepared the cavity for the filling with the fs laser instrument. In addition to the fact that we can obtain fast ablation of dental tissue with the FELS, the resulting dental cavity is clean of contaminants found on the tooth surface and the smear layer usually occurring with the conventional drilling is absent.

Micro-Raman spectra of dentin and enamel (figure 5) were dominated by the characteristic peaks of the P-O vibrational bands of the phosphate (PO_4^{3-}) group. The most intensive and narrow peak at 960 cm^{-1} was due to the ν_1 P-O symmetric stretching which is stronger in the enamel due to the higher content and higher crystallinity of the HA mineral. Peaks centered at 430 and 590 cm^{-1} were due to the ν_2 and ν_4 P-O bending vibrations. A doublet at 1045/1070 cm^{-1} originated from vibrations of the C-O bond in CO_3. The latter is characteristic of ν_3 C-O stretching and superimposes on the ν_3 P-O stretching. Peaks due to the amide I, II and III in collagen proteins were found at 1660, 1520 and 1247 cm^{-1}, respectively, with a higher intensity in the spectra of the dentin than those of the enamel which is equivalent to a higher concentration due to the superior organic content in the dentin tissue. Stretching vibration of the C=O bond in carbonyl groups originating from collagen proteins was observed at 1780

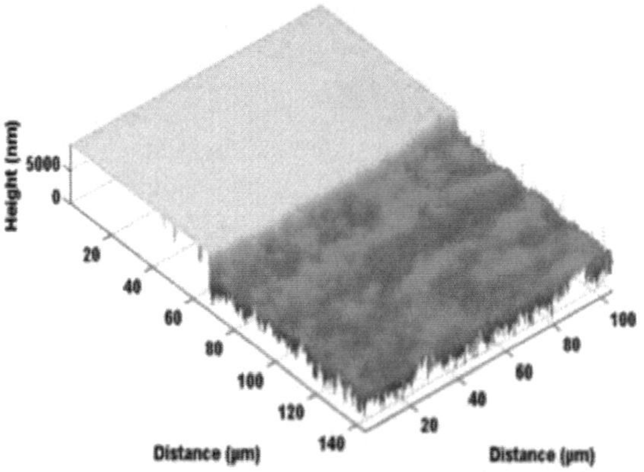

Figure 4. 3D image of the boundary between the smooth dentin surface and the ablated crater.

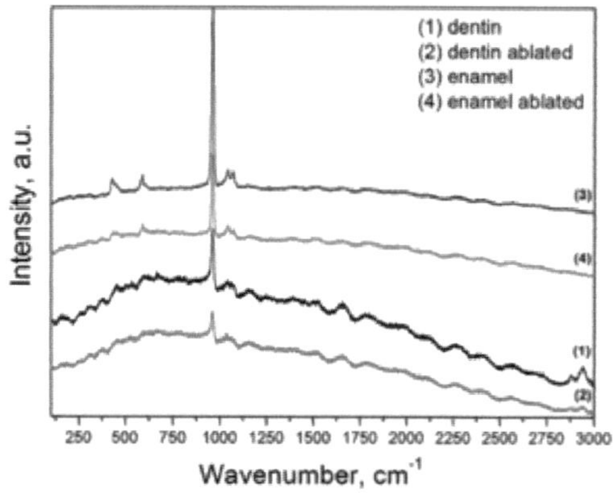

Figure 5. Raman spectra of intact and ablated enamel and dentin.

cm^{-1} in the spectra of intact and ablated dentin. Also in these two spectra, C-H bending at 1450 cm^{-1} and C-H stretching with doublet structure at 2882 and 2942 cm^{-1} due to organic groups were present. The ablation of enamel and dentin did not change their structure (spectra 2 and 4 in figure 5) and the characteristic features of the inorganic HA with weak peaks due to collagen proteins were preserved. This means that no overheating was obtained and the tooth nerves in the pulp cavity of a real tooth would not be damaged which is an additional advantage to the lack of collateral damage. The much lower peak intensity of the ablated materials is due to the fact that the ablated crater is deep enough (1 mm) and the signal coming back to the detector is weaker.

After having found the optimal parameters for the minimally invasive ablation of dentin and enamel, more research work will be performed for increasing the speed of ablation in order to surpass the speed of the conventional drilling instruments while using the benefits of the fs laser energy.

4. Conclusion

Craters with sizes of 1 mm^3 were ablated in enamel and dentin when the adequate parameters of laser ablation with the FELS were set after varying the laser pulse energy, pulse repetition rate and scanning speed. The craters in both materials were homogeneously machined, with well-defined precise vertical sides and edges. No collateral damages around the craters in either materials or overheating effects were observed and the material roughness was increased. Thus, the FELS used in this work is promising for an efficient and controlled cavity preparation in dentin and enamel.

Acknowledgements

The authors would like to acknowledge the Program for career development of young scientists, BAS for funding this work through a grant DFNP-159/2016.

References

[1] Cernavin I 1995 *Aust. Dent. J.* **40** 79
[2] Niemz M H 1998 *SPIE* **3255** 84
[3] Kim B, Feit M, Rubenchik A, Joslin E, Celliers P, Eichler J and Da Silva L 2001 *J. Biomed. Opt.* **6** 332
[4] Hairaye C, Mermet F, Engel T, Montgomery P and Fontaine J 2014 *J. Phys. Conf. Ser.* **558** 012063
[5] Gamaly E, Rode A, Tikhonchuk V, Luther-Davies B 2002 *Phys. Plasmas* **9** 949
[6] LeGeros R 1991 In HM Myers: *Calcium phosphates in oral biology and medicine* (Basel, Karger) vol 15

Photoresponsive azo-doped aerosil/7CB nematic nanocomposites: the effect from concentration of the azobenzene photoactive agent

G B Hadjichristov [1,*], Y G Marinov [1], A G Petrov [1] and S K Prasad [2]
[1] Georgi Nadjakov Institute of Solid State Physics, Bulgarian Academy of Sciences, 72 Tzarigradsko chaussee, Sofia 1784, Bulgaria
[2] Centre for Nano and Soft Matter Sciences, P.O.Box 1329, Jalahalli, Bangalore 560 013, India

E-mail: georgibh@issp.bas.bg

Abstract. We studied nanomaterials composed from 3 wt.% aerosil nanospheres and the room-temperature nematic liquid crystal 4-n-heptyl cyanobiphenyl (7CB), as doped with the photoactive liquid crystal 4-(4'-ethoxyphenylazo)phenyl hexanoate (EPH). The molecules of the azobenzene photoactive agent EPH were included at concentration ranging from 1 wt.% to 10 wt.%. The object of our interest is the effect of the EPH amount on the photosensitized electro-optical properties of thin films of aerosil/7CB/EPH nematic nanocomposites.

1. Introduction

Among the plenty of nanostructured materials with interesting properties and applications, the liquid crystals (LCs)-based nanocomposites, in particular those created by doping of LCs with nanoparticles (NPs) [1,2], have been attracted considerable research attention from the physics and engineering communities. At present, such nanomaterials are widely and intensely investigated because they have large potential to improve the performance of the LCs devices [3]. In this context, of special interest are the systems consisting of silica NPs doped into LCs [4]. Such nanostructured LCs have demonstrated advanced applications based on their valuable electric and electro-optic (EO) properties [1,4-6]. That is why, silica-nanostructured materials composed from aerosil NPs (ANPs) and nematic LCs have been extensively investigated [3,8,9].

Very recently, we have studied the EO properties of thin (25 μm) films of gel nanocomposites produced from nematic LC heptylcyanobiphenyl (7CB) and 3 wt.% hydrophilic ANPs [10,11]. Further, the photo-driven stimulation of the electro-optics of such nanomaterials doped with azobenzene LC 4-(4'-ethoxyphenylazo)phenyl hexanoate (EPH) is of research interest. Reasonably, the photo-stimulation of the electro-optics of EPH-doped ANPs/7CB films should be enhanced by use of higher intensity of light in reasonable limits. But also, the concentration of EPH photo-agent can affect their EO response. In the present work this was experimentally studied by varying the concentration of azobenzene EPH photoactive agent in the range 1 wt.% - 10 wt.%, by keeping fixed the UV light intensity.

* To whom any correspondence should be addressed.

IOP Conf. Series: Journal of Physics: Conf. Series **794** (2017) 012037 doi:10.1088/1742-6596/794/1/012037

2. Experimental

ANPs (hydrophilic Aerosil 300 nanospheres having a mean diameter of about 7 nm and capped with silanol groups) at a concentration of 3 wt.% were added to the well known room-temperature nematic LC 4'-heptyl-4-biphenylcarbonitrile (7CB). The formed nanocomposite was doped with the azo-containing LC compound 4-(4'-ethoxy phenylazo)phenyl hexanoate (EPH). Four different blends of EPH with aerosil/7CB were made, with the composition of EPH being X = 1 wt.%, 3 wt.% , 5 wt.% and 10 wt.%. The plan-parallel EPH-doped ANPs/7CB gel films with a thickness of 25 μm were formed in cells assembled from two ITO (indium tin oxide)-coated glass plates, to serve as electrodes.

The experimental set up and the optical layout for the experiments performed, have been described in details elsewhere [12]. Briefly, the EO response of EPH-doped ANPs/7CB films to an alternating-current (AC) electric field was investigated by use of a circularly polarized He-Ne laser beam (wavelength λ = 632.8 nm, 1 mW power on the samples) and computer-controlled lock-in amplifier (SR830 DSP, Stanford Research Systems) coupled with a photodiode. The UV continuous light source was a LED (λ = 375 nm). The UV light intensity on the sample was fixed at ~ 3 mW/cm2 and kept the same by all the measurements in this work.

The voltage-dependent transmittance of EPH-doped ANPs/7CB nanocomposite films were measured at 1 kHz frequency of the electric field applied on the films and with laser beam chopped at 90 Hz. The EO modulation characteristics of the films (the frequency spectra of the amplitude of the EO modulation of the laser beam passed through the films) were obtained in the range 1 Hz – 3 kHz at a fixed voltage of the electric field. In all cases, no polarizers were used. During the experiments, the temperature of the studied nanocomposite films was about 27° C.

3. Results and Discussion

Figure 1 reports UV-visible absorption spectra recorded for the EPH-doped aerosil/7CB nanogel samples we have prepared. The spectra show the absorbance change by variation of EPH concentration from 1 wt.% to 10 wt.%. The broad spectral band between 330 nm and 400 nm specific for EPH compound [13] (in fact, $\pi - \pi^*$ absorption band of the *trans*-azobenzenes at around 370 nm) is clearly seen. This implies that an efficient photoisomerization of EPH azo-molecules can occur by illumination with UV light at the wavelength of 375 nm that falls at the absorption band of EPH compound.

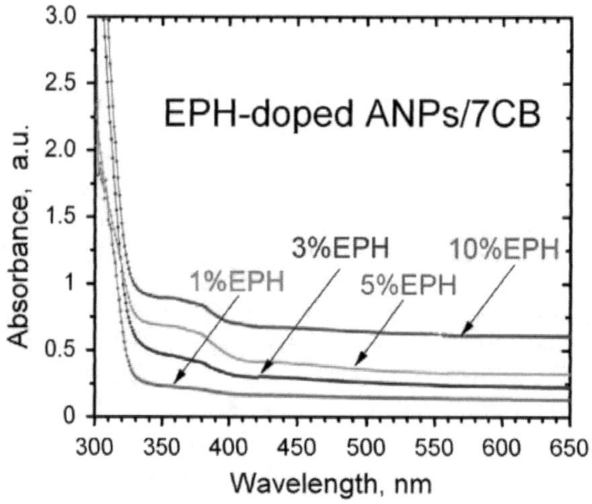

Figure 1. UV-Vis absorption spectra of ANPs/7CB nanocomposite films doped with azobenzene LC EPH molecules at various concentrations (denoted for each curve). All spectra were recorded under identical experimental conditions; the films were in cells assembled from quartz plates.

It should be noted that the shift of the baseline position of the spectra in figure 1 on the ordinate scale by increasing EPH concentration is mainly due to the increasing optical loss of EPH-doped aerosil/7CB films. This is resulting from strong light scattering (both specular and diffuse) by the films that takes place in the whole spectral range measured here. The light scattering problem has to be taken into consideration in the nematic state. Because of this, the optical transmission technique in our case provides qualitative results and atually the measured spectra may not represent the real absorption of the LC mixtures under study. Nevertheless, the effect from the azo-dopants is well distinguished.

It is well known that the strong light scattering by such a kind of nano-filled nematics is due to the formed gel structure [10,14,15]. Accordingly, the EO response of the examined three-component nanocomposites is determined from the gel network of nanoaerosil within the nematics, i.e. their EO response is directly related to the dispersed ANPs and their concentration. However, as seen from figure 1 the EPH dopants also strongly contribute to the light scattering even with no electric field applied on the films. Moreover, one can expect that the EPH doping does significantly change the structure of ANPs/7CB/EPH system where the azobenzene EPH molecules decorate the gel network of the nanostructured nematic.

By applying AC electric field, the nano(aerosil)-filled nematics scatter light due to electrically-controllable local inhomogeneities (ANPs-generated orientational defects in anisotropic LC matrix; the orientation of LC molecules in electric field leads to a disappearing the defects). By that, the electrically-induced changes do result from the electrically-driven reorientation of the LC molecules [11]. On the other hand, being exposed on UV light the ANPs/7CB/EPH nematic system should undergo changes due to *trans*-to-*cis* photoisomerization of azobenzene EPH LC molecules, i.e., from rod-shaped to bent-shaped conformation [12].

Figure 2 shows the light transmittance of ANPs/7CB/EPH films as a function of the applied AC electric field. The curves were obtained either without or with illumination of the films with UV light at 375 nm. For the sake of comparison, the voltage-dependent characteristics of the films doped with EPH at various concentrations were recorded keeping a constant intensity of the UV light. In general, the transparency of the ANPs/7CB films should be increased by the increase of the applied AC electric field [11].

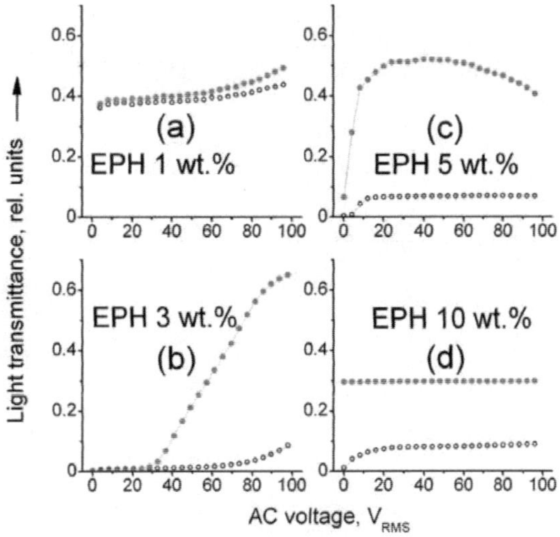

Figure 2. Voltage-dependent transmittance (at $\lambda = 633$ nm) of ANPs/7CB nanocomposite films doped with EPH at various concentration. All samples were in identical cells. Data obtained at a fixed UV-light intensity and under identical other experimental conditions. No UV light (open circles); with UV light (solid circles).

As seen from figure 2, the UV-light-induced changes of the EO response of the studied EPH-doped ANPs/7CB films are rather different as depending on the EPH concentration. So, instead of the smeared behavior of monotonic increase of light transmittance of the films at the lowest concentration (1 wt.%) of EPH we have used here (figure 2 a), a dependence with a well pronounced threshold of the electric field as well as an initiation of saturation of the film transparency, were present at the EPH concentration of 3 wt.% (figure 2 b). By increasing concentration of the EPH dopants (e.g., above 5 wt.%), a further UV-light-produced reduction of the voltage threshold takes place (figure 2 c). At higher concentration of EPH dopants, e.g., for the nanogel sample doped with 10 wt.% EPH, the voltage-dependent transmittance becomes threshold-less (figure 2 d), obviously because the sample is passed over its isotropic state at the given temperature (in our case 27° C). We have to point out that all the behaviors seen in figure 2 were consistent with our observations by optical microscopy of the samples.

Clearly, the observed UV-light-stimulated effect occurs due to *trans*-to-*cis* photoisomerization of azobenzene EPH LC molecules [13, 16-18] and is closely related to the UV-light-induced local Freedericksz transition of the examined nematic systems, as well as to their possible UV-light-induced transition to isotropic state at a given temperature. Furthermore, the observed UV-light-stimulated lowering of the threshold voltage of the local Freedericksz transition in the ANPs/7CB/EPH system can be attributed to an UV-light-induced increase of the characteristic size (δ_B) of the local bulk domains that undergo Freedericksz transition in this nano(aerosil)-filled nematic. Reasonably, the higher concentration of the EPH dopants that decorate the gel network, the larger δ_B.

In general, the enlargement of δ_B with the amount of EPH can be the reason for the changes observed in the amplitude-frequency EO modulation characteristics of the studied ANPs/7CB/EPH nanocomposite films by varying the EPH concentration. As an example, figure 3 compares the frequency spectra of second-harmonic (SH) modulation of light transmitted through such films doped with EPH at concentration 1 wt.% , 3 wt.% or 5 wt.%. The spectrum of SH amplitude-frequency EO modulation is relevant to the amplitude of the dielectric oscillations versus the field frequency, as measured for the LC films. These frequency characteristics were recorded without UV illumination, at a high intensity of the electric field applied (AC voltage of 100 V). As explained in [11], upon such electric-field strength the SH EO behaviors can be sensitive to the size of the bulk domains. It is seen from figure 3 that the larger the concentration of EPH dopants, the larger the slope of the SH EO modulation spectra. In accordance with [11], the results reported in figure 3 evidence the enlargement of the size δ_B of the active structural domains in the studied ANPs/7CB/EPH nanostructured nematic gel that scatter and modulate light. Note that these bulk domains are still smaller than those in pure (non-nanostructured and undoped) 7CB nematic LC. In the latter case, the slope of SH EO modulation spectra follows the reciprocal dependence on the field frequency [11].

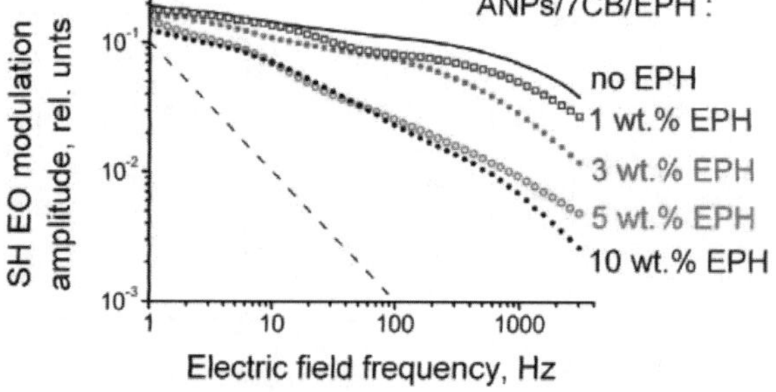

Figure 3. Frequency spectra of the amplitude of SH EO modulation of He-Ne laser beam transmitted through ANPs/7CB gel films doped with EPH at various concentrations. Data obtained by keeping the other experimental conditions the same; the driving AC voltage was 100 V_{RMS}. The dashed line represents the corresponding behavior relevant to pure 7CB nematic LC.

IOP Conf. Series: Journal of Physics: Conf. Series **794** (2017) 012037 doi:10.1088/1742-6596/794/1/012037

Figure 4 reports the UV light-induced change of SH EO modulation by ANPs/7CB films doped with EPH at various concentrations. For the sake of comparison, the spectra were scaled in identical manner. Again, the AC voltage applied on the films was fixed at 100 V_{RMS}, a voltage at which the UV-induced effect on the SH EO modulation behavior should be well pronounced. It is seen that even at the relatively low (1 wt.%) concentration of the azobenzene EPH dopants, the *trans*-to-*cis* photoisomerization of EPH is sufficient to photo-stimulate the electro-optics of EPH-doped ANPs/7CB films and to efficiently modify the SH EO modulation of intensity of the light transmitted through them. The illumination with UV light can considerably change both the amplitude of SH EO modulation and the slope of the SH EO modulation spectra of the films. Thus, the UV-light-modified SH EO modulation can become rather different than that achieved by the non-illuminated films. As seen, the UV-light-produced changes are enhanced at higher EPH-concentration (figure 4 c,d).

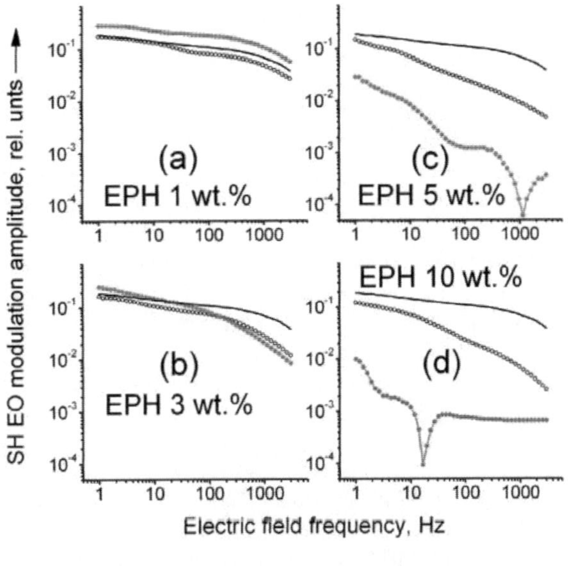

Figure 4. UV light-induced change of SH EO modulation response of ANPs/7CB/EPH films. The EPH concentration: 1 wt.% (a); 3 wt.% (b); 5 wt.% (c); 10 wt.% (d). The spectra were recorded either with (solid circles) or without (open circles) UV illumination (at the same level of the UV-light intensity as in the course of the experiments described above) by keeping the other experimental conditions identical; the driving AC voltage was 100 V_{RMS}. SH EO modulation spectrum achieved by undoped ANPs/7CB film is included as a reference (shown with lines).

Thus, at the presently used UV light intensity level, the UV illumination can increase the amplitude of SH EO modulation by the films doped with a small amount of azo-molecules (in our case, 1 wt.% of EPH, or at 3 wt.% of EPH - with a diminishing effect), or reduce the SH EO modulation amplitude (at 5 wt.% or 10 wt.% of EPH), as compared to the non-illuminated case. The UV-light-induced decrease of the SH EO modulation with increasing EPH concentration can be attributed to UV-light-produced enlargement of the bulk domains whose field-driven dielectric reorientation is heavily retarded. As a side effect, frequency dips can appear in the SH EO modulation spectra when the EPH concentration is high. Similar to the sharp single minima in the SH EO modulation spectra found for micro-confined nematic-based composites, such as micro-scaled polymer-dispersed LCs [19], the above mentioned dips upon UV illumination suggest that a strong scattering of light occurs at a given electric-field frequency, corresponding to a spatio-dielectro-optic coupling with the field-induced nematic director orientation deformations relevant to large-sized domains formed in the bulk of the UV-illuminated EPH-doped ANPs/7CB films.

It should be noted that at a low EPH-concentrations (like 1 wt.% and 3 wt.% in our experiments) an illumination with a blue light with intensity of about 30 mW/cm^2 restores the initial state of the films, i.e. SH EO modulation becomes close to that of the films non-exposed to UV light. This effect follows the back (*cis*-to-*trans*) transition of azobenzene molecules [20-22] after absorption of light in the blue spectral range (n-π* absorption band of the *cis*-azobenzenes at around 450 nm). In contrast, at higher EPH concentration the intensity of the continuous UV illumination at ~3 mW/cm^2 used here led to a change of the ANPs/7CB/EPH system that can not be restored by the same blue-light illumination. In our experiments this was clearly registered for EPH dopants at concentrations of 5 wt.% and 10 wt.%. Most probably, this is owing to UV-light-provoked formation of strongly-meshed agglomerate-like creatures-domains in the bulk of the azo-doped nano(aerosil)-filled nematic system under study.

4. Conclusion

We have studied the concentration effect of the of azobenzene photoactive molecules added to nano(aerosil)-filled nematic system, in particular EPH-photo-sensitized ANPs/7CB films, on the photo-driven stimulation of the electro-optics of such nanocomposite materials. It is demonstrated that at the relatively low-intensity (~ 3 mW/cm^2) continuous UV light, the UV-produced effect on both the EO transmittance and the SH EO modulation by 25 μm-thick EPH-doped ANPs/7CB films is well pronounced even at the relatively low (1 wt.%) concentration of the azobenzene EPH dopants. The UV-light-induced stimulation occurs due to *trans*-to-*cis* photoisomerization of EPH molecules that decorate the gel network of the nanostructured nematic and is enhanced at higher concentration of the EPH photoactive agent. Both factors (the optical density and the intensity of the activating UV light) should be considered and carefully balanced for proper photo-stimulated reversible function of the photoresponsive nematic nanocomposites such as the three-component system studied here. This balance is of importance for the practical application of the observed UV-stimulation effect in regular or specific scatter-based electro-optics of nematic LC systems doped with azo molecules.

Acknowledgments
The authors acknowledge the INERA EU project Research Potential "Research and Innovation Capacity Strengthening of ISSP-BAS in Multifunctional Nanostructures" (FP7 REGPOT-2012-2013-1) for their participation at 19th ISCMP “Advances in Nanostructured Condensed Matter: Research and Innovations”, Aug 28th – Sept 2nd, 2016, Varna, Bulgaria. Work supported by research project DFNI-TO2/18 “Nanostructured liquid crystals for tunable photonic devices” by the Ministry of Education and Science, National Science Fund of Bulgaria.

References

[1] Garbovskiy Y A and Glushchenko A V 2011 Liquid Crystalline Colloids of Nanoparticles: Preparation, Properties and Applications *Solid State Physics*, vol 62, eds R E Camley and R L Stamps (New York: Academic Press, Elsevier) p 1
[2] Liang H H and Lee J Y 2011 Enhanced Electro-Optical Properties of Liquid Crystals Devices by Doping with Ferroelectric Nanoparticles *Ferroelectrics - Material Aspects*, ed M Lallart (Rijeka, Croatia: InTech) chapter 10 p 193
[3] Q Li ed *Liquid Crystals Beyond Displays: Chemistry, Physics, and Applications* (Hoboken, NJ: Wiley & Sons)
[4] Iannacchione G S 2004 *Fluid Phase Equlibria* **222-223** 177
[5] Prasad S K, Sandhya K L, Nair G G, Hiremath U S, Yelamaggad C V and Sampath S 2006 *Liq. Cryst.* 2006 **33** 1121
[6] Chaudhary A, Malik P, Mehra R and Raina K K 2012 *Phase Transitions* **85** 244
[7] Jayalakshmi V, Nair G G and Prasad S K 2007 *J. Phys. Condens. Matter* **19** 226213
[8] Bapat P N, Rao D S S and Prasad S K 2009 *Thermochimica Acta* **495** 115
[9] Bapat P N, Rao D S S, Prasad S K and Yelamaggad C V 2010 *J. Phys. Chem. B* **114** 12825
[10] Kumar M V, Prasad S K, Marinov Y, Todorova L and Petrov A G 2015 *Mol. Cryst. Liq. Cryst.*

610 51

[11] Marinov Y G, Hadjichristov G B, Petrov A G and Prasad S K 2016 *Composites B: Engineering* **90** 471

[12] Hadjichristov G B, Marinov Y G and Yelamaggad C V 2014 *J. Phys. Conf. Ser.* **558** 012026

[13] Sridevi S, Hiremath U S, Yelamaggad C V, Prasad S K, Marinov Y G, Hadjichristov G B and Petrov AG 2011 *Mater. Chem. Phys.*. **130** 1329

[14] Iannacchione G S, Garland C W, Mang J T and Rieker T P 1998 *Phys. Rev. E* **58** 5966

[15] Lobo C V, Prasad S K and Yelamaggad C V 2006 *J. Phys. Condens. Matter* **18** 767

[16] Prasad S K, Nair G G, Hegde G, Sandhya K L, Rao D S S, Lobo C V and Yelamaggad C V 2005 *Phase Transitions* **78** 443

[17] Petrov A G, Marinov Y G, Hadjichristov G B, Sridevi S, Hiremath U S, Yelamaggad C V and Prasad S K 2011 *Mol. Cryst. Liq. Cryst.* **544** 3/[991]

[18] Yelamaggad C V, Prasad S K and Li Q 2012 Photo-Stimulated Phase Transformations in Liquid Crystals and Their Non-Display Applications *Liquid Crystals Beyond Displays: Chemistry, Physics, and Applications* ed Q Li (Hoboken, NJ: Wiley & Sons) chapter 5 p 157

[19] Petrov A G., Marinov Y, D'Elia S, Marino S, Versace C and Scaramuzza N 2007 *J. Optoelectr. Adv. Mater.* **9** 420

[20] Natansohn A and Rochon P 2002 *Chem. Rev.* **102** 4139

[21] Yager K G and Barrett C J 2009 Azobenzene Chromophores in Photo-Reversible Materials *Smart Light-Responsive Materials: Azobenzene-Containing Polymers and Liquid Crystals* eds Y Zhao and T Ikeda (Boca Raton, Florida: John Wiley and Sons) chapter 1 p 1

[22] Q A Acton ed 2013 *Azo Compounds-Advances in Research and Application*, Scholarly Brief (Atlanta, Georgia: Scholarly Editions)